SIR HENRY BESSEMER, F.R.S.

AN AUTOBIOGRAPHY.

WITH A CONCLUDING CHAPTER.

LONDON:

OFFICES OF "ENGINEERING," 35 AND 36, BEDFORD STREET, STRAND, W.C.

1905.

Book 451

Published in 1989 by
The Institute of Metals
1 Carlton House Terrace
London SW1Y 5DB
and
The Institute of Metals N American Publications Center
Old Post Road Brookfield VT05036 USA

British Library Cataloguing in Publication Data
Bessemer, Sir, Henry, 1813-1898
    Sir Henry Bessemer, F.R.S.
    1. Great Britain. Inventions. Bessemer,
    Sir Henry, 1813-1898
    I. Title
    609'.2'4
ISBN 0-901462-49-7

Library of Congress Cataloging in Publication Data
applied for

First printed at The Bedford Press, 20-21 Bedfordbury, Strand, London WC
This facsimile edition printed by Antony Rowe Ltd, Bumper's Farm, Chippenham, Wilts.

AGE 35

AGE 45

SIR HENRY BESSEMER IN HIS 80TH YEAR

AGE 56

AGE 70

Henry Bessemer

TO THE

PRESIDENT AND MEMBERS

OF THE

IRON AND STEEL INSTITUTE

OF

GREAT BRITAIN

THIS VOLUME IS RESPECTFULLY DEDICATED

BY THEIR OLD COLLEAGUE AND

PAST PRESIDENT

SIR HENRY BESSEMER

IN REMEMBRANCE

OF

TWENTY-FIVE YEARS

CORDIAL AND FRIENDLY INTERCOURSE

CONTENTS

CHAPTER I

EARLY DAYS

CHAPTER II

THE REWARD OF INVENTION

CHAPTER III

COMPRESSING PLUMBAGO DUST, CASTING TYPE, TYPE-COMPOSING MACHINE, ETC.

CHAPTER IV

UTRECHT VELVET

CHAPTER V

THE MANUFACTURE OF BRONZE POWDER

CHAPTER XIX

EBBW VALE

CHAPTER XX

THE BESSEMER SALOON STEAM-SHIP

CHAPTER XXI

CONCLUSION

LIST OF ILLUSTRATIONS

For Plates I-L, see plate section following page 169

FOREWORD

Bessemer's name became known throughout the world, among steel-makers and laypeople alike, for the invention of the converter process for making steel. Less generally known is that he was a prolific inventor in many other fields.

In early life he had access to his father's type foundry, and there he made castings for models which he had designed. He invented a die which would both emboss and perforate legal documents on parchment to prevent fraudulent re-use of the paper embossed stamps previously used (actually, the die was never used because Bessemer's fiancée pointed out that the addition of a dating mechanism to the existing die was a simpler matter, which solution of the problem he adopted). He devised a method of casting type for printing in water-cooled moulds and under vacuum to prevent porosity; also a machine for setting the type. Much more important was his invention of machinery to make bronze powder, previously produced extremely laboriously by hand.

A slightly comic note is struck by his foray into naval architecture. Bessemer suffered dreadfully from seasickness. He therefore ·experimented with the *SS Bessemer*, a cross-Channel steamship containing a saloon mounted in a hydraulic system which was intended to keep the saloon steady while the ship rolled. The scheme was not successful; the ship itself was quite unmanageable and, even with the saloon mechanism locked, it did a great deal of damage to Calais Harbour. The experiments were abandoned.

Before turning to the invention which made him famous, two very different projects are worthy of particular attention.

SUGAR CANE PROCESSING

The conventional method of extracting the juice from sugar cane at that time was to pass it between rollers. The yield was poor and a prize was offered for an improved method of extracting the sugar more efficiently. Bessemer won the prize, although he had no experience of the industry. His press consisted of a piston which cut off sections of sugar cane, ingeniously in such lengths that at least one end would be open, and subjected the cut pieces to prolonged compression in a perforated constricting chamber before ultimate release. It is to be supposed that others could have thought up such an idea but the energy with which he followed it up was, to say the least, unusual. He constructed an engine and had considerable shipments of sugar cane (the ends sealed in bitumen to avoid evaporation) imported from Madeira to prove its operation in London, whereby he convinced the sugar manufacturers of its worth. He showed his business acumen by protecting his invention by patent and also winning the prize for which he was congratulated by the Prince Consort.

Typically, his attention having been drawn to the then somewhat primitive methods of processing sugar from cane, he looked at many of the steps taken to produce the refined product from cane juice and took out no fewer than 13 patents on them.

OPTICAL GLASS

At one stage, Bessemer became interested in how to make defect-free optical glass. He wanted to make a solar furnace capable of melting several ounces of material in a crucible. At that time the only source of discs of glass suitable for making large lenses were those made in Germany by Frauenhofer; they were expensive and in poor supply. Bessemer studied the problems of mixing molten glass to eliminate concentration gradients and found that any attempt to mix by introducing a stirrer led to intractable gas bubble formation. He studied this by cold modelling the process using viscous castor oil and found by the addition of a dye as a tracer that very complete mixing could be achieved by rolling the container around in a particular way. His experiments in this field were very much ahead of their time. The work of, for example, the John Percy group in the Royal School of Mines during the past 20 years used similar cold modelling techniques. The tools available to that group included sophisticated devices such as laser Doppler anemometers and data loggers, but the principles are basically the same as Bessemer's! He developed a fine glass furnace in which a conically shaped crucible could be gently rolled around during the refining process. He included the idea of subjecting the molten glass to a vacuum 'through a platinum tube' to remove air bubbles. It is not clear whether this process was ever put into commercial scale operation but the inventor's mind, following the subject in his usual logical way, led on to developing methods of premixing the glass components by grinding them thoroughly together. This in turn led to his setting up a 12 cubic foot reverberatory melting furnace from which a stream of liquid glass was passed between rollers to make large sheets of glass. This particular process was taken up by Chance glassworks who bought the process and its patents.

THE BESSEMER PROCESS

At the time of the Crimean War spherical shot was being replaced by elongated projectiles, and Bessemer devised means of rotating them even though they were fired from the existing smooth-bore ordnance. He made longitudinal passages within or on the outside surface of the projectiles leading from the back forwards, and the front ends were turned tangentially to the longitudinal axis; when the gun was fired a small proportion of the propelling gases passed up these channels and, emerging as tangential jets, made the missile rotate.

It was thought that a stronger material than the cast iron then available should be used to make the guns, and Bessemer set out to produce a castable metal with properties comparable with those of wrought iron or steel. His first experiments were made in a specially designed reverberatory furnace, and, in order to consume completely the gases over the bath and increase the heat, air was blown in through perforations in the firebridge. One day he found on the side of the bath two shells of what proved to be decarburized iron; from this he concluded that air alone could convert grey pig iron into malleable iron without puddling. He therefore set out to melt some pig iron in a covered crucible in order to blow air into it through a pipe passing through the cover. Owing to bad chimney draught and the lack of metallurgical coke, he could scarcely melt the pig iron; however, a small puddle formed at the bottom of the crucible, so Bessemer decided to blow air onto it. On taking the lid off, half a minute later he was dumbfounded to find that all the iron had melted; on blowing for another seven or eight minutes the contents of the crucible were so hot that they put the furnace fuel in the shade. From this point on Bessemer ceased to use any external source of heat for his process.

In the *Journal of the Iron and Steel Institute* in 1890, Abraham S Hewitt wrote:

". . .the Bessemer invention takes rank with the great events which have changed the face of society since the Middle Ages. The invention of printing, the construction of the magnetic compass, the discovery of America and the introduction of the steam engine are the only capital events in modern history which belong to the same category as the Bessemer Process."

More recently, in 'The new science of strong materials' (Penguin Books, 2nd Edition, 1976), Professor J E Gordon stated:

"The cheapening and improvement of iron and steel during the eighteenth and nineteenth centuries was the most important event of its kind in history – or perhaps just the most important event in history."

Bessemer seems to have received no particular training as an engineer and certainly did not go on to any formal kind of higher education. He was, however, quite clearly self-confident, industrious, energetic, infinitely inquisitive and essentially intuitive. He lived to be 85 and was apparently active to the end. He was by no means modest, was capable of pique and, it has been said, sometimes ruthless in his treatment of others. The autobiography of such a man makes for intriguing and entertaining reading.

December 1988   *J H E Jeffes*
Emeritus Professor and Senior Research Fellow,
Imperial College

R B Wood
Deputy Secretary,
The Institute of Metals

PREFACE

IT is fifty years since Henry Bessemer made the great invention which has rendered his name famous, not only in English-speaking countries, but also in all civilised communities, and it is seven years since he died. If this Autobiography had dealt with the story of a lesser man, its appearance so long after his death might have reduced its interest and value so far as to render it scarcely worth while to place the narrative before the reader. But lapse of time cannot tarnish the lustre of Henry Bessemer's memory, nor can common and world-wide use of the great invention that crowned it, render uninteresting a story of the struggles through which he passed and the battles he had to fight before the world became enriched by his inventive genius.

The late Abram S. Hewitt, himself an engineer of universal reputation, and one of the pioneers of the Bessemer Process in the United States, speaking at the American meeting of the London Iron and Steel Institute, in 1890, said :—

A very few considerations will serve to show that the Bessemer invention takes its rank with the great events which have changed the face of society since the time of the Middle Ages. The invention of printing, the construction of the magnetic compass, the discovery of America, and the introduction of the steam-engine, are the only capital events in modern history which belong to the same category as the Bessemer process. They are all examples of the law and progress which evolve social and moral results from material discoveries and inventions. It is inconceivable to us how the world ever existed without the appliances of modern civilisation ; and it is quite certain that if we were deprived of the results of these inventions the greater portion of the human race would perish by starvation, and the remainder would relapse into barbarism. I know it is very high praise to class the invention of Bessemer with these great achievements, but I think a careful survey of the situation will lead us to the conclusion that no one of these has been more potent in preparing the way for the higher civilisation which awaits the coming century than the pneumatic process for the manufacture of steel . . . . The name of Bessemer will therefore be added to the honourable roll of men who have succeeded in spreading the gospel of "Peace on earth and goodwill toward men," which our Divine Master came on earth to teach and encourage.

The words of Abram S. Hewitt are frequently quoted in the following pages, always in the same spirit of appreciation of the great inventor ; but on no other occasion did he so justly and clearly crystallise his opinion of Bessemer as in the foregoing passage addressed to the Iron and Steel Institute, at a time when all the futile attempts that had been made to deprive Bessemer of the profit and glory of his great invention, had faded into almost forgotten history, and its practical outcome in the United States was measured by millions of tons of steel every year.

On an early page of this volume the author tells us he makes no claim to literary merit. He, certainly, was without training in the art of writing, but the happy gift, which characterised all his mechanical work, of instinctively selecting the simplest and best means of attaining a given end, did not desert him here. He wrote just as he talked, and infused into his writing the charm of his conversation. It was one of the great pleasures of his latter years to discuss with his old and valued friends—the proprietors of *Engineering*—the details of his Autobiography, and each printed page is more or less a reflection of the man himself in his varying moods. The eighty-five years of busy life which had been allotted him, had in no measure dimmed his memory, or even paled his enthusiasm : and in his Autobiography he lived over again the ambitions of youth, the struggles of manhood, the bitterness of injustice, the pleasure of appreciation, and the satisfaction of success. The world, as it recollects Bessemer, only knew him as the triumphant inventor, but in this volume we tread with him the thorny road to success, and more than that, he shows us the seamy side of the inventor's career.

Unfortunately, this Autobiography is not complete; even a Chapter of the history of the steel process is wanting—that recording its brilliant success in the United States. Sir Henry laid down his pen only a year before he died, but his self-told story goes no further than the episode of the Bessemer Saloon Steamer, in 1872. After that incident was closed he retired into private life, but not to a life of idleness. He had many occupations: the beautifying of his home; the installation of a large diamond-cutting and finishing plant; his telescope and observatory; his method of cutting and polishing optical lenses; his solar furnace; all these and other things kept him very busy, and formed not the least interesting part of his long life. It is unfortunate that he has left no consecutive record of this period; but he did leave many drawings, letters, and other documents referring to it, and from these has been prepared the supplementary Chapter which concludes the present volume.

CHAPTER I

EARLY DAYS

FOR many years past my most intimate friends have urged on me the desirability of giving to the world an authentic account of the origin and progress of the several inventions which together constitute what has, by common consent, been called the "Bessemer Steel Process;" thus tracing back to their earliest inception the various ideas and incidents which have led, by almost imperceptible degrees, to the development and practical working of that great steel industry, which, in so short a period, has spread itself over the whole of the continents of Europe and North America.

If we contemplate the rise and progress of almost all the great industries of the world, we find their origins lost in the mist of ages, with but few indications remaining of their gradual progress and development, or even of the names of those persons to whom we are indebted for their discovery.

This difficulty in tracing the origin of inventions is not less marked at the present day, when the increased rate of progress in all things brings about, in a few short years, a succession of changes, which, in olden times, centuries were required to effect; for the inventor of to-day is to-morrow overshadowed by the accumulated mass of improvements that follow in the wake of every new discovery.

I well remember how the world was startled by the great discovery of Daguerre;* how few minds could, at the first moment of its announcement, realise the wondrous fact that by the aid of chemistry combined with knowledge, he had seized upon and trapped the fleeting shadow on his silver plate and held it there immovable for ever.

* The production of Daguerreotype plates was announced on February 6th, 1839

The mind had scarce time to grasp the importance of this marvellous discovery before there commenced that ceaseless flow of inventive talent which, growing with years, has wholly submerged the original invention of Daguerre. Process succeeded process with immense rapidity. At every step new ground was covered; more beautiful and more permanent effects were almost daily produced by scientific investigators whose name was legion; until at last the glorious orb of day has taken over the business of the engraver, and daily produces its hundreds of deeply-etched blocks from which our common printing machines throw off their thousands of printed sheets with the same facility with which they print a page of common type. In the midst of these marvels of modern invention we look around and exclaim, "Where is now Daguerre?" and echo answers "Where?" Simply buried beneath the huge monument which, instead of being raised to his fame, has placed him out of sight and out of memory.

I have referred thus prominently to this great discovery ot Daguerre and its subsequent marvellous developments, not only because it made a deep impression on my youthful imagination at the time, but because I purpose making a somewhat extensive use of photography in illustrating the following pages, where its absolute truthfulness will afford indisputable evidence of some facts which would otherwise have been altogether omitted, rather than allow them to rest on the uncorroborated testimony of the writer. At the same time this beautiful art will serve to illustrate many existing objects, an equally realistic idea of which the most elaborate description would fail to impart.

It is to the rapid passing into oblivion of great inventions like that of Daguerre that I attribute the pressure of my kind friends who ask me to give them some account of my early life and its relation to the more immediate past, while yet the process which bears my name remains an existing fact among us, aud has not been engulfed in that ever-advancing tide of scientific knowledge and commercial enterprise which sweeps away the past and leaves us face to face only with the present.

So energetic in this matter was my friend, Mr. Price Williams, that some years ago he called on me with Mr. Samuel Smiles, LL.D., whose well-known talent as a biographer had all but tempted me to commit this task to him. We had a long consultation on the subject, but I could not feel that my life and its labours were a theme which could be treated in such a way as to make them interesting to the general reader, even when clothed in the beautiful language and charming style of that eminent writer. There were none of the exciting incidents of travel to relate : no hairbreadth escapes, no dangers by land and sea, to seize upon and captivate the imagination. Indeed, I could not help feeling that my daily pursuits were of too technical a character to supply the necessary materials to form an interesting book; and if the narrative were simply treated in the plain matter-of-fact style of which alone I was capable, I felt it would have inevitably failed to be of sufficient interest, either to the general reader or to the man of science. Thus the proposed biography was for the time abandoned.

Nevertheless, several of my friends have from time to time tried to induce me to write a concise account of my steel invention in my own quiet way. More especially was this view commended to my notice by my old friend Alexander Hollingsworth and his colleagues, the able editors of *Engineering*, William H. Maw and James Dredge. Thus it was in the year 1884 I found myself busily engaged in preparing large coloured drawings of the converting and other apparatus, and in the course of two or three months at least a dozen drawings were completed, from which photographic copies on a reduced scale were made on wood-blocks to illustrate the work I had just begun. At this time I was also engaged designing the whole of the machinery about to be erected by my grandson, William Bessemer Wright, at the new diamond mills in Clerkenwell ; and I became so deeply engrossed in working out the details of several experimental diamond-cutting machines which were in course of construction on my own premises at Denmark Hill, that by degrees my attention was gradually more and more drawn from the book I had commenced, and I became at last wholly absorbed in the more congenial work of construction going on every day in my workshop. Again the long-contemplated autobiography was laid aside, and I must confess that there always was in my mind an undercurrent of feeling averse to the task.

I have at all times keenly experienced the difficulty, which must necessarily confront an author when speaking of himself, and of what he has accomplished, of setting forth what I have done and what credit I am entitled to, without appearing to be self-assertive, and displaying a personal bias in relation to certain controversial matters into which I am obliged to enter. From this difficulty I see no way of escape without abandoning the work laid upon me by the importunity of my friends. I have, therefore, resolved to follow out rigidly the unenviable task of self-assertion, and not to shrink from fearlessly and truthfully claiming what is due to me, just as though I were speaking of some other person, whose advocate for the time I had constituted myself. And I shall, with equal candour, point out the persistent opposition and obstructive tactics to which my invention has been subjected in a few prominent cases ; while, on the other hand, I shall with pleasure place on record my grateful acknowledgments to those in the world of science who have honoured me by their kind appreciation : a gratitude which is also due from me to the many iron and steel manufacturers who have unreservedly acknowledged my patent-rights, and with rigid and scrupulous honour have fulfilled to the letter all their engagements with me.

Having thus entered upon a task so long deferred, I shall endeavour to make assured accuracy of historical detail take the place of literary ability, which I know but too well will be only conspicuous by its absence in these pages. Fortunately, I am in a position to review the past wholly uninfluenced by any mercantile considerations, having long ceased to possess pecuniary interests in the iron or steel manufacture; and having arrived at that late period of life when there is no desire for new worlds to conquer, and there are no strong ambitions to bias the mind and obscure the judgment.

The name of Bessemer does not sound like an English one, and has often given rise to doubts as to my nationality. I may therefore mention a few facts in relation to my father. He was born at No. 6, Old Broad Street, in the City of London, and at the age of eleven years was taken to Holland by his parents, who settled there. In due time he was articled to a mechanical engineer, and during his apprenticeship assisted in erecting the first steam-engine in Holland, this engine being employed in draining the turf pits near Haarlem.

After arriving at the age of twenty-one, my father went to Paris, and there commenced a career which did him much honour. At the early age of twenty-six he was made a member of the Academy of Sciences, as a reward for a great improvement he had effected in the microscope. He was at that period engaged in the Paris Mint, and while there invented that very simple and beautiful machine now known as the Portrait Lathe, by means of which medallion dies of any desired size can be engraved in steel from an enlarged model.

He was still residing in Paris at the time of the great French Revolution, and, as an active member of the Commissariat Department, he had to distribute a certain dole of bread and rice to the starving thousands, who formed a long *queue* for many hours every morning before the municipal bakery was opened. Everyone in Paris at that time felt the pinch for food. My father had a small estate some twenty miles out of town, and when he saw the probability of a famine, he had a few sacks of wheat taken to his house in Paris, and there secretly stowed away; for a knowledge of their presence would have brought the hungry mob upon him. It was my mother's task at night, when the household had retired to rest, to grind some of this wheat in a coffee mill, so that cakes might be made for the morrow's breakfast; and thus in secret my parents enjoyed the luxury of whole-meal bread of their own manufacture.

My father was most anxious to return to England, but it was very difficult to get away. He could obtain nothing from his bankers but the paper money then well known as *Assignats*, which were issued for amounts as low as fifty sous, or about two shillings in English value.

Fortunately a short lull occurred in those stormy times, and, taking advantage of the opportunity, my parents escaped to England, bringing with them about £6,000 in nominal value in *Assignats*, and only a very small sum in cash.

Arrived in London, my father had to begin the world over again; so availing himself of his intimate knowledge of the use of the stamping-press and dies, and the working of gold, he commenced the manufacture of gold chains of a novel and beautiful description. By using gold of a high standard of quality, and with the assistance of finely-executed steel dies for stamping each link, a splendid chain was produced, which appeared very massive while in reality it was very light. These chains were bought by the retail jewellers as rapidly as they could be made.

While this new branch of trade was going on satisfactorily, a great panic was created in London by a report that Napoleon was about to invade England in flat-bottomed boats, which were said to be then at Boulogne, prepared for the expedition. My father, who had lost all in Paris, was determined at this juncture to secure some solid property in his own country, and at once dispatched his traveller to collect all the money he could from his various customers. With this money he purchased a small landed estate in the village of Charlton, near Hitchen in Hertfordshire, to which he shortly afterwards retired, and where I was born on the 19th January, 1813.

My father's active business habits did not permit him to lead a life of idleness, and, after a year or two of quiet retirement, he commenced to cut letter-punches for Mr. Henry Caslon, the proprietor of the well-known Caslon type-foundry of London. The eminence my father had acquired in this art, while in the Paris Mint, enabled him to produce specimens of typography far more beautiful than any others that could be met with at that time. An immense accession of trade to the Caslon foundry resulted, and Mr. Henry Caslon became a frequent visitor at my father's house at Charlton; where, on one of these occasions, he acted as my godfather, and gave me the name of Henry.

Some years later, my father was joined in business by a former partner of Mr. Caslon's, and a type-foundry was built on our estate at Charlton. The knowledge of metal work which I acquired in this foundry, assisted, I doubt not, in fostering and developing that taste for casting and other metallurgical works in which, as an amateur, I took so deep and abiding an interest.

After leaving school, I begged my father to let me remain at home, and learn something of practical engineering. This he acceded to, and as a preliminary step he bought me one of those beautiful small slide-rest lathes, made by Messrs. Holtzapffel of London, and which are still produced in all their original excellence by that eminent firm.

After a year or two at the vice and lathe, and other practical mechanical work, my father allowed me to employ myself in making working models of any of the too-numerous schemes which the vivid imagination of youth suggested. Among these, I well remember, was a machine for making bricks, which was one of the most successful of my early attempts, producing pretty little model bricks in white pipeclay. I always had access to molten type-metal, which I used for casting wheels, pulleys, and other parts of mechanical models where strength was not much required. Hence arose various devices for moulding different forms, a matter that caused me very little trouble, for by some intuitive instinct modelling came to me unsought and unstudied. Often during my evening walks round the fields, with a favourite dog, I would take a small lump of yellow clay from the roadside, and fashion it into some grotesque head or natural object, from which I would afterwards make a mould and cast it in type-metal.

In this quiet village life there was a break every two months, when the large melting-furnace was used to make type-metal, in which proceeding a great secret was involved. In spite of injunctions to the contrary, I would, by some means or other, find my way into the melting-house, where large masses of antimony were broken up to form the alloy with lead. The dust arising from the powdered antimony, on more that one occasion, caused me severe sickness, and betrayed my clandestine visits to the melting-house, where I discovered that the addition of tin and copper, in small quantities, to the ordinary alloy, was the secret by which my father's type lasted so much longer than that produced by other typefounders.

There was, however, one other attraction in the village, which played a not-unimportant part in moulding my ideas at this very early period. I was very fond of machinery, and of watching it when in motion; and if ever I was absent from meals, I could probably have been found at the flour mill at the other end of the village, where I passed many hours, gazing with pleasure upon the broad sheet of water falling into the ever-receding buckets of the great overshot water-wheel; or, perhaps, I might have been watching, with a feeling almost of awe, the huge wooden spur-wheel which brought up the speed, and was one of the

wonders of the millwright's craft in those days. Its massive oak shaft
and polished horn-beam cogs have long since passed away, and yielded
to their successor, cast iron, which in its turn is now being rapidly
replaced by the stronger metal, steel, thus keeping up that ever-changing
cycle of advancement in the arts which is carrying us forward to
discoveries that may change every phase of civilised life, if the exhaustion
of our coal does not land us again into a state of barbarism.

I had now arrived at my seventeenth year, and had attained my
full height, a fraction over six feet. I was well endowed with youthful
energy, and was of an extremely sanguine temperament. At this period
of life all things seem possible if you have once made up your mind to
conquer, and not to allow any temporary disappointments to weaken
your resolution. The opportunity to put this beautiful theory to the
proof was about to be afforded to me, for my father had resolved to
remove his business to London, when I should have to change my
solitary country life, which had so many irresistible charms, for a totally
different one. I should see for the first time the great metropolis, about
which I had heard so much but knew so little.

On March 4th, 1830, I arrived in London, where a new world seemed
opened to me. I was overwhelmed with wonder and astonishment; all
the ideal scenes in the "Arabian Nights," which had held me spellbound
in my native village, were as nothing to the ceaseless panorama which
London presented, with its thousands of vehicles and pedestrians, its
gorgeous shops and stately buildings, and its endless miles of streets
and numerous squares. I was never tired of walking about, for every
turn presented some new object to rivet my attention; and in this way
I passed my first week's residence in London. I usually returned home
in the evening, greatly tired and worn out, only to go forth on the
morrow to make new explorations and again lose myself in those endless
labyrinths of streets; and yet, with all the delight inspired by the
novelty of the scene, there was one thing strange to me, and sadly
wanting. I felt that I was alone; no one knew me. I never met, in
all this excited rush, one human countenance that I could recognise, or
a friendly face to smile and give a passing salutation as in my old
home : where the little children on their way to school would drop a

curtsy and leave me the best side of the path, while the farm-labourer at his cottage door would give me "Good morning, Master Henry!" All this had passed away for ever, and here amidst the countless thousands I stood alone, as much uncared for as the lamp-post beside me. How often I thought, in those early days in London, "Shall I ever be known here? Shall I ever have the pleasure of seeing a smile of recognition light up the face of any person in these ceaseless streams of unsympathetic strangers?" The thought made me very sad, and at times sigh for the old home; but it has been truly said that "hope springs eternal in the human breast," and so I found the advantage of my sanguine temperament. "Why," I asked myself, "instead of pining after the old associations of my native village, should I not strive to make a name for myself, even in this mighty London? It is not impossible, for many others have done it, and I at least will make the effort." Such reflections as these enabled me to settle down again, and resume my old home occupations.

I knew full well that I laboured under the great disadvantage of not having been brought up to any regular trade or profession, but, on the other hand, I felt a consciousness that Nature had endowed me with an inventive turn of mind, and perhaps more than the usual amount of persistent perseverance, which I thought I might be able to use to advantage.

In the course of my ramblings I had met with an Italian, who had shown me several boxes full of plaster casts of the most beautiful medallions; real gems of art at one penny a-piece. I selected a number of them with the intention of casting them in metal, an occupation in which I took a deep interest at that time. But the moulding and casting of more intricate objects had even a much greater charm, and I began to try my hand on the reproduction in metal of natural objects, both vegetable and animal. For this purpose the article to be cast was immersed in a semi-fluid composition, of which plaster-of-Paris formed the base. The mould was gradually dried and then made red-hot, and the object was thus destroyed. An opening into the mould on one side allowed the ashes to be removed, and gave entrance for the metal of which the object was to be formed. In this way a rosebud or other

c

flower, with its stalk and leaves, could be produced; but, alas! the whole of those thin, delicate leaves were destroyed in attempting to break away the mould. Some of the fragments were exquisitely beautiful, but no entire cast could be obtained.

All sorts of schemes were tried, and tried in vain, until, when on the eve of abandoning the whole affair as impossible, I hit upon the happy idea of using unburned blue-lias limestone ground to a fine powder. This, and the dust of Flanders brick, with a small quantity of plaster-of-Paris, formed the mould; the destruction of the succulent vegetable, by making the mould red-hot, had also the effect of burning the limestone portion of the composition, while the brickdust served to destroy much of the cohesive strength of the plaster-of-Paris, the hardness of which had proved so great an obstacle in extricating the casting. When the mould had cooled down, all that was required to get out the casting was to apply cold water to it, when the burnt lime slaked, became hot, and fell away from the cast. A sharp jet of water from a tap on the main service sufficed to wash out all the small particles from the deep recesses, and liberate the casting perfect and unbroken. I prepared for this purpose an alloy of antimony, iron, bismuth, and tin, and in all cases made the mould with a very tall gate or runner, keeping it red-hot for half an hour after the metal was poured into it. In this way the static pressure of the metal which remained fluid forced the air slowly but surely through the pores of the mould, and occupied every minute cavity; so that the fine pile on the back of a leaf and the tiny prickles on the stem of a rose were all produced as sharp as needle-points.

The love of improvement, however, knows no bounds or finality. Beautiful as these representations of nature were, there was one great drawback which I still desired to surmount. They were only white metal, and were sometimes looked upon as merely "lead castings."* I therefore attempted to cast them in brass or yellow metal, but this I found was impossible. I then conceived the idea of coating them with a deposit of copper from an acid solution of that metal. Many were the trials

---

* *Vide* Dr. Ure's *Dictionary of Arts, Manufactures, and Mining*

and failures in these attempts, but after a time I made more suitable solutions, and found out how to cleanse the surfaces of the delicate objects without injuring them ; and finally I succeeded in getting a beautiful thin coating of copper on every part of the surface. The castings were simply laid on the bottom of a shallow zinc tray, and a saturated solution of sulphate and of nitrate of copper, in certain proportions, was poured into the bath, which resulted in producing a thin coating of bright metallic copper over the entire surface of the castings, so that no suspicion could be entertained as to the metal of which they were really formed. In the case of medallions I sometimes put into the solution some crystals of distilled verdigris, which produced a good imitation of antique bronze. Several specimens of these bronzed medals and copper-coated castings of natural objects were exhibited by me at Topliss' Museum of Arts and Manufactures, which at that time occupied the present site of the National Gallery in Trafalgar Square. Among the things I exhibited there were a basso-relievo of one of the cartoons of Raphael, a large medallion head of St. Peter, and several smaller casts of medals. I also exhibited a group of three prawns lying on a large grape-vine leaf, a moss-rose bud with leaves, and a beautiful piece of Scotch kale, the intricate convolutions of which appeared to all who saw it a thing impossible either to mould or cast, but which was nevertheless a comparatively easy one, because this vegetable leaf is very thick and succulent, and consequently leaves scarcely any ash in the mould when burned.

I may mention that various devices were tried to get rid of the fine ash resulting from the burned vegetable matter.

Sometimes small passages open to the outer air were left in the mould, and into these a blast of air was blown to assist the combustion and destruction of the vegetable matter while still in a red-hot state. At other times the mould, when cooled down, was filled with a strong solution of nitre, which saturated the dried vegetable matter. The remainder of the fluid was then poured out and the mould again made red-hot, when the nitre, causing complete combustion, reduced the contents to a fine white ash. When the mould had again cooled down, the ash so formed was floated out of it, by pouring mercury in and well shaking it.

In fact, the treatment resorted to for cleansing the mould had to be adapted, in each case, to the nature of the object to be destroyed and got rid of.

I had a strong belief that the mode I have described, of reproducing the most delicate and, at the same time, the most intricate vegetable forms, might be utilised by botanists and other collectors, in remote or solitary places, from whence the transmission of such objects in their natural state would be impossible. It would be perfectly easy for the botanist to take abroad with him a few tin cans filled with the dry powdered materials required for his moulds, ready to be mixed with water at a moment's notice. A number of small cardboard boxes, painted in oil colour so as to render them waterproof, and fitting inside each other, would enable him to choose one suitable in size, for any particular specimen to be moulded in. He would have nothing to do but to mix with water a small quantity of his prepared plaster, place the delicate fungus, lichen, or other specimen, in the bottom of the box, and pour in the semi-fluid mixture, filling the box, and gently tapping its sides and bottom to ensure the penetration of the fluid matter into every interstice of the specimen. In less than a quarter of an hour, he would find in his fragile little box a hard, solid, square mass, in which the specimen would be safely embedded, where it might remain uninjured for any necessary period, and then be burnt out, and the object reproduced in metal. An absolutely perfect copy of nature's most beautiful work, in an indestructible material, would thereby be obtained by a minimum of labour and cost.

I made many attempts to impress the importance of these facts on some of the managers of the British Museum, with whom I had several interviews, but all to no purpose; and so the whole thing dropped, and I had all my trouble in vain.

Returning from this digression, I may state that the site occupied by Topliss' Museum was required for the erection of the present National Gallery.* A museum was, however, erected in Leicester Square, where the Panopticon was subsequently built, and it was to this new home that my specimens of casting were removed.

* Opened April, 1838

About the year 1836 I made the acquaintance of Dr. Ure, author of the well-known *Dictionary of Arts, Manufactures, and Mining*, and of him I can only speak with affectionate regard. I had sought his assistance in some analyses, and after several interesting interviews he furnished me with the information I desired, and seemed to take so much interest in me that I was induced to give him a short sketch of my isolated country life. I described to him my love of experiments in casting, and how I made use of the continued statical pressure of a high column of metal, retained in a fluid state for a considerable time, in a red-hot mould. He was much interested in my founding operations, and generously declined to take any fee for the analysis he had made for me, saying, if I wanted any further analyses he would be happy to do them for me. Needless to say how pleased and grateful I felt for his disinterested kindness, and the encouragement which his appreciation gave me.

With the exception of three or four samples still remaining in my possession, all these beautiful castings have been sold or given away many years ago.

Among those I still have is one of the cartoons of Raphael, representing the "Woman taken in Adultery," a plaster-of-Paris copy of which I bought for a few pence from an itinerant Italian. It was fairly sharp and perfect in detail, and I planed some strips of type-metal into an ornamental moulding, so as to form a frame around it. I then made a mould from it as framed, and took a cast in a white metal alloy, which I afterwards coated with a thin film of copper, in the manner already described.

It is now in much the same condition as it was when cast, with the exception of the loss of the more prominent features, caused by the continued rubbing and dusting of the housemaid—or rather, I may say, of a succession of housemaids, who during the last sixty-four years have gradually wiped away, not only the copper film from the projecting parts, but a noticeable quantity also of the soft metal of which it is made : as will be at once seen on examining the photographic reproduction, Fig. 1, Plate I., which is a full-sized representation of it.

I also give, in Fig. 2, Plate II., a copy of an oval medallion, the numerous figures on which were originally very sharp and perfect, but it has suffered somewhat by rubbing and dusting from time to time during these many years.

These medallions were, as Dr. Ure says in his article on Electro-Metallurgy, simply used as "mantel-piece ornaments," and I doubt not that this oval medallion was one of those I had shown to him in 1836, and had made in 1832 when I was about nineteen years of age. This perfect casting was, like the cartoon, coated with a thin film of copper, giving it the appearance of being cast in that metal.

ELECTRO-METALLURGY.   629

rounded by a cylinder of zinc, and then introduced into another vessel (a wooden tub for instance) containing dilute sulphuric acid. The earthen vessel is intended to contain the solution of gold or silver, and is furnished with a web of copper wire, which is made to communicate with the zinc by means of one or more conducting wires. The objects to be gilt or silvered are placed upon the net-work. The earthen vessel containing a zinc cylinder, and some hydrochloric acid, is introduced into another vessel, containing the solution of gold or silver, placed in the centre of a wire web partition, which communicates with the zinc cylinder by means of a conducting wire. In the first case, the articles which are to receive the thickest coating are placed nearest the outer sides of the apparatus; in the second, nearest to the earthen vessel : in both cases it is advisable to shift their position occasionally. By combining these different arrangements, the deposit obtained is more abundant, and more equally distributed upon the surface to be gilded or to be silvered. For this purpose an opening is made in the centre of the web in which the zinc cylinder is inserted, with connecting wires to the web. When the articles to be operated upon can be easily suspended from a given point, the web of the apparatus may be made with wider meshes, and the articles suspended vertically between them. Dr. Philipp prefers a single galvanic arrangement to a battery, as it affords more solid deposition.

ELECTRO-METALLURGY. By this elegant art perfectly exact copies of any object can be made in copper, silver, gold, and some other metals, through the agency of voltaic electricity. The earliest application of this kind seems to have been practised about 16 years ago, by Mr. Bessemer, of Camden Town, London, who deposited a coating of copper on lead castings, so as to produce antique heads in relief, about 3 or 4 inches in size. He contented himself with forming a few such ornaments for his mantelpiece ; and though he made no secret of his purpose, he published nothing upon the subject. A letter of the 22d of May, 1839, written by Mr. J. C. Jordan, which appeared in the *Mechanics' Mag.* for June 8, following, contains the first printed notice of the manipulation requisite for obtaining electro-metallic casts; and to this gentleman, therefore, the world is indebted for the first discovery of this new and important application of science to the uses of life. It appears that Mr. Jordan had made his experiments in the preceding summer, and having become otherwise busily occupied, did not think of publishing till he observed a vague statement in the Journals, that Professor Jacobi, of St. Petersburg, had done something of the same kind. Mr. Jordan's apparatus consisted

FIG. 3. EXTRACT FROM DR. URE'S DICTIONARY, "ELECTRO-METALLURGY"

When the Doctor published a supplement to his *Dictionary* in 1846, he referred to these medallion castings, under the head of "Electro-Metallurgy," as having to his knowledge been cast in " lead " *(sic)* and coated with copper, about ten years previously : that is, about five or six years prior to the discovery of the electrotype process, by Jacobi of St. Petersburg, Jordan of London, and Spencer of Liverpool. This process was afterwards perfected by Dr. Wright, and Messrs Elkington, of Birmingham, to whose joint labours we owe the practical development of the beautiful art of Electro-Metallurgy, to which I had "approached within measurable distance," but of which I had nevertheless failed to recognise the full importance, excepting as a means of producing artificial bronzes by coating the cheaper white metal castings, as is now largely practised in France, in imitation of bronze for clocks, etc. In order that there shall be no misapprehension as to what Dr. Ure has said on the subject, I give on the opposite page a photographic reproduction of the upper part of page 629 (Electro-Metallurgy), in the Fourth Edition of his *Dictionary*, published by Longman, Brown, Green and Longman, the title-page of which bears date 1853. This extract clearly shows that I practised the art of depositing copper from an acid solution of that metal on the surface of ornamental castings, *several years before* we had any known or published account of that process.

It is quite true, as the Doctor states, that I kept some of these medallions as ornaments on my mantel-piece, where three of them may be found at this very day; but Dr. Ure appears to have forgotten, or possibly was unaware of the existence of, the beautiful specimens of natural objects cast in white metal and coated with copper, which I exhibited in Trafalgar Square, and afterwards in Leicester Square, and which had the effect of bringing me in contact with several large business firms, and in one case resulted in the development of an entirely new and important branch of the Utrecht velvet manufacture, to which I shall have occasion hereafter to refer.

There is yet another description of casting known as the "lost wax" process, which was at that time practised in France; and I was anxious, if possible, to acquire this art, as it seemed to offer greater facility for obtaining white metal casts of busts and statuettes from wax models that were cast in plaster-of-Paris moulds. To the metal casting so obtained, the appearance of real bronze could be imparted by depositing a green copper coating thereon. By doing this the "lost wax" process need not have been confined, as it then was, to the production of original works of art modelled in wax; and the effect would have been to immensely facilitate the multiplication of copies of the highest examples of classic art, by simply obtaining thin wax casts of them, in lieu of the plaster

casts sold so cheaply in the streets by itinerant Italians. The loss
of the wax model so produced forms only a very small part of the cost
of the process.

All persons conversant with the ordinary mode of casting a bust
or statuette must be aware that the mould is formed of a great number
of small pieces, more or less perfectly fitted together, and that the metal
in casting will run into all the minute joints or cracks which lie between
the numerous parts of which the mould is composed, forming little ribs
or "fins," which cross the face and other portions whereon the talent
of the artist who prepares the model is chiefly expended. The removal
of these ribs or fins, with chisels or files, must be done by a workman
after the cast is made, and this greatly interferes with the delicate
touches of the artist, and not unfrequently mars his work. In the
"lost wax" process, the artist may put on his finest touches, may
"undercut" as much as he pleases, or form intricate hollows to any
extent, which he could not do if the mould had to be made in pieces;
when the model is complete in wax, a mould can be made over it in one
piece, after which the wax original is melted out or "lost." Molten
metal is then poured into the space previously occupied by the wax,
and a cast is produced, absolutely identical with the model. Every
delicate touch of the artist is there, free from the fins and ribs so
inseparable from casts made in a mould which is built up in pieces. But
here lies the difficulty and great risk of this process. The artist's model
is irretrievably destroyed, and if a bad casting should result, all his
labour is lost.

How well I remember the heartbreaking disappointments that beset
all my early attempts to cast from the lost wax in plaster-of-Paris
moulds. For this was the very great desideratum for which I was
striving. The plan was to first carefully dry the massive plaster envelope
in which the wax model was embedded, and then to put it into a
stove heated sufficiently to melt the wax, which, if the mould was
inverted, would run out.

The difficulty in obtaining a good casting arose from the plaster
mould absorbing a small portion of the wax during the melting process,
so that when the molten metal was poured into the mould, the wax so

absorbed and retained in its pores was converted into gas, which bubbled up through the metal, and made a most unsound and imperfect casting.

Over and over again I essayed to prevent this result, but all to no purpose, and I almost gave it up in despair. I pondered over many schemes to remedy this defect, when at last it occurred to me that there was only one way that must succeed. Plaster-of-Paris very quickly sets, and gets hard and firm while it is quite saturated with water ; and it seemed probable that if I kept the mould in this saturated condition instead of drying it, the melted wax could not be absorbed by it. After several trials I found it advisable to render the wax a little more fusible by the addition to it of a small quantity of animal fat, and then it was quite easy to melt out the wax by simply immersing the mould in a caldron filled with boiling water. The wax was melted by the heat of the water, and floated up to the surface, allowing the water to take its place. As soon as the melted wax ceased to rise to the surface the mould was taken out of the bath, emptied of the water which had taken the place of the melted wax, and slowly dried ; after this its temperature was raised to a point sufficiently high to allow the metal to be retained in it in a fluid state for some time, thus ensuring a perfect cast. By this simple device there was no absorption of the wax, and consequently no gas produced in the mould, and no longer any fear that the model would be lost without getting a perfect cast in return.

A fine bust of Shakespeare which I produced in this way, and coated with copper, was purchased of me by an eminent sculptor, who saw in this simple plan a means of getting faithful copies of his works uninjured by the chipping and filing of a mere mechanic. Indeed, it was finally arranged that I should cast for him a bust of the Hon. George Canning, for which he had received a commission ; but, unfortunately, while engaged in modelling this bust, he was seized with a sudden illness which terminated fatally, and the work was never completed.

My attention was at this period directed to the production of castings suitable for stamping ornamental scroll-work, medallions, and *basso relievos* in cardboard. This was a much more difficult subject to deal with than casting in white-metal alloys, and required moulds of quite a

D

different character. It was, however, mainly a question of mixing metals so as to produce great hardness, while absolutely free from brittleness, and in this manner to obtain an alloy that would melt at a comparatively low temperature, and run very fluid in the mould. All of these conditions to their fullest extent could not be combined in any one alloy ; but after a few months expended in making a systematic series of experiments, I succeeded in obtaining a die metal that pretty closely approximated to all the desired requirements, and I also found a "facing" for the moulds, which stood the heat of these hard alloys without suffering any destructive action or the formation of surface cracks. I could thus form moulds capable of taking an impression of the finest and most delicate lines, and in these I succeeded in casting many works of art in brass.

After a certain amount of practice, I produced a great many very beautiful dies, from which thousands of fine sharp impressions were made. I erected a powerful "fly-press" for stamping impressions from these dies, and thus achieved what was in reality my first commercial work. It will be easy to imagine my delight on securing a first order for 500 copies on buff-coloured cardboard of a beautiful *basso-relievo* of one of the cartoons of Raphael, from Messrs. Ackerman, the well-known art publishers. These impressions cost me only threepence each, including the material, and I found a ready sale for them at half-a-crown, when taken in wholesale quantities. They must still exist in many families, for hundreds were stamped in leather on the covers of a beautifully got-up quarto edition of the Bible, the cartoon chosen being the one in which Raphael represents Our Saviour giving the keys to St. Peter. I also made a great many dies in this way for bookbinders, cardboard-box manufacturers, etc., thus turning to commercial account the art of "fine-casting," which I had heretofore only practised as an amusement.

A year or so previously, 1831, I made the acquaintance of a Mr. Richard Cull, then a youth about my own age, who, when I last heard of him, had become a noted philologist and a member of the Antiquarian Society. My friend Cull was a great admirer of these beautiful cast dies, and we very nearly entered into a deed of partnership, with the intention of carrying on this business on a greatly extended scale ; but for some reason or other this intention was never carried out.

CHAPTER II

THE REWARD OF INVENTION

WHILE this die-making and stamping business was going on, I had discovered another and distinctly different mode of making, from an embossed paper stamp, dies which were capable of reproducing thousands of facsimile impressions. I at once saw to what a dangerous result this discovery might lead if made known to unscrupulous persons, and hence I carefully guarded the secret, which was, in fact quite useless to me, and might soon have been forgotten had not my attention been directed by some accidental circumstance to the fact that the forgery of stamps to an alarming extent was known by the Government to have been practised.

One of these sources of fraud was the removal from old and useless parchment deeds of stamps, which were again stuck on to new skins of parchment. Thinking over this subject, it struck me that a stamp might be made which it would be impossible to transfer from one deed to another, and at the same time would be much more difficult to produce by the stamping press; while it would be impracticable to obtain from it a die that would be capable of reproducing the stamp.

This appeared to me to be a most important invention, and one that I conceived it would be impolitic for the Government to reject; I supposed that I should be handsomely rewarded if I brought it under the notice of the authorities. I felt the more certain of success because I was able to show that their ordinary receipt and bill stamps, as well as the blue paper adhesive stamps on parchment deeds, could be forged by any office-boy, who could make a die from a paper stamp for a few pence, wholly without talent or technical knowledge.

Thus confident of success, I set to work to make a die for parchment deeds on my new plan, for the time putting aside and neglecting everything else, for this grand project was to make my fortune at once. After providing myself with a suitable press and experimenting with different forms of cutting punches, I decided on a plan. Having worked for some months, making long days which not unfrequently extended to the early hours of the morning, the task was finished, and I prepared some specimens to take with me to Somerset House.

With the idea of showing that there was no escape from the adoption of my new plan, I thought it advisable to make a die from a genuine Government stamp. For this purpose, I obtained a dozen ordinary embossed bill stamps, and from one of them made a die, and stamped about as many impressions with it as I had real stamps. In order that I might be able to prove that these were forged ones, I stamped the impressions on a large sheet of paper, and then cut out a slip from it with a slightly indented edge, but otherwise of the same form and size as those I had purchased. I may mention that the Stamp Office presses were so constructed that they could not put a stamp in the middle of a large sheet of paper, and hence I was enabled to prove that these particular stamps, with their slightly-indented edges, did not emanate from the Stamp Office. I made up a small parcel containing six genuine stamps and six of those I had myself made, and also the sheets of paper from the centre of which they had been cut. With these I also enclosed a few impressions of my new parchment stamp. The old form of Government stamps is illustrated in Fig. 4, Plate III., while Fig. 5 illustrates my perforated stamp. Below is shown a system introduced a few years ago for cancelling cheques and other documents.

Full of hope and high expectation, I started off one morning to call on Sir Charles Presley, the then President of the Stamp Office. I had, up to this moment, kept all my plans and what I was doing a profound secret. The whole affair seemed to my overwrought imagination almost like a skilful plot, such as we see depicted on the stage or read of in a sensational novel; and I had, like the hero of the piece, only to walk into Somerset House and accept unconditional surrender.

On my way to the scene of my intended conquest I passed up Farringdon Street, and went into a fruiterer's shop at the corner of the New Market to buy an orange. How vividly I still remember this trifling incident in all its details. I ate my orange as I went jauntily up Fleet Street, thinking of nothing but how I should introduce the subject, what they would say, and how I should go through the ordeal I had to face, on the results of which depended all my dearest earthly hopes. Had I not in the silent hours of night, when I was pursuing my experiments, and wearily working at these new dies, told myself triumphantly : " A few more weeks will seal the fate of my whole life. If I succeed in saving the Government so much revenue, they must liberally reward me. I shall then establish myself in a new home, and marry the young lady to whom I have for two years been engaged." I had needed no stronger incentive to urge me onward as the lonely hours of night found me engaged in the laborious work of making these dies. I now felt that the task was over, and that I was well on the road to my reward; but suddenly my day-dream came to an end, for just as I approached Temple Bar I discovered that I was not in possession of my parcel of stamps. I was staggered for a moment, and a cold perspiration seemed to break out all over me. I felt faint and alarmed, for in a second I began to fully realise the fact that I had actually been possessed of forged stamps, and had left them on the counter of the shop where I had bought the orange. The little paper parcel was not sealed. What if curiosity had caused it to be looked into and handed over to the police ? It was but a momentary hesitation, for I knew well that I was innocent of all intentional wrong, though perhaps not technically so, and I hastened back with all speed to the fruiterer's shop.

"Did you," I asked, " see a small parcel left here by me half an hour ago ?"

"Oh, yes, sir," was the reply. "I put it on the shelf, thinking you would come back for it."

How gladly I once more grasped it, and felt that I was now safe, even from a momentary suspicion. I own that I was a little crestfallen and unnerved ; but a sharp walk soon restored my confidence, and I entered Somerset House with a firm step and full faith that I should succeed in my mission. I was admitted into the private office of Sir Charles Presley, and said that I desired him to tell me if a dozen receipt stamps, which I handed him, were genuine. He looked at them attentively with a large magnifying glass, and laid two aside which he thought were not genuine. As far as I can remember exactly what passed, I said there were more forgeries among them, when he enquired, "How do you know that?"

I answered: "Simply because I forged them myself."

I could not quite suppress a smile as I said this somewhat triumphantly, and I distinctly remember his severe frown, as he said: "Young man, you treat this subject with a great deal of levity."

I at once apologised, and assured him that my object was solely to prevent all future forgery of stamps, and that I had ventured to test his experienced eye in order that he might himself appreciate the full danger to the State if my system were publicly known; unless, indeed, some remedy could be suggested for the prevention of further forgery.

As my scheme was unfolded he gradually relaxed that severe expression of countenance which plainly evinced that he felt annoyed at being tricked by a youth in so bold a manner, and the importance he evidently attached to my communication was manifested by his request that I would call again in a few days.

I may here briefly state that one of the plans I brought before the Stamp Office authorities was adopted by them, and has been to this day employed as a security against forgery on every stamp issued by the Stamp Office during the last half century; but I was nevertheless pushed from pillar to post, and denied all remuneration for the important services I had rendered. I was too busy making my way in life at this period to press any legal claims on the Government. I had no friend at Court, and had to bear this shameful treatment as best I could; and so, this matter of the stamps sunk gradually into oblivion until the year 1878, when my angry feelings against the Government were again excited by their refusal to allow me to accept the Grand Cross of the Legion of Honour, which the French Government desired to present me with, provided that the British Government would permit me to wear it. The failure of all attempts to get this permission aroused my just indignation, and I, as so many aggrieved persons have done before me, and doubtless will do again, wrote a letter to *The Times*. As this letter has played a not-unimportant part in my life's history, I think it desirable to insert it in this place, although it is not in the chronological order of events.

I may, however, say that I no sooner saw this letter in print than it occurred to me that an *ex parte* statement of so grave a character against the Government in general, and some of its officials in particular, demanded at my hands some documentary or other proof of the truth of the statements thus publicly made, and that I ought to lay the whole matter before the Government of the day in justice to myself. With this object I determined to address myself to our then Prime Minister, Lord Beaconsfield, and also to furnish printed copies of this communication to each of the other Ministers of State. The following is a *verbatim* copy of a portion of these communications, as well as of my letter to *The Times* :—

To the Right Hon. the Earl of Beaconsfield.

Denmark Hill,

November 16th, 1878

My Lord,

Under a feeling of some irritation, excited by recent events in connection with the Paris Exhibition, I felt impelled to relieve my mind of a long-suppressed grievance which my excessive dislike to controversy has hitherto prevented me from making public.

Under these circumstances I addressed a letter to *The Times* on the "Reward of Invention," which was published in that journal on the 1st November, 1878, a *verbatim* copy of which is embodied in this communication, and, as you will see, brings a very grave charge against some of the executive of a former Government; and, after perusing it in print, I saw at once that it was due to my own honour, and but fair to the Government, that I should bring forward some evidence in corroboration of the serious allegations therein contained, the more so as the public press have warmly espoused my cause, and commented in not very measured terms on the treatment I had received at the hands of the Government of that day.

No sooner, however, did the desirability of such corroborative evidence present itself to my mind than I took the necessary measures to acquire it ; and notwithstanding the length of time that has elapsed since these events took place, I have succeeded in obtaining the most unimpeachable testimony in support of the charge brought by me against the British Stamp Office ; but prior to bringing these proofs before the public, I have deemed it a duty which I owe, alike to myself and to the State, to bring the whole subject under the individual attention of each one of Her Majesty's present Cabinet Ministers ; hence I have forwarded a copy of this letter separately addressed to each of them.

As far as my experience of the great commercial transactions of this country extends, I have found that in every instance where a firm takes in a new partner, and in every change of the directors of a railway, bank, or other public institution, all those who have been elected to administer these great establishments have ever held inviolate the engagements of those whose position they have been called upon to occupy; nor can I for one moment doubt but that Her Majesty's Ministers will feel themselves equally bound in honour, if not to carry out the letter of the engagements entered into with me by their predecessors, at least to make such reparation and acknowledgment of my services to the State as will be both satisfactory to me and honourable to themselves, for I cannot believe it possible that my just claims will be repudiated by the British Government, and that its present Ministers will plead the Statute of Limitations as a sufficient bar to them; for this, after all, would be but to reduce it to a simple debt of honour, a form of obligation which it has ever been the pride of Englishmen to regard as their most sacred bond; and you will, I hope, pardon me when I confess that I cannot but coincide in the opinion so pithily expressed at the close of a leader in an influential journal,* viz., that " The Rulers of the State at the present day must be held to have inherited the responsibility of rendering to Mr. Bessemer the reward of the services by which they and the country have so largely profited."

In order that you may fully understand and appreciate the value of the evidence which I have the honour to lay before you, I must beg the favour of your perusal of my letter on the " Reward of Invention," in which you will find a detailed account of my transactions with the Stamp Office, and on which my present claims are based.

The following is a *verbatim* copy of that letter :—

The Reward of Invention.
To the Editor of *The Times*.

Sir,

The letter which you favoured me by publishing last week in relation to the refusal of our Government to allow the Grand Cross to be accepted by our countrymen, has elicited many kindly and sympathising expressions from private correspondents; but to the mind of one gentleman I appear to have written "with some bitterness." Now, I may plead guilty to such feeling whenever my memory is driven back by force of circumstances to a period when the Government of this country inflicted on me a great and grievous injustice in exchange for a great and permanent benefit conferred by me on the State.

Perhaps nothing would tend so much to dispel this morbid feeling as a brief recital of the circumstances to which I refer.

The facts are briefly these :—At the age of seventeen, I came to London from a small country village, knowing no one, and myself unknown, a mere cypher in this vast sea of human enterprise. My studious habits and love of invention soon gained for me a footing, and at twenty I found myself pursuing a mode I had invented of taking copies from antique and modern *basso-relievos* in a manner that enabled me to stamp them on cardboard, thus producing thousands of embossed copies of the highest works of art at a small cost. The facility with which I could make a permanent die, even from a thin paper original, capable of producing a thousand copies, would have opened a wide door to successful fraud if my process had been known to unscrupulous persons; for there is not a Government stamp or the paper seal of any corporate body that every common office-clerk could not forge in a few minutes at the office of his employer or at his own home. The production of a die from a common paper stamp is the work of only ten minutes; the materials cost less than a penny. No sort of technical skill is necessary, and a common copying-press or letter-stamp yields most successful copies. There is no need for the would-be forger to associate himself

* *The Times*

with a skilful die-sinker capable of making a good imitation in steel of the original, for the merest tyro could make an absolute copy on the first attempt. The public knowledge of such a means of forging would, at that time, have shattered the whole system of the British Stamp Office, had I been so incautious as to allow a knowledge of my method to escape. The secret has, however, been carefully guarded to this day.

No sooner, however, had this fact dawned on me than I began to consider if some new sort of stamp could be devised to prevent so serious a mischief. During the time I was engaged in studying this question, I was informed that the Government were themselves cognisant of the fact that they were losers to a great amount annually by the transfer of stamps from old and useless deeds to new skins of parchment, thus making the stamps do duty a second or third time, to the serious loss of the Revenue. At a later date, this fact was confirmed by Sir Charles Presley, of the Stamp Office, who told me that he believed they were defrauded in this way to the extent of probably £100,000 per annum. To fully appreciate the importance of this fact, and realise the facility afforded for this species of fraud by the system then in use, it must be understood that the ordinary impressed or embossed stamp, such as is employed on all bills of exchange, if impressed directly on a skin of parchment, would be entirely obliterated if the deed be exposed for a few months to a damp atmosphere. The deed would thus appear as if unstamped, and therefore invalid. To prevent this, it has been the practice as far back as the reign of Queen Anne, to gum a small piece of blue paper on to the parchment; and to render it still more secure a strip of metal foil is passed through it, and another piece of paper with the printed initials of the Sovereign is gummed over the loose ends of the foil at the back. The stamp is then impressed on the blue paper, which, unlike parchment, is incapable of losing the impression by exposure to a damp atmosphere. But, practically, it has been found that a little piece of moistened blotting-paper applied for a whole night so softens the gum that the two pieces of paper and the slip of foil can be removed from the old deed most easily and applied to a new skin of parchment, and thus be made to do duty a second or third time. Thus the expensive stamps on thousands of old deeds of partnership, leases and other documents, when no longer of value, offered a rich harvest to those who were dishonest enough to use them.

With a knowledge of these facts I was enabled to fully appreciate the importance of any system of stamps that would effectually prevent so great a loss to the Government; nor did I for one moment doubt but that Government would amply reward me if I were successful in so doing. After some months of study and experiment—which I cheerfully undertook, although it interfered considerably with the pursuit of my regular business, inasmuch as it was necessary to carry on the experiments with the strictest secrecy, and to do all the work myself during the night after my people had left work—at last I succeeded in making a stamp that satisfied all the necessary conditions. It was impossible to remove it from one deed and transfer it to another. No amount of damp, or even saturation with water, could obliterate it, and it was impossible to take any impression from it capable of producing a duplicate.

I knew nothing of patents or patent law in those days, and if I had for a moment thought it necessary to make any preliminary conditions with Government, I should have at once scouted the idea as one utterly unworthy. Dealing direct with Government, I argued, must render my interest absolutely secure; and in this full confidence, I wended my way one fine morning to Somerset House, and was ushered into the presence of the chief, Sir Charles Presley. I explained the object of my call, and showed him numerous proofs in my possession: how easily all his stamps could be forged, and also my mode of prevention. He was greatly astonished at what I had communicated and shown to him, and asked me to call again in

E

a few days, which I did, and after further conversation on the subject he suggested that I should work out the principle of my invention more fully. This I was only too anxious to do ; and some five or six weeks later, I called on him again with a newly-designed stamp, which greatly pleased him. The design was circular, about $2\frac{1}{2}$ inches in diameter, and consisted of the garter, with the motto in capital letters surrounded by a crown. Within the Garter was a shield, with the words "Five Pounds." The space between the shield and the Garter was filled with network in imitation of lace.* The die had been executed in steel, which had pierced the parchment with more than four hundred holes, each one of the necessary form to produce its special portion of the design. Since that period, perforated paper has been largely employed for valentines and other ornamental purposes, but was previously unknown. It was at once obvious that the transfer of such a stamp was impossible. It was equally clear that mere dampness could not obliterate it ; nor was it possible to take any impression from it capable of perforating another skin of parchment.

The design gave great satisfaction, and everything went on smoothly ; Sir Charles again consulted Lord Althorp, and the Stamp Office authorities determined to adopt it. I was then asked if, instead of receiving a sum of money from the Treasury, I should be satisfied with the position of Superintendent of Stamps, at some £600 or £800 per annum. This was all I could desire, and great was my rejoicing at the prospect before me, for I was at that time engaged to be married, and my future position in life seemed now assured. A few days after affairs had assumed this satisfactory position, I called on the young lady to whom I was engaged (now Mrs. Bessemer), and showed her the pretty piece of network which constituted my new parchment stamp. I explained to her how it could never be removed from the parchment and used again, mentioning the fact that old deeds with stamps on them dated as far back as the reign of Queen Anne could be fraudulently used, when she at once said, "Yes, I understand this ; but surely, if all the stamps had a date put on them they could not at a future time be used again without detection ?" This was, indeed, a new light, and I confess greatly startled me, but I at once said the steel dies used for this purpose can have but one date engraved upon them. But after a little consideration I saw that moveable dates were by no means impossible ; and shortly afterwards it came into my mind that this could easily be effected by drilling three holes of about a quarter of an inch in diameter in the steel die, and fitting into each of these openings a steel plug or type with sunk figures engraved on their ends, giving on one the day of the month, on the next the month of the year, and on the third circular steel type the last two figures of the year. I saw clearly that this plan would be most simple and efficient, would take less time and money to inaugurate than the elaborate plan I had devised ; but I must confess that while I felt pleased and proud at the clever and simple suggestion of the young lady, I saw also that all my more elaborate system of piercing dies, the result of months of study, and the toil of many a weary and lonely night, was shattered to pieces by it, and I more than half feared to disturb the decision that Sir Charles Presley had come to as to the adoption of my perforated stamp ; but with my strong conviction of the advantages of my new plan I felt in honour bound not to suppress it, whatever might be the result. Thus it was that I soon found myself again closeted with Sir Charles at Somerset House, discussing the new scheme, which he much preferred, because he said all the old dies, old presses, and old workmen could be employed, and there would be but little change in the Office ; so little, in fact, that no new Superintendent of Stamps was required, which the then unknown art of making and using piercing dies would have rendered absolutely necessary. After due consideration my first

* See engraving of this stamp, Fig. 5, Plate III

plan was definitely abandoned by the Office in favour of the dated stamps, with which everyone is now familiar. In six or eight weeks from this time, an Act of Parliament was passed calling in the private stock of stamps dispersed throughout the country, and authorising the issue of the new dated ones.

Thus was inaugurated a system that has been in operation some forty-five years,* successfully preventing that source of fraud from which the Revenue had so severely suffered. If anything like Sir Charles Presley's estimate of £100,000 per annum was correct, this saving must now amount to some millions sterling; but whatever the varying amount might have been, it is certain that so important and long-established a system as that in use at the Stamp Office would never have been voluntarily broken up by its own officials except under the strongest conviction that their losses were very great, and that the new order of things would prove an effectual barrier to future fraud.

During all the bustle of this great change, no steps had been taken to instal me in the office. Lord Althorp had resigned, and no one seemed to have any authority to do anything for me; all sorts of half promises and excuses followed each other with long delays between, and I gradually saw the whole thing sliding out of my grasp. Instead of holding fast to my first plan, which they could not have executed without my aid and the special knowledge I had acquired, I had in all the trustfulness of youthful inexperience shown them another so simple that they could put it in operation without any assistance from me. I had no patent to fall back upon. I could not go to law, even if I wished to do so, for I was reminded when pressing for mere money out of pocket, that I had done all the work voluntarily and of my own accord. Wearied and disgusted, I at last ceased to waste time in calling at the Stamp Office, for time was precious to me in those days, and I felt that nothing but increased exertions could make up for the loss of some nine months of toil and expenditure. Thus, sad and dispirited, and with a burning sense of injustice overpowering all other feelings, I went my way from the Stamp Office, too proud to ask as a favour that which was indubitably my just right; and up to this hour I have never received one shilling or any kind of acknowledgement from the British Government. Such has been my reward.

I am, Sir,

Your obedient Servant,

(Signed) HENRY BESSEMER.

Denmark Hill,
29th October, 1878.

In all the early stages of the development of my invention for piercing designs on parchment, I had depended entirely on my own hands; but when I was desired by the Stamp Office authorities to show how I proposed practically to carry out the invention, I designed the form of stamp described in my letter to *The Times*, and which is faithfully represented by an impression on the fly-leaf at the commencement of this letter; the execution of the somewhat elaborate design in steel, represented by this impression, was entrusted by me to Messrs. Porter and Son, die-sinkers of some eminence, at that time carrying on business in Percival Street, Clerkenwell, and whom I had frequently before employed to re-touch the cast-metal dies used by me for stamping works of art in relief on cardboard.

Now, in order to obtain positive evidence in corroboration of my letter to *The Times* of November 1st, it was of paramount importance that I should find Mr. Porter, if still

* Now over seventy years ago

alive; I had strong hopes of doing so, as I had both seen and conversed with him twice within the last eight or ten years, but had no knowledge of his present residence; failing to obtain this information, I resorted to an advertisement in the second column of *The Times*, on November 6th and six following days, which happily resulted in Mr. Porter communicating with me. He knew me well as an old customer of his firm, and reminded me of some of the more important dies re-touched by him; in consequence of the extremely novel character of the piercing die referred to, and the unusually difficult and laborious nature of the work, consequent on the extreme depth of the engraving, it had been fully impressed on his memory, and he was enabled at once to recognise the impression given on the fly-leaf of this letter as a faithful (though somewhat less artistically finished) copy of the piercing die executed by his firm for me in 1833.*

In order to secure permanently this important evidence, Mr. Porter made, at my request, a statutory declaration to that effect, his identity being witnessed by a gentleman of position who had known him intimately for the last thirty-five years. A *verbatim* copy of this declaration is appended hereto.

The advertisement referred to induced many persons to whom I was known to tender such information as they might happen to possess in reference to Mr. Porter; one of these letters was from a Mr. Richard Cull, a gentleman with whom I became personally and intimately acquainted soon after my first arrival in London, about the year 1831. Being a man of taste and superior education, he took great interest in my invention for cheaply reproducing works of art in *bas-relief*, and during our intimacy of that period he proposed to join me as partner in the commercial carrying-out of my invention, but this proposition was never carried into effect.

Many years had elapsed since I had seen Mr. Cull, during which time he had risen to the highest eminence as a philologist and a prominent member of the Society of Antiquarians. Most fortunately, he had happened to see my advertisement for Mr. Porter in *The Times*, and having been acquainted in early life with Mr. Porter and his family, he at once wrote to me on the subject; he had also seen my letter in *The Times* on the "Reward of Invention," and it is to that circumstance, no doubt, that I owe the closing remarks of his letter, of which the following is a *verbatim* copy, omitting only some irrelevant family matters relative to Mr. Porter:—

> 12, Tavistock Street,
> Bedford Square,
> November 9th, 1878.

DEAR MR. BESSEMER,

It is some time since we met, but seeing your advertisement for Mr. Porter, the die-sinker, I determined to write to inform you what I can on the subject. He was a very good artist, but he failed in business and took a situation in the City; I knew him very well, and his family, including his father, mother, and sister. . . . . I think he must now be dead, as I have not met him for fourteen or fifteen years, and he never said where he lived after leaving Percival Street.

I remember SEVERAL CONVERSATIONS with you concerning Sir C. Presley and your

* See Fig. 5, Plate III

invention, AT THE TIME OF YOUR INTERVIEWS WITH HIM. I well remember the unfavourable opinions I formed of that official.

I am,

Yours very truly,

(Signed) R. CULL.

This letter from a gentleman I had for so many years lost sight of was a most unexpected and spontaneous confirmation of the fact that I was at the time mentioned in constant communication with Sir Charles Presley on the subject of my newly-invented stamps, and also that our conversations at the time had impressed Mr. Cull with "an unfavourable opinion of that official." I have no doubt but that in our frequent and friendly intercourse I had complained loudly of the constant evasions with which my claims were met at the Stamp Office, which must have given rise to this unfavourable impression in the mind of my friend, and which, it appears, was strongly enough imprinted to survive for so many years, although the precise reasons for it are no longer distinctly remembered. At my suggestion, Mr. Cull unhesitatingly made a statutory declaration on the 15th of November embodying these facts, a *verbatim* copy of which is appended.

In my letter on the "Reward of Invention," I stated that I was twenty years of age when my experiments for the prevention of forgery were commenced. Now, I was born on the 19th January, 1813, hence I had arrived at twenty years of age in January, 1833. I have also stated that after some months of study and experiment, I succeeded in producing a stamp which satisfied all the necessary conditions; then follow the intervals between my several interviews with Sir Charles Presley, and also the five or six weeks occupied by Mr. Porter in engraving the die, which was accepted by the Stamp Office authorities; and then came the application to Parliament for an Act to empower the Commissioners of Stamps to call in all the old stamps and issue new ones in lieu of them. This Act of Parliament, if I correctly understood Sir Charles Presley, was hurried through the House in six or eight weeks; it was, in fact, as I now find, passed on August 29th, 1833, or just seven months and ten days after I was twenty years of age; thus, proving how accurate I was in my statement of the period when these transactions took place, and which family matters had impressed indelibly on the memory.

I mentioned also in my letter to *The Times* that an Act of Parliament was passed calling in all stocks of stamps dispersed throughout the country, and authorising the issue of the new dated ones. I did not know of my own knowledge that such an Act had been passed, but I perfectly well remember being told so by Sir Charles Presley, because it was an absolute assurance to me that my plans would be adopted; but I relied solely on Sir Charles Presley's statement to that effect. Hence, when it occurred to me that this Act of Parliament would form a most important link in the chain of evidence I desired to establish, I must confess to some trepidation lest Sir Charles had misinformed me, or had spoken only of an Act in the course of passing through Parliament, but which might have been thrown out and never passed at all; thus, when I applied to my solicitor to obtain, if possible, a copy of the Act in question, I was greatly pleased to find that not only was the statement of Sir Charles Presley (repeated by me in *The Times*) confirmed, but I found that this Act of Parliament* in its preamble admitted the fact that "the laws

---

* This Act is the 3rd and 4th of William IV., Chapter 97, dated August 29th, 1833

heretofore enacted, and now in force in Great Britain, have been FOUND INSUFFICIENT TO PREVENT THE SELLING AND UTTERING OF FORGED STAMPS ON VELLUM, PARCHMENT, AND PAPER.

Powers are given under the different sections of this Act to BUY UP AND DESTROY ALL STAMPS AND STAMPED PARCHMENTS then in possession of all vendors of stamps throughout the country. Full powers are also given to the Commissioners to DISCONTINUE the use of ALL DIES HERETOFORE USED in the Stamp Office, and authorising the employment of ANY NEW DIE OR DIES, with such DEVICE OR DEVICES as the Commissioners MAY THINK FIT.

The Act also declares that after three months from that date all stamps previously issued, or any deeds stamped therewith, shall be deemed to be illegal.

Then follows a most stringent clause (Section 12), making it felony punishable by TRANSPORTATION FOR LIFE BEYOND SEAS, for any person to JOIN, FIX, OR PLACE UPON any vellum parchment or paper, any stamp, mark or impression, which shall have been CUT, TORN, OR GOTTEN OFF, OR REMOVED from any vellum, parchment or paper, etc.

The object of this last clause is clearly to add, by the terrors of a most sweeping and stringent penal law, to the security which the new stamp was calculated to afford against the heavy losses which the Government had for so many years sustained by the transfer of stamps from one deed to another; and bears evidence, as indeed does the whole document, of the perfect state of panic into which the Stamp Office was thrown when they fully realised the extreme facility which my method of making composition dies from any paper impressions afforded for successfully forging every description of embossed stamp.

It is almost impossible to realise the spectacle afforded by one of the most conservative of all the institutions of the State—one which has stood its ground for generations—suddenly and without the smallest reserve flinging over every tradition of the past, repudiating all its former issues, and buying back again from the public all the stamps it could lay hands upon, for no better purpose than their destruction, and proclaiming by advertisement that their use, if not brought back, would be illegal; thus suddenly waking up, as it were, from a long period of fancied security, and seeking in hot haste powers from the legislature to protect them by the most severe of all penal laws next to that of death, and asking at the same time full powers to search all domiciles, shops, warehouses or places, under the mere suspicion that forged stamps may be concealed there, and to seize all stamps suspected of being forged; thus showing a not unnatural dread lest the secret of my method of reproducing embossed impressions might become known, and result in flooding the country with spurious stamps.

Thus does it sometimes happen that the stern realities of life overstep the boldest flights of imagination. Who in his wildest dreams could have supposed that one of the oldest departments of the State would be thrown into utter confusion, requiring immediate legislative action for its security; that the loss of vast sums annually to the Revenue would be prevented, and that a great temptation and incentive to crime would find a perfect remedy at the hands of a mere boy and girl? Such things are of themselves strange enough, but it is still more extraordinary that the Government of a country which prides itself more than any other in the civilised world on its simple justice and inviolable honour should have received so great a boon at the hands of a youth who was struggling hard to create for himself a position in the world, and who, in the fulness of his unbounded faith in their honour and integrity, placed unreservedly in their hands the power of doing all this, without retaining the smallest check on them for his own protection ; and who up to this hour has never received one iota of the remuneration held out to him as an inducement to persevere with his invention, or even one word of thanks or acknowledgment of the great and lasting benefits he has conferred upon the State.

Such, then, are the circumstances under which I now come forward to vindicate my honour, by proving the truth of the statements publicly made through *The Times;* and to claim, at the hands of Her Majesty's present Ministers, such payment or acknowledgment of my past services as may be consistent with the honour and dignity of the State, and at the same time acceptable to myself.

I scarcely need say that I shall at any time be happy to give personally any further facts or explanations that may be desired in relation to this matter; and I may further add that Mrs. Bessemer as well as myself, has a perfect remembrance of the circumstances connected with her suggestions of the dating on stamps, and which has for more than half a generation been a sort of tradition in the family, perfectly well known and fully understood by more than a dozen of its members.

I have mentioned all these facts most unreservedly, that you might be in a position to judge if I have not had substantial grounds for dissatisfaction with the administrators of former Governments.

But the one and only claim I now make has reference to the engagements entered into with me by the Stamp Office, and in this case I merely ask that a simple act of common justice may be done, such as in private life the law would compel, and individual character would render imperative; nor do I doubt for one moment that Her Majesty's present Ministers, who have so nobly maintained untarnished the honour of the British nation in every part of the world, will (now they are aware of the fact) most gladly blot out from the page of history the deep stain on the nation's honour which has been so long recorded in the annals of the British Stamp Office.

In conclusion, allow me to apologise for the length to which I have extended this letter, and to offer you my most grateful thanks for your kind perusal of it; and further allow me the honour to subscribe myself—

Your most obedient, humble Servant,

(Signed) HENRY BESSEMER.

I need not say how anxious I was to receive a reply from Lord Beaconsfield to this rather bold assertion of my claims on the Government, but I felt well assured that every enquiry among the still existing officials at Somerset House could not fail in establishing the justice of my demands. These printed letters to Her Majesty's Cabinet Ministers were posted on May 5th, 1879, and were most courteously acknowledged, and resulted in an investigation being instituted. On May 29th, I was honoured by an autograph reply from Lord Beaconsfield, of which a photographic copy is here given (Fig. 6, Plate IV.); and which clearly shows that both he and his colleagues were not only satisfied of the truth of the charges I had made, but were honourable enough to offer such compensation as they had in their power to bestow, and which I cordially accepted as a full acknowledgment of the services rendered. The form taken was the more satisfactory to me, inasmuch as it was a reward in which Mrs. Bessemer would take an equal share with myself, as she had already done in the invention which had been of such signal service to the State.

On the 21st June I received an intimation from the Right Honourable R. A. Cross that Her Most Gracious Majesty had been pleased to signify her intention of conferring on me the honour of Knighthood, after the Council which would be held at Windsor Castle, on Thursday, the 26th instant ensuing. I accordingly repaired to Windsor on that day. One of the royal carriages awaited my arrival at the station, and conveyed me to the Castle, where I had the honour of passing through the quaint and interesting ceremony of kneeling on one knee before Her Majesty, and receiving a gentle blow across the shoulder from a light and beautifully jewelled sword, and was commanded to express my gratitude by kissing the hand of Her Most Gracious Majesty. I afterwards took lunch at the Castle, and then returned to London.

CHAPTER IV

UTRECHT VELVET

AMONGST the many persons who had seen my castings from Nature coated with copper, at the Museum, was a member of the old-established firm of decorators, Messrs. Pratt, of Bond Street; and being at that time in search of someone to carry out an idea of his own, he sought an interview with me. He explained his object, and asked me if I thought it possible to produce an imitation of a particular material which he required, showing me at the same time some splendid old specimens of figured Genoa velvet with a satin ground. Mr. Pratt's idea was to produce an imitation of this beautiful fabric on Utrecht velvet, woven plain, and to have the desired patterns produced thereon by stamping, after the manner of the embossed cotton velvet so much in fashion at that time. He told me that various qualities of Utrecht velvet had been tried for him by the best manufacturers of embossed cotton velvet, but all attempts to produce a permanent effect on this stubborn material had utterly failed, and he had abandoned the idea of getting it made, until he had by chance seen the metal castings from Nature before referred to. Explaining this circumstance to me, he complimented me by saying that the idea at once struck him that the man who had found out how to produce such marvellous castings would, in all probability, soon discover how to emboss Utrecht velvet.

The result of this interview was that Mr. Pratt left with me a specimen of his woven Genoa velvet, a copy of which I undertook to try and produce by heat and presssure on a plain fabric. This Utrecht velvet is a long-piled, very harsh and stubborn worsted material, as, indeed, every one would at once recognise who had seen chairs

Just before I had embarked on my luckless Stamp Office enterprise, I become aware of some curious facts relative to the manufacture of black lead pencils. The only mine in Great Britain which yields plumbago, or black lead as it is called, suitable for pencil-making, is situated in one of the mountains at Borrowdale, in Cumberland, and is about 1000 ft. deep. This rare and very valuable mineral substance became the subject of continued robbery about one hundred and forty years ago, and is said to have enriched many persons resident in the neighbourhood. It was strongly guarded by the proprietors, but they were more than once overpowered by an infuriated mob, and possession of the mines was held for a considerable time by the desperadoes. When the owners again got possession, their carts, which conveyed the produce of the mine to Keswick, were always guarded by soldiers.

The entrance to the mine was afterwards protected by a strong building, consisting of a well-appointed guard room and three other apartments on the ground floor, in one of which was an opening into the mine, secured by a trap-door, through which alone the miners could enter. In another of these apartments, called the dressing-room, the miners changed their ordinary clothes for a working dress, and after six hours' work in the mine they had again to change their dress under inspection, lest some of this valuable substance might be concealed about them.

The plumbago, when perfectly cleaned, was packed up in casks and despatched to London, and there disposed of at monthly sales by auction, at the offices of the proprietors, in Thames Street, where it realised from thirty-five to forty-five shillings per pound, the annual sales ranging in value from £30,000 to £40,000 sterling.

Plumbago is found in small irregular nodules about the size and shape of a potato, and consists of carbon in a peculiar state of aggregation, with a small impregnation of iron.

The trade in pencil-making at the time of which I am speaking—about 1838—was chiefly in the hands of the Jews, and one important branch of it consisted in sawing these little nodules of plumbago into slices of about one-sixteenth of an inch in thickness. This art of sawing the plumbago was a most difficult one to acquire, and hitherto all efforts to replace hand labour by machinery had failed; hence it remained a monopoly in the hands of the Jewish workmen, who were paid as much as a guinea per pound for sawing the material. The difficulty of cutting it into slices without breaking them was very great, while the rounded shape of the nodules and their slippery surface rendered it most troublesome to hold them firmly during the sawing operation; moreover, the thin slices thus obtained were so brittle as to be easily broken by the accumulation of sawdust in the bottom of the saw-cut. Another difficulty arose from the presence of minute sparks of black diamond dispersed here and there throughout the mass; whenever the saw struck against one of them the slice was broken. The hand-saw used by the workmen had what is called a "wide-set"; that is, the teeth were bent right and left so as to well relieve it from the pressure of accumulated sawdust; the consequence being that the saw-cut was nearly as wide as the slice of plumbago produced, and hence each pound was reduced to about nine ounces of slices and seven ounces of dust. The result was that the price of the slices, augmented by twenty shillings for the labour of sawing, was brought up in value from about forty shillings to nearly £4 10s. per pound.

On enquiry into this matter, I found that I could purchase the sawdust for about half-a-crown per pound. These facts held out promises of a very profitable manufacture, if I could only succeed either in making a sawing-machine that would be less wasteful of the material, or in finding some means of consolidating this large quantity of dust, without such an admixture of extraneous matter as would prevent its being used in the manufacture of the best pencils.

I first tried the sawing-machine, which I constructed with great care. The principal features of novelty in this machine related to the saws; these were made from the main-springs of watches which had been broken while in use; they were extremely thin, and of a beautifully fine quality of steel. The "set" on the tooth was made especially small, and consequently the saw-cut was so narrow as to waste only a very little of the material in the form of dust. I entirely avoided the clogging of the saw in these narrow cuts, and the consequent splitting-off of the slice, by putting the teeth of the saw uppermost, and bringing the piece of blacklead to be cut downward upon it by the slow motion of a fine screw. By this means the dust fell freely downwards out of the saw-cut, and never clogged the saw or broke a slice of the material.

I was also successful in getting over the difficulty caused by striking against the little black diamond sparks, by the use of a spring friction clutch on the connecting-rod which reciprocated the saw-frame. This delicately-adjusted clutch was tightened up just sufficiently to overcome the usual resistance to the saw; but whenever that resistance was increased by contact with a diamond spark, the friction clutch simply yielded and the saw was rendered motionless, although the machine continued to work until it was thrown out of gear. It was in this way almost impossible to break a slice in the process of cutting; whenever the machine was thus rendered inactive, the diamond was searched for and removed in the usual way, when the sawing process was resumed.

Having thus succeeded in making a machine capable of saving a large quantity of the plumbago which had hitherto been wasted as dust in the ordinary process of sawing by hand, I considered it advisable to bring my invention under the notice of the eminent pencil-makers, Messrs. Mordan and Co., offering to saw their plumbago at a mere nominal cost, and share with them the value of the material saved. Every offer was rejected by them, under the plea that the firm could not suffer their "prepared plumbago" to leave their premises; they, in fact, wished me to put up my machine and work it in their manufactory; but this I declined to do, and consequently I laid the machine aside for the moment in deep disgust at this unexpected rebuff.

I then determined to try and utilise the plumbago dust which at that time could be obtained so cheaply, and after several preliminary trials I obtained leave from a city firm to use in private their powerful hydraulic press, a machine capable, if necessary, of exerting a pressure of 400 tons on the plunger of my experimental mould, which was simply a cylindrical mass of iron, having an internal diameter of three inches. This cylinder was half filled with the plumbago sawdust in a pure state, and the short ram or plunger occupied the other half; it projected above the surface as shown in Figs. 7 and 8, where A, Fig. 7, represents a section of the cylinder in which the plunger, B, is fitted, and c shows a recessed plate of iron on which the cylinder rests. The powder to be pressed is shown at D; in this state the apparatus was placed in a furnace and heated to redness, after which it was removed to the hydraulic press, and the plunger forced down with a pressure of about five tons to the square inch; the pressure being continued until the whole had cooled down, and the powder had formed into a solid mass. The cylinder was then placed over a hollow block of iron as shown at E, in Fig. 8, when pressure was again applied to the plunger and the cylindrical mass of plumbago was forced out, after which it was found to be in every way suitable for making the very best lead pencils.

A young friend of mine to whom I showed some of this compressed

FIGS. 7 AND 8. METHOD OF COMPRESSING PLUMBAGO POWDER

plumbago, offered to purchase the invention, at the same time saying that he could not risk more than £200 on the venture; I, remembering the rebuff with the sawing-machine, accepted his offer without further consideration; my friend then went off to Cumberland, and made arrangements with the Plumbago Company. At the present day we find that the best lead pencils in the market are made by crushing the small lumps and odd pieces of plumbago, then washing and floating the powder, by that means getting entirely rid of the little black diamonds, and producing various grades of hardness by different degrees of heat and pressure.

I fear this little episode does not speak very favourably for my business capacity in those early days, for I certainly ought to have made much more than I did by this really important invention.

When I was experimenting with plumbago (about 1838) I was engaged in designing a new system of casting types by machinery, some features of which are of sufficient interest to be recorded. The moulds in this machine were entirely composed of hardened and tempered steel, shaped by laps, as the metal could be neither planed nor filed. From fifty-five to sixty types were cast per minute in each of the two compartments of the mould ; and in order that the solidification of the metal should take place in the extremely small interval of time allowed for that purpose, the moulds were cooled by a constant flow of cold water through suitable passages made in them, in close proximity to those parts where the fluid metal came in contact. Another special feature of this mode of casting was the employment of a force pump placed within the bath of melted metal, by means of which the latter was injected into the mould at the proper moment, the pressure of the injected fluid being under the perfect control of a loaded valve. It will be readily understood that a sharp jet of fluid metal would propel with it an induced current of air, and consequently produce a bubbly and spongy casting, which would have been wholly valueless. The short space of time occupied in its solidification afforded no opportunity for the escape of air in the usual way by floating in bubbles upward, as in the case of castings where the metal is retained in its molten state in the mould for several minutes.

I found an absolute cure for this apparently insuperable difficulty, by forming a vacuum in the mould at the very instant at which the injection of metal took place ; and so successful was this system of exhausting the moulds, that one might break a hundred types in succession without finding a single blowhole in any one of them.

The iron or brass founder, whose slow and tedious operations are performed by quietly pouring his molten metal into the mould with a ladle, will at once see what a new departure in the art of founding this machine presented. Firstly, there was the same mould producing fifty-five to sixty castings per minute, instead of being broken up and destroyed after one cast : then pouring the metal from a ladle was replaced by injecting it with a force-pump, the mould itself having a continuous stream of cold water running through suitable passages formed in it so as to cool every part of its surface in contact with the fluid metal; and, finally, instead of the mould being composed of porous materials through which the confined air gradually escaped, there was an almost indestructible mould, wholly free from pores, from which all the contained air was withdrawn in the fraction of a second by its sudden connection with an exhausted vessel at the moment when the metal was injected.

The valve through which the metal was injected into the mould being extremely small, required to be fitted very closely to prevent its leaking; it was found that after it had been opened and closed some six or seven thousand times, a portion of the fluid metal would, by friction against the sides of the valve, be rubbed into powder, and more or less obstruct its action. Otherwise, the really beautiful mechanism of this casting machine performed all its functions with perfect precision, and formed the bodies of the type so parallel and so perfect in other respects, that it soon began to create much jealous feeling and opposition among the type-founders, whose occupation was threatened by it. For this reason, Messrs. Wilson, the well-known type-founders, of Edinburgh, to whom I had sold my invention, preferred to make no further efforts to improve the valve arrangements, and allowed the whole matter to sink quietly into oblivion rather than face the storm they saw was brewing.

About this period my attention was directed to the art of engine-turning, which was a very profitable one to the few who had sufficient originality of thought to work out those marvellous combinations of interlacing lines, such as we see at the present time on the coupons of many foreign bonds. I was a most enthusiastic admirer of these productions, especially those of that greatest of all engine-turners, Jacob Perkins, the well-known American engineer. I felt certain that I could employ one of these beautiful machines to advantage, and I was fortunate enough to purchase a very good one for £65.

How well I remember its being delivered at my premises one afternoon; I had it placed in my private office, close to the window. I knew pretty well nearly every detail of its construction, but I commenced by taking it all to pieces, the better to impress my mind with the smallest detail. Having put it together again, and taken my evening meal, I lit my large argand lamp, and, with my back to the window, I sat facing the Rose engine, and commenced my first essay on some odd pieces of brass which I had mounted on the straight-line chuck. I found myself rather awkward at first, but I soon began to manipulate more successfully, and in a short time became deeply absorbed in my work. I was ruling some very fine waved lines, which I could not see so clearly as I wished, when, looking round to the window on which my back had so long been turned, I was surprised to find the grey morning light stealing quietly in, and rendering my lamp useless. I had no idea that I had been sitting up all night, so imperceptibly had the time glided by. I was, however, well satisfied with the progress I was making, and was much delighted with my Rose engine, additions to which I never seemed tired of devising, and thus obtaining the infinity of beautiful effects which simple interlaced curved lines were capable of producing. Nor was this delightful work unaccompanied by a substantial reward, for almost fabulous prices were sometimes paid for unique specimens of the art, applicable as patent medicine labels, coupons, and for other purposes where it was desirable to render fraudulent imitation impossible.

On this machine I engraved many rollers for paper-embossing and printing for Messrs. De la Rue, and for the firm of Vizetelly and Co., etc. In cutting deeply-incised lines in metal for surface printing, there was always a tendency in curves to drag or blur the surface of the metal block. A little study of the subject convinced me that this defect was owing to the quality of the metal employed, and after several attempts I succeeded in making an alloy of tin and bismuth, which answered admirably. It made a sharp creaking sound as the tool glided over it, cutting very crisp and raising no burr on the sides of the line cut. Indeed, so perfectly did this alloy remove a serious practical difficulty, that I used to manufacture blocks of the metal for the trade. This was the case also with another alloy, of equal parts of tin and zinc, to which were added 8 per cent. of copper and 3 per cent. of antimony. The metals forming this alloy have a tendency to solidify in the order of their fusibility, and the alloy has the peculiar property of passing from the fluid to the solid state so slowly that it may be used at an intermediate stage, when it is neither liquid nor solid; in this state it lends itself admirably to the formation of what are called "forcers," used in embossing leather or cards. This raised impression, or "forcer," is made by pouring the melted alloy into an open frame laid on the edges of the die; when the metal has attained a state of partial solidification, a beautiful impression of the die may be obtained by gentle pressure, and the alloy, when quite cold, is hard enough to stand the wear and tear of stamping in a most remarkable manner. The sale of these alloys to the trade was a welcome source of profit to me, and by no one was their usefulness more appreciated than by the late Mr. Thomas De la Rue, the talented founder of that well-known firm of fancy stationers, whom I had the advantage of knowing intimately and numbering among my best customers.

Thus, one branch of trade seemed to lead imperceptibly to another; but I was always waiting and looking forward to the establishment of the one large and steady branch of business that I hoped would some day allow me to drop the many schemes which my versatile mind so easily created, seized upon, and engrafted on the business I was carrying on; but this one great branch of trade, so earnestly desired, had not yet manifested itself. I was accordingly content in the meantime to hold on to everything that fairly paid for the time and capital employed in its production.

My life at this time was pretty much one of hard work and steady attention to business, from which I could only snatch short intervals. Late in the evening I would drop in and have a chat with my father, then advanced in years, but ever anxious to hear of my progress, and desirous to see the latest specimens of Rose engine work, or to discuss with me some of the many new schemes that occupied my thoughts. At that time my two sisters kept house for my father, and in this little family circle I spent many a quiet evening. There was another house, however, to which my steps were involuntarily wont to lead me. My friend, Mr. Richard Allen, had a fair daughter, to whom I had for some time been engaged; thus, between the two families

G

all my leisure hours were spent in friendly intercourse and quiet meetings, without even a desire on my part to mix in any of those gaieties which the world calls Society. Pleasant and delightful as were these evenings, replete with all the charm of unrestricted social amenities, they were, nevertheless, only steps to one great end and aim of all my earthly aspirations: for above all things I desired to exchange my lonely bachelor's apartments for a home of my own. I did not see the wisdom of waiting for an indefinite time on "fickle fortune," so as my betrothed was willing to share my lot in life, we were married. We settled down quietly in Northampton Square, close to my place of business, and I am happy to say that in all the changes and vicissitudes of the sixty-four years that have passed since that happy event, I have never had reason to regret a step which I had taken in the full confidence of youth that I should, in time, be able to carve out for myself a name and a position in the world worthy of her to whom my life was henceforth to be devoted.

The white metal medallions and casts of natural objects, coated with a film of copper and exhibited by me at the Museum of Arts and Manufactures, in Leicester Square, attracted the attention of a gentleman who had in his possession a great many of the beautiful dies that had been engraved in the French mint, the impressions from which are generally known as the "Napoleon Medals." Some of them were engraved in steel, others were cut in brass, and all were of the most exquisite workmanship. I made arrangements with the owner of these dies to produce a great quantity of bronzed impressions of them at prices which were highly renumerative. For this purpose, I devised a simple apparatus for rapidly stamping the impressions in semi-fluid metal, the only mode by which perfect impressions could be obtained from those dies that were engraved in brass. After some considerable trouble, I produced an alloy of tin and other metals, which differed from the alloy named before in having no zinc in it, though it nevertheless passed so slowly and so gradually from the fluid to the solid state, that the most perfect impressions were obtained with unerring certainty. The shower of splashes inseparable from stamping semi-fluid metal was received in the case surrounding the dies, and this was automatically closed as the press descended. Immense quantities of these fine medallions were made, and beautifully bronzed without impairing their sharpness. I still possess a few of them, more or less damaged by time; and as an example of their general character, I give photographic reproductions of some of them in the figures on Plates V. and VI., each being the same size as the original. Those I have selected include the famous "double - head," Napoleon and Josephine (Fig. 9, Plate V.), said to be the finest portrait medals of the Emperor ever produced. Fig. 10, Plate VI., is another of these Napoleon medals, and Fig. 11 is a medallion of the head of Minerva.

One day I was called upon by a gentleman, a Mr. James Young, who presented a card of introduction from a barrister to whom I was well known. His object was to obtain the assistance of a mechanician to devise, or construct, a machine for setting up printing type. I had a long and pleasant conversation with this most agreeable client; indeed, our frequent meetings and friendly discussions resulted in a close friendship, terminating only with his death, which occurred several years later. My friend Young, who was a silk merchant at Lille, had persuaded himself that by playing on keys, arranged somewhat after the style of a pianoforte, all the letters required in a printed page could be mechanically arranged in lines and columns more quickly than by hand; but as he was personally wholly unacquainted with mechanism, he desired someone to elaborate all the details of such a machine, and asked me if I would professionally study the subject for him, and prepare models to illustrate each proposition. The matter seemed a very difficult one at first sight, and I said that it would be impossible for me to devote more than a portion of each day to its consideration. It was then arranged that I should give as much thought to the subject as I could, consistent with due attention to my general business, and to these terms was attached a guinea per day as a consulting fee.

The general idea on which the machine was based was the arranging of the respective letters in long narrow boxes, from which a touch of the key referring to any particular letter would detach the type required; this, when set at liberty, was to slide down an inclined plane to a terminal point, where other mechanism was to divide the letters so received, into lines if required, and thus build up a page of matter, such as a column in a newspaper, etc.

It will be at once understood that this was not a very simple matter, in consequence of the many signs required. We have first the twenty-six small letters of the alphabet, and the double letters, such as *fi*, *fl*, *ff*, *ffi*, *ffl*; then we have the points, or punctuations, signs of reference, etc.; there are also the ten figures and the twenty-six capital letters and their respective double letters, as well as blank types, called "spaces," of different thicknesses, required to divide separate words from each other, etc. Now, as a primary necessity, these numerous letters, when wanted, must, of course, come from different places, and all must descend grooves in the inclined planes in precisely equal times. The time of the whole journey down the incline, say, 2 ft. long, must not occupy any one type more that one-hundredth of a second more or less than the one before or behind it, or its arrival will be too soon or too late, and the word will be wrongly spelt. Thus, suppose the word ACT is required, and the keys A, C, and T, are touched rapidly in succession. If the letter C should arrive first instead of A, the word would not be "ACT" but "CAT," and so for every word. A type that is less than 1 in. in length must never, on its journey, arrive its own length in advance or in the rear of the others that are simultaneously rushing down the inclined plane to the same terminus.

The difficulty that this fact presented was almost beyond belief. Many models were made and much study devoted to it. Thus, suppose a type detached at the point A in the accompanying diagram (Fig. 12) is required to slide down the inclined plane to c, and another one from the point B is immediately to follow, it will be seen that not only is the road to be travelled by A much longer than that by B, but B also has the advantage of coming straight down the inclined surface, encountering friction only on the one surface on which it rests; while A has not only got a longer journey to perform, but it lays its whole weight on the inclined surface, and rubs also against the inclined side of its groove, thus causing additional friction, so lessening the speed of its descent, and resulting in the arrival of B at its destination before, instead of after, A.

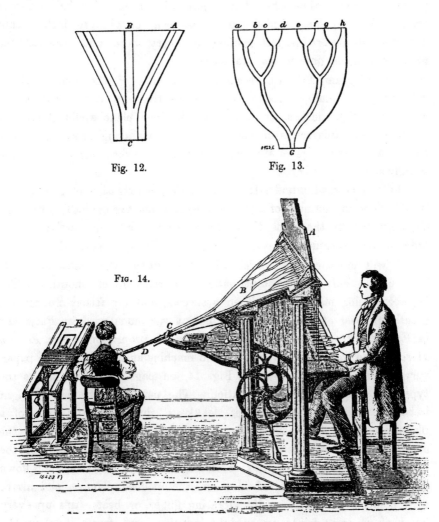

Fig. 12.

Fig. 13.

FIG. 14.

FIGS. 12 TO 14.   YOUNG'S COMPOSING MACHINE

The result of studying this part of the question forced on my mind the important fact that the grooves on the surface of the inclined plane would have to be all of precisely the same length, and every letter, in descending, would have to encounter exactly the same amount of sideway rubbing surface.   This knotty point was at last settled in so simple and perfect a manner, that when I had accomplished it I felt half ashamed that it had so long eluded me.   The form of grooved incline thus indicated ensured a perfect spelling of every word, and removed the greatest obstacle on the way to success.

The diagram, Fig. 13, represents a portion of the inclined plane, with its small shallow grooves so arranged that any one of the letters a, b, c, d, e, f, g, and h, at the top of the inclined plane would, if allowed to slide down this series of curved grooves, pass along precisely similar paths, and travel precisely equal distances, before arriving at the terminus c.

It will be readily understood that a simple extension of this system would allow any number of letters arranged along the upper line to reach the terminus in the same time ; hence each one would arrive in the order of its departure and every word would be spelt correctly.

I will not tire the reader with the many other difficult points surmounted, only by constant patience, during fifteen months.   The type-composing machine was then a success, and my friend Young was greatly pleased at the result.   His patent was much used in Paris, and in England it was employed by the spirited proprietor of the *Family Herald*, who gave an engraving of the machine at the head of the paper, very similar to the illustration, Fig. 14, on page 45, which shows the type-composing machine in operation.   The person shown on the right is seated before a double set of flat keys, similar to the keys of a pianoforte, each key having its proper letter marked thereon ; the depression of a key detaches its corresponding type from one of the numerous partitions in the box or case A; this type will then slide down the series of grooves allotted to it on the inclined plane B, and arrive at a point, c, where a rapidly vibrating finger or beater tips up every letter as it arrives into an upright position, and forces it along the channel D.   These rows of letters are moved laterally, forming one line of the intended page.   The boy on the left hand divides the words with a hyphen if necessary, or he so spaces them as to fill one complete line ; this operation he can complete while another line is forming in the channel D. In this way he makes line after line until part of a page is set up, when he moves on the galley E, shown at his left hand. Thus a page or a long column of matter was produced with the greatest ease, and in a very short space of time.

In the ordinary way of composing types, each letter is picked up by hand from one of the numerous small divisions of a shallow box, or "case," as it is called, and the letters are then arranged in their right positions in a small frame held in the left hand of the compositor. About 1700 or 1800 letters per hour can be formed into lines and columns by a dexterous compositor, while as many as 6000 types per hour could be set by the composing machine. A young lady in the office of the *Family Herald* undertook the following task at the suggestion of the proprietor of *The Times*, viz.: she was to set up not less than 5000 types per hour for ten consecutive hours, on six consecutive days; giving a total of 300,000 letters in the week. This she easily accomplished, and was then presented with a £5 note by Mr. Walter.

This mode of composing types by playing on keys arranged precisely like the keys of a pianoforte would have formed an excellent occupation for women; but it did not find favour with the lords of creation, who strongly objected to such successful competition by female labour, and so the machine eventually died a natural death.

CHAPTER III

COMPRESSING PLUMBAGO DUST; CASTING TYPE; TYPE-COMPOSING MACHINE, ETC.

AFTER this long digression I must retrace my steps, forget for a time all the great doings of the 26th June, 1879, and remember only, so far as this little personal history is concerned, that I was at the time of which I am writing, simply Henry Bessemer, an unknown youth struggling to get a footing in the world by working with hand and brain for many hours every day, a task most cheerfully performed. In those days I had one great and paramount object always before me; one bright guiding star that kept me from falling into the almost irresistible temptations which the pleasures and gaieties of London hold out to every youth of a sanguine temperament who, like myself, happens to be sole master of his own actions. With no friendly voice to give counsel, or to guide and regulate my hours of leisure, or check my wanderings, that one silent but ever-present irresistible control which the desire to be worthy of, and united to, a beloved object, ever exercised over me, kept me in the straight path, made my labour sweet, and almost converted it into an amusement.

At this period the enthusiasm of the amateur was fast giving way to a more steady commercial instinct, and I let no opportunity slip of improving my position, but I felt that I was still labouring under the disadvantage of not having acquired some technical profession. With the exception of my card-embossing and die-making business, I had nothing to depend upon, and I but too readily allowed my attention to be directed to new subjects which always exercised a sort of fascination over me; this tendency I found difficult to control, but I invariably made myself believe that as soon as I could strike some "good vein" I should work it to its full capacity, and never again be tempted to turn aside after mere novelties.

F

covered with it, and sat upon for years, without the pile being flattened down.

I provided myself with a flat brass die, or plate, engraved nearly a quarter of an inch deep, each of the parts sunk in it having vertical sides and a flat bottom, so that the pile at certain parts was left wholly untouched by the die, and therefore in its normal state; while those parts which came in contact with the plate were crushed down. All this was perfect enough, as far as it went; but I, like others, failed to produce a permanent effect, for in two or three days the pile so pressed down would partially rise again, and the pattern almost disappear. Many things were tried, but neither hot water nor steaming, nor the application of alkaline solutions, were of any avail, and I began to fear that I should be no more successful than others had been in dealing with this material. Further consideration, however, and a little study of the nature and properties of hair and wool, led to the idea that these substances were really of the nature of horn; and this material, I knew, was capable of semi-fusion at high temperature, and was, in that condition, suitable for being moulded into various ornamental shapes, which permanently retained, when cold, the forms thus impressed upon them in a heated state. I now felt that I was on the right scent, and believed that if I could rapidly submit the material to a very high temperature, and then move it away as quickly, a partial fusion of the part in contact with the hot surface of the die would take place, and produce a glossy surface like satin, which would never again stand up as pile.

I had no sooner got this view of the subject than I took measures to put it to a practical test. The result went to show that by maintaining the metal surface, which was in contact with the velvet, at a very high temperature for a short and definite period, and acting under a carefully-regulated amount of pressure, the process could be made a perfect success. These experiments also proved that the temperature must be so high as to produce a semi-fusion of the wool, and that if continued for a fraction of a minute too long the fabric would be destroyed.

The next step was to devise a machine in which these very critical conditions could be practically carried out on a commercial scale. This

I undertook to do at my own cost, in consideration of the very liberal
price per yard offered me for embossing the velvet. I erected, on my
own premises, the machine I had designed, and personally regulated
its operations. The apparatus consisted mainly of a massive iron frame,
in which was mounted a very deeply-engraved hollow roller of cast
iron, having a plain or unindented paper roller running in contact
with its under-side. The iron roller was not heated by steam, as the
temperature absolutely necessary was too high for that mode of heating;
so I had to apply a powerful Bunsen gas-burner, extending the whole
length of the interior of the open-ended, hollow-engraved roller, and
by that means I kept it at a constant temperature just short of what
would be destructive to the fabric. Now, a cast-iron roller working in
the open air is not a thing to which one can apply the glass bulb of a
thermometer, and ascertain the precise temperature of its external
surface; consequently, the accurate control of the temperature of the
roller presented many difficulties; but, after some study of the question,
I found a most satisfactory way of ascertaining this all-important fact.

I was aware that metallic lead fuses at a temperature of 640 deg.
Fahr., and by additions to that metal of tin and bismuth, in varying
proportions, its melting temperature can be lowered until the alloy will
fuse at the boiling point of water, viz., 212 deg. With these facts before
me, I had simply to form a standard alloy, fusible at, say, 450 deg. Fahr.,
that being the required temperature of the roll. This we may call alloy B;
another alloy, A, was made that would fuse at 10 deg, lower than B, and
a third, C, was made whose melting temperature was 10 deg. Fahr. higher
than B. These three alloys were made into rods about the length and
size of a black-lead pencil. Their use was extremely simple. When
commencing to heat up the roller for working, one end of the most
fusible rod, A, was pressed against the hot iron roller as it revolved, and
as soon as the first symptom of the fusion of the end of the rod manifested
itself, it was known that the roller was within 10 deg. of its proper working
heat. Care was then taken to gradually regulate the gas supply, and
when the end of the standard or working rod B was found to fuse on being
pressed against the roller, the machine was put in motion at the exactly-
ascertained speed, thus producing with certainty a beautiful figured fabric

that twenty years after would be found in much the same condition, less the amount of wear and tear to which it had been subjected.

The first practical working of this new process was upon a beautiful design, for which Messrs. Pratt had obtained an order for furnishing a suite of apartments at Windsor Castle, so that the new material, under so favourable an introduction, was certain to become fashionable.

In those palmy days of Utrecht velvet embossing, I was paid six shillings a yard for putting fabric through the rolls; but gradually this very high price was reduced, and when it came down to a shilling per yard immense quantities were embossed. Prices were still on the decline, when my machines and the stock of engraved rollers were purchased from me by Messrs. Gillett, Lees and Company, the well-known Utrecht velvet weavers of Banbury. A general taste for this material soon afterwards set in ; prices for embossing were lowered ultimately to one penny per yard, and many persons may still remember, some forty years ago, seeing the cushions of cabs and omnibuses covered with this decorative fabric. It is curious that the present fashion for antiquated furniture has again brought it into use, and it may now be seen in many of the best houses.

The original specimen of figured Genoa Velvet brought me by Mr. Pratt had what is called a narrow edging of " Terry," or uncut velvet, forming a series of little ribs which surrounded each leaf or scroll in the design, and made a sort of natural shading between the dark untouched pile of velvet, and the bright and satiny pressed-down surface of the ground on which the design was formed. A very beautiful specimen of my imitation of this " Terry" edging came into the possession of my niece, Mrs. Ada Allen, of Wingerworth Hall, and this she kindly presented to me; it has some historic interest, being the design of Mr. Pugin for covering the benches in the House of Lords; the roller was engraved by myself, and it was the first attempt to produce an imitation of the " Terry " edging in this new fabric. A photographic reproduction of this old specimen is given in Fig. 15, Plate VII., and it shows that after the many years' use of the fabric the design still retains marked tracing of the " Terry" edging.

In the early part of these pages, I referred to the fact that the origin of many important inventions and manufactures was lost in the

" mist of ages," but here we have an example of one that has passed
out of memory whilst its originator is still living : for I venture to say
that few indeed of the thousands who daily lounge in their easy chairs
on embossed Utrecht velvet would ever suspect that this material issued
from the same room in " Baxter House " in which all my first steel
experiments were made ; and that the same hand which regulated and
controlled the fiery steel converter also drew the first few hundred yards
of that very beautiful material through the rolls.

CHAPTER V

THE MANUFACTURE OF BRONZE POWDER

MY eldest sister was a very clever painter in water colours, and in her early life, in the little village of Charlton, she had ample opportunities of indulging her taste for flower-painting. My father had lived too long in Holland not to have imbibed a love of the beautiful Dutch tulips, for which there was a great rage in his young days, so much so, indeed, that a single bulb would sometimes realise a fabulous price. At Charlton, my father grew his beloved tulips, and my sister used to paint all the finest specimens he produced. I also well remember the many beautifully-coloured chrysanthemums we there cultivated, although none of the magnificent varieties since introduced from Japan were then known. My sister had accumulated a great collection of charming groups of these and other flowers, and had, with much ingenuity made a most tastefully-decorated portfolio for their reception. She wished to have the words—

STUDIES OF FLOWERS

FROM NATURE,

BY

MISS BESSEMER,

written in bold printing letters within a wreath of acorns and oak leaves which she had painted on the outside of the portfolio; as I was somewhat of an expert in writing ornamental characters, she asked me to do this for her, and handed me the portfolio to take home with me for that purpose.

How trivial and how very unimportant this incident must appear to my readers. It was, nevertheless, fraught with the most momentous consequences to me; in fact, it changed the whole current of my life, and rendered possible that still greater change which the iron and steel industry of the world has undergone, and with it the fortunes of hundreds of persons who have been directly, or indirectly, affected by it.

The portfolio was so prettily finished that I did not like to write the desired inscription in common ink ; and as I had seen, on one occasion, some gold powder used by japanners, it struck me that this would be a very appropriate material for the lettering I had undertaken.

How distinctly I remember going to the shop of a Mr. Clark, a colourman in St. John Street, Clerkenwell, to purchase this " Gold Powder." He showed me samples of two colours, which I approved. The material was not called " gold," but " bronze " powder, and I ordered an ounce of each shade of colour, for which I was to call on the following day. I did so, and was greatly astonished to find that I had to pay seven shillings per ounce for it.

On my way home, I could not help asking myself, over and over again, " How can this simple metallic powder cost so much money ? " for there cannot be gold enough in it, even at that price, to give it this beautiful rich colour. It is, probably, only a better sort of brass ; and for brass in almost any conceivable form, seven shillings per ounce is a marvellous price."

I hurried home, and submitted a portion of both samples to the action of dilute sulphuric acid, and satisfied myself that no gold was present. I still remember with what impatience I watched the solution of the powder, and how forcibly I was struck with the immense advantage it offered as a manufacture, if skilled labour could be superseded by steam power. Here was powdered brass selling retail at £5 12s. per pound, while the raw material from which it was made cost probably no more than sixpence. " It must, surely," I thought, " be made slowly and laboriously, by some old-fashioned hand process ; and if so, it offers a splendid opportunity for any mechanic who can devise a machine capable of producing it simply by power."

I adopted this view of the case with that eagerness for novel inventions which my surroundings had so strongly favoured, and I plunged headlong into this new and deeply-interesting subject.

At first, I endeavoured to ascertain how the powder was then made, but no one could tell me. At last I found that it was made chiefly at

Nüremberg, and its mode of manufacture was kept a profound secret. I hunted up many old books and encyclopædias, and in one which I found at the British Museum, the powder was described as being made of various copper alloys beaten into thin leaves, after the manner of making gold leaf, in books of parchment and gold-beaters' skin. The delicate thin leaves so made were ground by hand labour to powder on a marble slab with a stone muller, and mixed with a thick solution of gum arabic to form a stiff paste and facilitate the grinding process. The gum so added was afterwards got rid of by successive washings in hot water.

It thus became evident to me that the great cost of bronze powder was due to this slow and most expensive mode of manufacture, and it was equally evident that if I could devise some means of producing it from a solid lump of brass, by steam power, the profits would be very considerable. With these convictions I at once set to work. I had at that time a two-horse power engine, partly made by myself, which I finished and erected in a small private room at the back of my own house, for there I could make my experiments in secret.

Then came the all-important question, from what point was I to attack the new problem? An attempt to imitate the old process by any sort of automatic mechanism seemed to present insurmountable obstacles—the thousands of delicate skins to be manipulated, the fragile leaves of metal that would be carried away by the smallest current of air from a revolving drum or a strap in motion, and the large amount of power which must of necessity be employed to reduce the metal in whatever way it was treated. This necessity for delicate handling combined with great mechanical force, gave a direct negative to any hopes of producing the powder in a way analogous to the one in use.

How could I then proceed? A mass of solid brass did not appear to be a likely thing to fall to powder under treatment by a pestle and mortar. Then came the question: Can the metal be rendered brittle, and so facilitate its reduction? No, it cannot be made brittle except by alloying it with such other metals as will destroy its beautiful gold colour. Then there was the question of solution of the metal in acid, and its precipitation in the form of powder. These and many other plans were thought of, only to be again put aside as theoretically improbable or impracticable schemes.

The first idea which presented itself to my mind as a possible mode of reducing a piece of hard, tough brass to extremely minute, brilliant particles, was based on the principles of the common turning-lathe, with which I made my first attempt on a circular disc of brass, one-quarter of an inch in thickness, and four inches in diameter. This was mounted on a suitable mandril, and made to revolve at a speed of 200 revolutions per minute. The revolving brass disc was tightly pressed between two small steel rollers, having fine but very sharp diagonal grooves formed on their surfaces, sloping to the left on one of them and to the right on the other; the effect of this was to impress diagonal lines crossing each other on the periphery of the brass disc, and to form on it a series of minute squares. If the reader examines the milled edge of a sovereign, he will see just such indented lines

FIG. 16.   DIAGRAM SHOWING BASE OF PYRAMIDS FOR BRONZE POWDER

running across its periphery, but in the experiment described the lines impressed on the brass disc were V-shaped. A flat-faced turning tool mounted on a slide-rest was slowly advanced in the direction of the disc, so as to shave off an extremely thin film of metal from the apex of every one of the truncated pyramids formed on the periphery of the disc. The actual size of the base of each of these pyramids is shown in Fig. 16, where a surface of 1 in. square is divided into a hundred lines to the inch, and is crossed at right angles by another series of lines of similar pitch, forming, of course, 10,000 small squares, which represent the base of each pyramid; hence it will be seen that if the small square upper surface of each pyramid is one-half the width of its base, its area will be one-fourth that of the base, or only one 40,000th of a square inch, and this will be the uniform shape and size of each particle of the powder so produced. Thus, if the area of the periphery of the disc is equal to four square inches, and is revolving at the very moderate speed of 200 revolutions per minute, we shall have 40,000 by 200, or just 8,000,000 small particles of brass cut off per minute, every one of exactly the same form and size, the continued pressure of the steel rollers renewing the depth of the grooves as fast as the cutter pares them down.

From this it will be obvious that in a machine closely resembling a lathe, discs of much larger diameter and much thicker than my small 4-in. experimental disc, could be employed; and, further, that ten or a dozen such discs could be put at a small distance apart on the same mandril. Thus, large quantities of solid brass could, in a short space of time, be made into powder by this simple device. It will also be understood that the cutting tool could be advanced so slowly by a fine screw properly geared, that a mere film of brass would be taken off the summit of each pyramid, and so very fine powder would be produced.

Such then was the theory on which I relied in my first attempt to produce a bronze powder direct from solid brass. My experimental apparatus was made very accurately in all its working parts, and it was with much anxiety that I awaited the time necessary to get the first results of this novel scheme, which I may say at once were very unsatisfactory. It is true that the machine worked admirably, and minute particles of brass were produced and thrown up like a little fountain of yellow dust as the disc spun round; but, alas! neither to the touch nor to the eye did it resemble the bronze powder of commerce. I was, I may freely own, deeply disappointed at this failure, because the promise was so large. The direct production of powder, worth sixty shillings to eighty shillings per pound wholesale, from brass plates costing only ninepence per pound, was, to use a common phrase, "too good to be true," and so I found it; and I well remember that at the time it required all my philosophy to persuade myself that I must look forward to such disappointments as the natural result of trying so many novel schemes. It was not the first castle I had built, only to see it topple over. Fortunately, my sanguine temperament soon enabled me to forget this failure, and to again quietly pursue my usual avocations.

About a year after the incidents I have just related, I happened to be talking to the elder Mr. De La Rue, when he mentioned to me

I

a matter in which he was at that moment greatly interested; indeed,
I may say, he was very justly irritated with a merchant who sold him
arrowroot largely adulterated with potato starch, which had spoiled
a considerable amount of valuable work for which the pure starch
of arrowroot was required. He had, he said, just found out a mode
by which he could accurately ascertain the percentage of potato starch
present; he added that chemically these substances were so much alike
in their constituents that he could not rely on simple analysis as a proof
of fraud. He told me that by putting, say 100 granules of the adulterated
starch, in the form of powder, under the microscope, he could see that
there were present granules of two distinct shapes. The genuine arrowroot
consisted of oval granules, while the potato-starch granules were perfectly
spherical; and by simply counting the number of each shape in any given
quantity he could ascertain beyond question the percentage of adulteration.

I was a good deal struck by this ingenious mode of detecting adultera
tion; and a few days later, when thinking it over, it occurred to me that
possibly the microscope might throw some light on the cause of the
failure of my then almost forgotten attempts to produce bronze powder.
I submitted some of the brass powder I had made, and some of the
ordinary bronze powder of commerce, to microscopic examination, and saw
in a moment the cause of my failure. The ordinary bronze powder is, as
before mentioned, made from an exceedingly thin leaf of beaten metal,
resembling an ordinary leaf of gold. Now, such a thin flake, rubbed
or torn to fragments, will, on a smaller scale, resemble a sheet of paper
torn into minute pieces; and if such fragments of paper were allowed
to fall on a varnished or adhesive surface, they would not stand up on
edge, but would lie flat down, and when pressed open would represent
a continuous surface of white paper. So it was with the bronze powder
of commerce; when applied to an adhesive surface, the small flat
fragments of leaf (for such they are) present a continuous bright surface,
and reflect light as from a polished metal plane. But the particles
of metal made from my machine, minute as they were, presented a
perfectly different appearance, and under a high magnifying power they
were found to be little curled-up pieces, one side being bright and the
other rough and corrugated, and destitute of any brilliancy; while on

being applied to an adhesive surface they arranged themselves, without order, like grains of sand or other amorphous bodies, and reflected scarcely any light to the eye. The reason of my failure was thus rendered perfectly obvious.

This critical examination, and the evidence it afforded me of what was really necessary to constitute bronze powder, began to excite my imagination; for to make a pound of brass in an hour, by machinery, equal in value to an ounce of gold, was too seductive a problem to be easily relinquished. Again the idea and the hope of its realisation took possession of me. "Was this to be, after all," I asked myself, "the one great success I had so long hoped for, which was to wipe away all my other pursuits in life, and land me in the lap of luxury, if not of absolute wealth?"

I studied the whole question over and over again, from every point of view, and week after week I became more and more certain that I was on the right track. At length I came to an absolute decision. "Yes," I said, "I will throw myself into it again."

I then went systematically to work, and drew out the detailed plans for the different machines that were necessary to test my idea thoroughly. I purchased a four horse-power steam engine, and erected it in close connection with my dwelling-house. I made part of the machinery in my own workshops, and personally erected the whole of it in a room into which no one was ever allowed to enter but myself. At last, after months of labour, the great day of trial once more arrived, and I had to submit the raw material to the inexorable test. I watched the operations with a beating heart, and saw the iron monster do its appointed work, not to perfection, but so far well as to constitute an actual commercial success. I felt that on the result of that hour's trial hung the whole of my future life's history, and so it did, as the sequel will clearly show.

I now became most anxious to have my views confirmed by some of the importers of German bronze. With this object, I tried for a week or so to improve the working of the machinery, and then produced a very fair sample of my new material, which I put into the small ounce packages common in the trade, and with it called on a Jewish importer.

This worthy individual looked critically at my samples, and when I requested him to purchase some he was very curious, asking me many fishing questions, for his practised eye had at once shown him that the powder differed slightly in appearance from the usual make of bronze. He, however, made a distinct offer of twenty shillings per pound for all I could manufacture.

Such an offer from a Jewish importer of bronze convinced me at once that the sample I had shown him was worth much more than the price he had named; and this view was still further confirmed by a long conversation, which terminated in an offer to give me £500 per annum for the sole use of the machinery I had invented. This proposal I could not for a moment entertain, for I could no longer doubt that my new mode of producing bronze powder was destined to be a great commercial success.

As I have already explained, I had become intimately acquainted with Mr. Young, the inventor of the type-composing machine. I told him all that I had achieved, and showed him some of the powder I had produced. He was of opinion that I ought to build a large works, and make bronze powder for "all the world." Then arose the question of capital, and this he proposed to supply, and to share with me the profits of the venture, an offer which I eventually accepted; but we had several knotty points to settle before a single step could be taken. Up to this juncture the details of my invention and the nature of the several machines used in the process were an absolute secret, and I feared to patent these inventions: firstly, because they might be modified or improved by others, but chiefly because secret machinery could be erected abroad, and the article smuggled into this country without fear of detection, because powder cannot be identified as having been made by any special machinery. Thus, a patent would have afforded no protection whatever to me. Then came the difficult question of continued secrecy; there were powerful machines of many tons in weight to be made; some of them were necessarily very complicated, and somebody must know for whom they were. Also the people who tended the machine must know all about it; and I had still to find out how all the various alloys were made, and the way in which such varied colours as the trade required were produced. The result of a review of all these difficulties was this :—

Firstly, we both agreed that if brass were still to be sold at a higher price than silver, it would be impossible for us to maintain this price if all the details of my system were shown and described in a patent blue-book, which anyone could buy for sixpence. This fact absolutely decided me not to patent the invention.

Secondly, how could we trust workpeople who could have a thousand pounds or so given them at any time for an hour or two's talk with a rival manufacturer ? This difficulty we proposed to meet by engaging, at high salaries, my wife's three young brothers, on whom we felt we could entirely rely ; so this point was satisfactorily arranged.*

Thirdly, how about making these massive machines ? What engineers could we trust ?—for any engineer must have such work done in his workshops open to the eyes of all his men.

Fortunately, here I was enabled to step in. I could undertake personally to make, not only all the general plans, but also each of the working drawings, to a large scale, for each of the machines required ; and when I had thus devised and settled every machine as a whole, I undertook to dissect it and make separate drawings of each part, accurately figured for dimensions, and to take these separate parts of the several machines and get them made : some in Manchester, some in Glasgow, some in Liverpool, and some in London, so that no engineer could ever guess what these parts of machines were intended to be used for. Of course, I was able to undertake the proper fitting together of all these detached parts after they had arrived in London.

All this was plain sailing, but it imposed on me one great difficulty. I proposed to do the work of seventy or eighty men, and I wanted this carried out by my three relatives without much labour or trouble to any of them. It simply meant this : I must design each class of machine to be what is called a "self-acting machine"; that is, a machine that could take care of itself ; and when a certain quantity of raw material had been put in place it must deal with it without a skilled attendant, do its appointed work with unerring certainty, and throw itself out of gear

* This important secret was kept inviolably for more than forty years

when its task was accomplished, to prevent injury to itself. This I also took upon myself to do, notwithstanding that one of the most powerful machines in the series would sometimes stop the career of a 20 horse-power engine, and pull it up dead, while others were performing noiselessly the most delicate operations conceivable.

Fourthly, there came the question of making the various alloys necessary to give, by oxidation, the almost endless variety of tints required in the trade. I had previously done a great deal in making alloys of copper, tin, bismuth, and other metals, and this matter we both agreed to leave for future development. My friend Young, who had acquired great confidence in my inventive faculties, remarked, "Oh, you will be certain to do it when the time comes." Relying thus with implicit faith on me, he agreed to enter into this new manufacture.

It was, indeed, no light matter, and I felt the great responsibility I was assuming. It is true I had been successful on a small scale in overcoming one of the main difficulties in the new process, but there was still much to invent, and much that at that period I necessarily knew nothing about. There were, in fact, the hundred-and-one little secrets of the trade which the ingenuity of many men and long practice had built up and accumulated around the ancient art of bronze-powder making. All of these were still kept absolutely secret by the German manufacturers, whom I proposed to rival and beat in the open markets of the world by a series of processes, absolutely new, and bearing not the faintest resemblance to any of the methods then in use. In my process, the power of steam, acting through delicate and complicated mechanism, was intended to replace the skill and well-trained muscular efforts and intelligent manipulation of the practised workman, and to imitate in every detail the ordinary commercial article. Self-reliance, and the power of readily discriminating between the first crude and imperfectly-formed ideas that strike the mind, in contradistinction to the well-considered theory on which any novel scheme really rests, allowed me deliberately, and with full confidence, to enter on this new undertaking, even though it entailed, to a large extent, the sacrifice of a small but increasing business that had been laboriously built up during several years of close application to it.

If not with a light heart, at least with a stolid and unflinching resolution, I applied myself to the task thus deliberately self-imposed. Firstly, I had to reconsider all my rough plans; I had to arrange every detail of the six different classes of machines necessary to prepare, to manufacture, and to polish and colour the bronze; all had to be made automatic and self-controlling; and when all these details had been arranged from hand sketches and figured dimensions, the labour of making the different working drawings of each machine to an accurate scale, was begun. I had, of course, to make all the necessary calculations of the strength requisite in the parts subjected to strain; of the best speed of working each machine so as to secure the highest results; then the size and proportions of each of the six machines had to be estimated, so that each one could do its part in the day's production, neither lagging behind nor doing too much. This furnished me with laborious work at the drawing-board for several months; and when all was done, each of the machines had to be dissected, and I had to commence making complete—nay, even elaborate—drawings, in detail, of every different piece required in in each of these varied machines, and to so divide the work between several engineers resident in different towns, that each had certain shaped pieces to make which he supposed were individual parts of one machine, whereas they were separate sections of several different machines, all drawn to the same scale, and sometimes represented on the same sheet of drawings. Elaborate specifications were thus rendered necessary, because neither master nor workman could use his judgment, as he would have done in the execution of any machine for a known and well-understood purpose, the full details of which are usually embodied in a complete drawing of the whole.

After much personal labour and study, this part of the undertaking was accomplished, and the making of all the machines was commenced. Meanwhile, I sought for quiet, unobtrusive premises, with sufficient land to build a factory and engine-house, and on which there was also a dwelling-house for myself and family: for such premises must not be left unguarded either by day or night. In the quiet suburb of St. Pancras I found just what I wanted, viz., an old-fashioned, unostentatious, but comfortable house, lying some distance back from the high road, and having a large garden in the rear. Such was old "Baxter House," the scene of so many experiments, and the birthplace of several entirely new manufactures.

The ground for the factory having been chosen, and a long lease of the premises obtained, I had next to plan the necessary buildings. One or two cardinal points were first determined. A substantial wall was to separate the engine- and boiler-house from the factory proper, into which the engine-driver could have no access or connection whatever, except in so far that the shafting from the 20 horse-power engine passed through a stuffing-box in the wall of separation. Access to the engine-house and coal-store was confined to a back entrance leading into another street.

The factory proper was to have but one external door, opening into a large hall, from which all the other rooms were separated by locked doors; there were no windows, except to this one outer room, all light being obtained by means of double skylights, through which no one could look; and these were further secured by impregnable inside sliding shutters. Adjoining the entrance-hall was a washing- and dressing-room, as a change of clothes on going in and coming out was imperative.

Then came other important provisions rendered necessary by the fact that the machinery was massive and very heavy, and no labourers or other workmen could be admitted to assist in putting it together and erecting it in its destined place. Concrete foundations and iron bed-plates had been put in wherever necessary, with bolts inserted therein corresponding with bolt-holes in the machine framing then being made. Heavy beams were fixed on the walls crossing over the several places where the weighty machines were to be erected, each beam having stout eye-bolts inserted in it for the purpose of attaching a block-and-tackle for hoisting. In order to facilitate the erection of all this machinery by myself and my three unpractised assistants, I had so divided the large frame castings that no single piece would weigh over ten or fifteen hundredweight.

All the smaller shafts and driving-drums were put in place, the gas and water laid on, and Chubb's safety-locks were affixed to every door before any of the machinery had arrived. The last workman had already departed, and silence reigned supreme in the empty building, into which, from that day forward, for probably twenty years, only five persons ever passed. In such a case secrecy must be absolute to be effective, and although mere vague curiosity induced many persons of my intimate acquaintance to ask to be allowed to just go in and have a peep, I never admitted anyone. Even my own sons were rigidly excluded until they were grown up. When mere lads, if they teased me to let them in, I would sometimes say, "No, you will find much more amusement at the theatres, and to-night you may go if you wish." I need scarcely say that this was greatly preferred.

Meanwhile, two steam engines and all other requisite appliances had been erected in the engine-house, where the heavy gearing was also located ; this communicated with the factory proper by two lines of 7-in. diameter shafting, which passed through the party wall.

A new phase in the undertaking was soon in active progress.

From day to day, at odd times, one of Pickford's vans would bring detached portions of the machinery, carefully packed in large wooden cases, which were delivered into the entrance of the factory by ordinary labourers, and there left to be further dealt with by ourselves alone.

The work, as a whole, had been admirably executed, and we succeeded in putting together the several parts sooner than I expected. It was with no small degree of satisfaction that we found this laborious part of the undertaking completed, and the machines ready for work.

But with the cessation of bodily labour, I entered on a period of deep and almost painful anxiety, for I felt that my position in life for many years to come was at that moment about to be determined. A few days would show if all these elaborate contrivances were based on sound mechanical principles, and whether the mass of novel machinery, occupying several large rooms, would perform its allotted task and carry forward, step by step, the successive changes necessary to convert in a single day a hundredweight of solid brass into countless millions of shining, delicate particles known as bronze powder; or whether, on the contrary, several thousand pounds, a year's increasing mental strain,

K

and much laborious physical exertion, had been cast away and thrown to the winds, leaving nothing behind but professional discredit, crushed hopes, and the inevitable regret that waits on failure of every kind.

I had, indeed, much reason for anxiety, for this was no simple test of a modification of an old and well-known machine, but the trial of a whole series of absolutely new mechanical inventions, each performing entirely new processes, following on and dependent on each other, all of which must succeed or the whole would prove a failure. But I may truly say that my hopes of success and my confidence in the whole scheme had never been shaken, although a full appreciation of the importance of the issue about to be tried necessarily caused me to feel anxious and excited. While standing alone in the silent factory, face to face with the giant whom, like Frankenstein, I had created, cold and motionless in all its grim reality, I knew that on the morrow I should, as it were, breathe into its nostrils the breath of life, by simply turning on the steam, when all those varied combinations of mechanism would be instinct with motion, and essay the task of superseding human labour and intelligence in the production of a material which, for hundreds of years, both in China and Japan, as well as in Germany, had been wholly dependent on human skill and intellect for its marvellous delicacy and beauty.

Well, the time of trial came at last, and one by one the different machines were tested. There were little hitches here and there, which took some time to rectify, but gradually each machine was got to work, and before the close of that eventful day absolute proof had been obtained of the soundness and success of the whole scheme. It was an immense relief from the severe mental strain of the few previous days, such as those only can feel who have lived on hope for more than a whole year, with a full knowledge that the time was approaching, day by day, when all their cherished expectations were to be realised or utterly destroyed.

The next thing of importance to the successful working of all this machinery was to keep inviolate the secret of its character and mode of action. Each different machine worked by itself in a room, the door of which was secured by a Chubb's detector lock; and, in addition to this precaution, each machine was itself concealed in a complete case, or covering, so that, without breaking open this case, no one could see or understand either its internal structure or its mode of operation.

It has often been remarked that the unforeseen is always sure to happen, and thus it was in reference to the intense and ceaseless noise in No. 2 Room, where thirty pieces of solid brass were being simultaneously operated upon at a very high speed, each piece throwing off from its respective surface some 2000 or 3000 fine needle-like filaments per minute. These fell in a continuous shower, and became so felted and interlaced that it was not safe to attempt to lift any portion of the accumulated mass by the naked hand, for with the slightest pressure the hand was pierced, and dozens of these fine pieces, three-eighths of an inch in length, entered the skin, and were found sticking to the fingers in every direction, like the spines on a prickly pear, or the thorns on the stem of a rose. These needle-like pieces owed their form to the intense vibration of the machine, and each one of the millions of filaments, as it was forcibly severed from the parent mass, uttered its shrill protest, and helped to swell the fearful chorus. Let those who have, happily, never heard this machine in motion, imagine the screech of a hundred discordant fiddles, accompanied by the piercing screams of as many locomotives, all bottled up in a small room, their shrill sounds echoing and reverberating from wall to wall and from floor to ceiling, until the very atmosphere seemed thick with the ceaseless roar, and the human voice at its highest pitch was wholly lost and inaudible. This was a result I might reasonably have anticipated, knowing, as I did, what the machine had to do, but in reality it never crossed my mind. Double doors covered with baize were found necessary to deaden the sound, and prevent its penetrating into the main building, while the machine itself was doomed thenceforward to work in absolute solitude.

These little filaments of brass were mechanically fed in succession into two differently constructed self-acting laminating machines consisting of highly-polished chilled-iron rolls, 12 in. in diameter and 18 in. in length, the brasses on the axes of which were pressed upon by massive spiral springs, each of which required a force of three tons to compress it half an inch. This stream of filaments was conducted between the rolls matted and felted together in inextricable confusion, and in this state they had a strong tendency to unite and so weld themselves together under pressure as to issue from the rolls with a smooth, continuous surface, resembling an ordinary sheet of solid brass. This would soon have become too compact to separate and break up again, but the tendency to unite was entirely overcome by putting about three drops of olive oil to each pound of filaments, thus not only preventing too strong an adhesion from taking place, but allowing all contiguous surfaces to slide over each other, and become more or less polished. The continuous passing and repassing through the rolls thus extended the surfaces of the filaments, and made them gradually thinner and thinner, until the whole charge under operation became soft and pliable, and was finally reduced to a leafy, flaky powder of varying degrees of fineness, the largest particles passing freely through a wire-gauze sieve having 10,000 meshes to the square inch, so that no sifting operations could possibly divide them into the ten different standard degrees of fineness required by the trade.

The crude powder, after passing through each of these two laminating machines, was polished in an apparatus, into which it was perpetually poured from a height of five or six feet, thus falling heavily on to a quantity of bronze which occupied the lower part of the receiver, but which in its turn was also lifted up and allowed to fall many thousands of times. When falling in large quantities this stream of metallic powder behaved very much like a heavy fluid, falling with considerable force, and rebounding in powerful jets; and thus by the friction of its own particles rushing among each other, their surfaces became highly polished and much smoother to the touch.

The material so far manufactured was then taken to the sorting-room, where its separation into different grades of fineness was effected.

What a remarkable contrast this room presented to the noisy cutting-room, for in this there was not a sound to attract the ear or to disturb the thoughts! Quietly and noiselessly the separation took place; just as the snowflakes silently fall and by a gentle breeze arrange themselves in a beautifully-formed snow-drift, so this apparatus did its appointed work, separating microscopic particles, inconceivably minute, from those next them in size, and so on to the coarsest powder, which was only used for inferior kinds of work.

As this mode of separating powder into various grades may be useful for many other purposes, I will here give such a description of it in detail as will make its action readily understood.

The arrangement consisted of a table about 40 ft. in length and about 2 ft. 6 in. in width, covered with black varnished cloth, on which the powder was slowly deposited; a long mahogany box, or tunnel, was inverted over the table, but was capable of being partially lifted on hinges at one side, thus giving access for the removal of the powder. At one end of the table a sheet-iron drum, or churn, was supported on hollow axes or trunnions, both of which were left open. The interior of the drum was provided with inclined shelves. Rotatory motion was given to the drum by a belt passing round it; the effect of this slow rotation of the drum was to lift the powder, and allow it again to fall in a thinly-divided shower on those shelves which occupied the lower part of the drum. A gentle current of air was caused to enter the outer end of the drum's axis, and, passing through the falling shower of powder, it emerged through the opposite axis, and quietly flowed along the tunnel already mentioned, carrying with it an almost imperceptible cloud of fine particles, which were slowly and gradually deposited upon the varnished cloth covering of the table. The largest and heaviest deposited themselves quite near the entrance of the tunnel, and others of smaller size fell farther away, the very finest reaching the distant end of the tunnel, where there was a raised box, or cupboard, in which were two cylindrical bags made of very closely-woven silk, their lower ends open to the tunnel and their upper ends closed. A blowing fan of ordinary construction was used to exhaust air from the cupboard, causing the silk bags to become inflated, and the air in the interior of the tunnel to pass through them; this was effected so gently through some 50 square-feet surface of silk as to detain in the interior any minute particles of bronze which had not fallen on to the table, while a very light current of air was steadily maintained, the force of which was accurately controlled by a large and very lightly-balanced valve in connection with the cupboard.

It is difficult to imagine the beauty of this golden snowdrift of 40 ft. in length, varying at every foot in appearance, and ranging from pieces too coarse for use, and which required further lamination, to the extremely minute particles arrested by the silk surfaces, and which, between the fingers, felt like the dust of pure plumbago, or some other wonderfully smooth lubricant. The contents of these silk bags were called No. 2000, and have been sold as high as one hundred shillings per pound. Pure copper powder so produced was supplied by me for many years to Messrs. Elkington, of Birmingham, for metallising the surfaces of elastic non-metallic moulds employed by them in the production of works of art by the electro-deposition of metals.

Thus far I have described the manufacture of raw uncoloured bronze, in which state it was used for many of the paler shades. But an almost endless variety of different colours may be produced by varying degrees of oxidation, the colour being in part dependent on the nature and quantity of the other metals with which copper is alloyed, and in part on the length of time and on the degree of temperature to which the powder is exposed, while in a heated state, to the action of the air.

One of the great difficulties in producing a beautiful uniform tint in bronze arises from the fact that almost all tints, more or less perfect, can be obtained by varying degrees of oxidation, even of pure copper ; a slight oxidation gives it a pale red-gold colour, which soon becomes richer and more golden, and passes on to citron, orange, and, in a short time, to crimson, from which it changes rapidly into claret, purple, green, pale-green, green-gold, and then still paler, until it is almost white; it then passes again to gold, and through all the series of colours, but less perfect than the first time. Now, it will be readily understood that every one of the countless millions of particles in 20 lb. of bronze powder should, as far as possible, receive precisely the same temperature for the same length of time, and be equally exposed to the current of air : then a beautiful uniform tint of colour will necessarily result. But if some parts of the mass are made hotter than others, or are longer exposed to heat or to a more perfect current of air, the powder may consist of a mixture of almost every imaginable shade of colour, be really of no standard colour at all, and thus be rendered worthless.

This delicate colouring operation was performed with unerring certainty in a gun-metal revolving vessel, mounted on trunnions, somewhat similar to a steel converter, for the purpose of discharging its contents rapidly at the right moment; this vessel was heated by an easily-controllable Bunsen burner of large size, and was provided with a means of taking out a small sample every minute for examination without interrupting its action. This important and most delicate and difficult operation was thus performed mechanically, and the device was, perhaps, one of the most perfect machines it has ever been my good fortune to design.

Still there was one more tedious task to perform. I had to justify the faith of my friend Young that "when the time comes you will be sure to find out all the proper alloys." One of a range of small buildings at Baxter House was fitted up for this purpose with a powerful air-furnace, for actual commercial working; and a smaller one for the necessary series of experiments in the production of alloys that would, when oxidised, produce the desired colours, but that must, nevertheless, be tough and ductile. There was already known to metallurgists a series of copper alloys passing under different names, and more or less resembling gold in colour; thus we had "Pinchbeck," "Mannheim Gold," "Red Tomback," "Dutch Pan Metal," "Mosaic Gold," etc., the nature of all of which had to be investigated. Then came the question of the best source of pure copper as the base of all the alloys to be made. I tried best English copper, red Japan copper, and Russian charcoal copper, made into coin. I may say that I have, since then, melted scores of barrels of Russian kopeks, on account of their purity. Dutch pan metal is, as the admirers of some of our old Dutch paintings may easily imagine, a beautiful gold-coloured brass, which I have used extensively, and which owes its beauty to its purity and mode of production. One of the ores of zinc, "Lapis Calaminares," is put into the lower part of a large crucible; small fragments of broken crucibles are laid upon it, and on this is placed granulated or shot copper (produced by pouring molten copper into water); the crucible is then covered over, the zinc contained in the ore is volatilised by heat, and, passing up through the stratum of broken pieces of crucible, is absorbed by the copper, which becomes a beautiful gold-coloured brass. Those impurities in the zinc ore which are not volatile remain at the bottom, and do not contaminate the gold-coloured alloy, which is afterwards melted in another crucible.

The production of a new tint of colour was the aim of the trade, and, with this view, a whole series of alloys were made with copper as the base. Alloys with bismuth, nickel, tungsten, molybdenum, tin, cadmium, and silver, were tried, the latter in the proportion of three of silver to seven of copper; this made a most beautiful cream-coloured bronze in its natural state, and a brilliant peacock purple when fully oxidised.

One of the most successful novelties was a margarate of copper, obtained by using animal fat in the oxidising process, producing margarate acid, and making a superb green: large quantities of this bronze found a ready sale amongst French clockmakers in Paris.

Some of the rare metals referred to were extremely difficult to reduce from their ores or oxides; but as they were not wanted in a pure state, but merely for the purpose of alloying, I found it much easier to reduce their refractory oxides with oxides of copper. In this way the oxide of molybdenum was easily reduced in combination with oxide of copper intimately blended with a black flux, consisting simply of resin in a melted state mixed with charcoal powder. The mineral wolfram readily yielded an alloy when mixed with fine granulated copper, or with copper oxide, but alone it proved very refractory.

I was quite unable to make any white metal alloy hard enough to be made into powder by my machinery. All the soft tin alloys welded by pressure into a perfectly indivisible mass, whilst the harder alloys, such as German silver, Chinese tutencg, and other nickel compounds, were not white enough to take the place of the so-called "silver powder" produced in the old mode of manufacture by the further beating of thin tin foil. I was much annoyed at being unable to execute orders for "silver bronze," and had to make an exchange with the German importers, giving them gold bronze for their cheaper white powder. This changing "old lamps for new ones" annoyed me very much ; but knowing that brass pins are whitened by a film of tin deposited on them by boiling them in a bath of tartaric acid and tin shavings, I determined to try if this system could be employed to whiten the brass powder, which we could make so easily and cheaply. There were two great obstacles in the way which threatened to render the scheme impossible, viz., the probability that these minute particles of brass would, in the act of being coated with tin, become united and stick together, and also that the tin deposit, being naturally dull, like "frosted silver," would fail in being sufficiently bright.

However, after due consideration, I planned a machine which I had reason to hope would overcome both these difficulties ; it consisted of a brass churn with a steam-jacket, so as to enable it to boil any water contained in its interior. Into this churn was put a strong solution of carbonate of soda—not tartaric acid as usual ; about 20 lb. of bright brass powder was then put in, to which was added 12 lb. of small spherical shot, formed of *pure tin* by pouring molten tin into oil. The churn was then put in action, so that the tin shot not only provided the necessary metal for solution, but by their continuous motion, as the churn revolved, counteracted any tendency of the bronze particles to become matted together by the deposited tin ; while the friction of all these rubbing surfaces in constant motion entirely prevented the dull " frosted" deposit from taking place, but on the contrary gave a beautiful polished surface to the bronze. This process was a great success, and white bronze so produced was freely purchased by the trade.

This apparatus suggested the deposition of real gold on the surface of the bronze. Some few costly experiments were made with this object, but were not successful. Probably at some future time a method of carrying out this idea may be discovered, and a large and profitable trade secured to the fortunate inventor.

While all these investigations were going on, I had taken offices in London Wall, and commenced the actual sale of bronze to the trade ; a traveller was engaged, and he sent in his first small order

L

for two pounds of pale-gold bronze for the Coalbrookdale Iron Company at eighty shillings per pound net.

The new bronze caused quite a stir in the trade. The locality of its origin and its mode of manufacture were kept a profound secret. Many consumers gladly purchased it on the favourable terms offered; while others could under no circumstances whatever be prevailed upon to give it a trial, even long after our trade was well established. As an example, I may mention one case in which my traveller made many unsuccessful attempts to do business with a very large consumer of bronze in the City, who used it in the manufacture of paper-hangings, and who said that he obtained his bronze from a descendant of Baron Scheller, an old German, who happened to be a large customer of ours, and who, for more than two years, had purchased a particular quality of our bronze, which we afterwards found that he supplied to this manufacturer at twenty shillings per pound above the price we charged to him. The old German died rather suddenly, and the paper manufacturer was informed that for years he had been using our bronze, improved (in price only) by passing through the old German's hands; he looked very crestfallen at the discovery, but kept on using the same quality, which, he told my traveller, no one in the trade but Scheller could equal.

The sharp competition with the German importers was going on pretty fiercely, when one day I was asked to receive a deputation from the trade, who came to expostulate with me for "spoiling the business, and ruining the trade and myself at the same time." I told them that they were labouring under a great mistake: that if I could maintain existing prices, it would make my fortune. They asked in all seriousness, "Can you really sell bronze at your present price without absolute loss?" I replied that I could do so, and that if they chose to deal with me, and supply my article to the consumer instead of importing it, I would allow them a discount of 25 per cent. on present prices; that I would withdraw my traveller; and in future supply no consumer below their retail prices. They took time to confer with their brethren, and finally accepted my terms, and from that time I became exclusively a wholesale manufacturer.

I was anxious to find new outlets for the bronze, and saw clearly that if I could use it as a "paint," it would answer for a great variety of purposes where a loose powder could not be applied; for instance, it could be used for gilding the raised stucco patterns on the ceilings of rooms, for temporary theatrical decorations, etc.; but quick-drying turpentine varnishes all destroyed the bronze, and turned it black. After much trouble and study of the subject, I found that the succinic acid in spirits of turpentine, and some other acids found in resinous gums and in burnt oil, could be neutralised by mixing the varnish with dry lime, and I devised a novel system of filtration, whereby all the lime, after neutralising the acid, was perfectly removed. Thus, my new "gold paint" was brought out, and those who knew how to use it, and what substances it could be successfully used upon, were delighted with it; while the attempts of others were a complete fiasco, and it was by them condemned as a failure, notwithstanding which as many as 80,000 bottles of it have been sold in the course of a year. Among its various uses, a very odd one was due to the 'cuteness of a Birmingham manufacturer of "coffin furniture." Instead of stamping in brass the variety of ornaments used on the sides of coffins, he stamped them in the cheaper metal zinc, and made them beautiful with gold paint; they lasted much longer than was necessary for the purpose, and only turned black after some time.

On one occasion, when giving an order for varnish at the factory of Messrs. Hayward and Sons, they asked if I would like to go through the works; and as I always take an interest in any manufacture that I am unacquainted with, I accepted their kind offer, and passed a very interesting hour or two. Everything was shown to me and lucidly explained; but there was one thing which seemed to stand out from all the rest, which, I thought, was a wasteful and unnecessary source of expense, and so I expressed myself to Mr. Hayward at the time. It is only another of the many proofs I have had of the very different impressions which the same facts make on differently constituted minds; here was an important fact, presented to me for the first time, but which my friend Mr. Hayward, during forty years of practical experience, had had every day before him, but had never seen, at least from my point of view. I said to him: "Why do you not do so-and-so, and save this great cost?" He was much struck with the idea; and when we returned to the offices to partake of a biscuit and a glass of sherry, I said: "If you will give me a sheet of paper, I will draw you a sketch of a simple apparatus which, I doubt not, will have the effect I have described." I made the sketch, which my friend received in a very kindly spirit, albeit with a full share of doubt as to the possibility of its effecting so great a desideratum by such simple means. His son, Mr. Sharp Hayward, whose more recent chemical studies gave him an advantage in forming an opinion, unbiassed by long routine practice, said: "I will see this tried as soon as possible"; and so the matter passed, and was soon quite forgotten by me. Some two or three months later, however, when I was sitting at breakfast at Baxter House, I saw a horse and cart stop at my front garden gate, and the driver bring a letter up to the door. It was from my friend the varnish manufacturer; he told me briefly that they had tried the method I suggested to him on the occasion of my visit to his works; it was, he said, a perfect success, and that I should greatly add to the obligation conferred if, in speaking of the circumstance at any future time, I omitted to mention the nature of the improvement I had suggested. The letter went on to say that one of his sons was a wine-grower in Madeira, and, having had a splendid vintage, he had sent his father a pipe of Madeira as a present; "and," said my friend Mr. Hayward, "it at once struck me that it was a fortunate opportunity, accidentally placed in my way, of acknowledging my indebtedness to you; will you, therefore, oblige me by accepting it as a souvenir of your visit to our varnish manufactory, which has been of so much advantage to me." Of course, I accepted with great pleasure this most welcome gift. I had the wine bottled, and in due time it turned out to be of excellent quality, and I may safely say that I have never drank of wine which gave me so much pleasure as this did; it was treasured up, and always reserved for special occasions, and I believe that at this time of writing there are still some few bottles remaining, safely stowed away in my cellar. I shall have occasion to refer again to this incident later on.

The bronze business was now progressing most satisfactorily. I had given up many of my former employments, and felt that I might indulge in some luxuries from which I had hitherto carefully abstained. I thought that a brougham would be very useful to me, and, at the same time, a source of much convenience and pleasure to my wife and children; but I had no suitable place for it at Baxter House. I imagined that I needed a meadow for a horse, but it is most probable that it was really for myself that I felt the need of "pastures new"; for the instinct of the village boy was evidently in the ascendant, and I sighed for the large kitchen garden, and the poultry-yard, and other rural delights, the very thoughts of which had long slumbered and been forgotten. The result of all these aspirations was the taking on, a fourteen years' lease, of a house, the grounds of which abutted on the beautifully-wooded domain of Lady Burdett Coutts, at Highgate; and here I built a large conservatory, kept my cows and Shetland ponies, played at cricket or quoits on summer evenings, and could sometimes, in my quiet walks round my own meadows, almost fancy myself at my dear old birthplace, Charlton, and myself again a village boy. I had given the name "Charlton House" to my residence at Highgate, and while living there I used to go down to Baxter House every morning to business, which, as far as the bronze powder was concerned, was conducted almost entirely without my assistance; so that I had ample time to devote to the many new and interesting subjects that seemed for ever to present themselves to my mind and demand investigation.

I had a good light drawing-office fitted up at Baxter House, and was always at work there on some novel invention, for which patents were being taken out; in some cases experiments were made on the premises, and all sorts of machinery and furnaces were erected to put the ideas to the test of practice. So much did the work at the drawing-board increase, that on one occasion, when much pressed, I applied to my friend, Mr. Bunning, the City Architect, for the loan of an assistant draughtsman to finish some patent drawings. "Well," he said, "I think I can let you have a pupil of mine who is just out of his time; he is a clever architect, an expert at the drawing-board, and is a gentlemanly young fellow, in whom you can place implicit confidence." He then called the young man into the office to see me, and this was my first introduction to my friend and partner, and afterwards my brother-in-law, Mr. Robert Longsdon. We soon arranged terms, and he came to Baxter House to assist me for a while with my drawings ; we worked side by side in the same room for many months, during which time I gained something in architectural taste and knowledge, and he gained from me, and from his daily occupation, a further insight into engineering. It is not surprising, under these circumstances, that a real and solid friendship should spring up between us ; after a time I proposed that we should take more convenient offices in the City, and do something jointly in the way of architecture and engineering, while I was still to devote myself chiefly to my inventions.

We fixed on No. 4, Queen Street Place, for our City offices, and it was from there that so many of my patented inventions were dated. I had now, for the most part, discontinued my labours at Baxter House, except for the erection of experimental machinery or furnaces. On one of these occasions, while busily engaged there, our local policeman called in to see me on a private matter that had exercised his mind very much for the previous two days. He told me that he thought my house was going to be robbed, for it had been watched from early morning until late at night by a person stationed at one of the windows of a public-house that commanded a view of the front door of Baxter House. He said that the man was of gentlemanly appearance, but he did not think he was a member of the "swell mob"; and, in fact, it was to him quite a mystery. I asked, "Do you think he is a German?" "Probably so," he said. "At any rate he is a foreigner." I commended the officer for his vigilance, and giving him a small gratuity, I told him to let me know if anything further occurred.

I at once formed the opinion that the person referred to was watching to see some of the numerous workpeople, who, he might naturally suppose, were employed in my bronze factory, and of whom he might try to obtain information as to my secret process. Now, it so happened that, with the exception of my engine-driver, there were no operatives employed, but only my three relatives, who never left the

office all at one time, and when they did leave might well be taken for
office clerks, who would know nothing of the manufacture; and so the
foreigner watched in vain for an opportunity of bribing some of my
imaginary workmen.

I was very desirous of probing this mystery, however, for which
purpose I called into my office my engine-driver, a steady, honest
Scotchman, who had long been in my employ. I told him what the
police officer had communicated to me, and arranged that he should go
just as he was, with his shirt-sleeves tucked up (the very beau-ideal of a
British workman), over to the public-house, leaving by my front door,
so as to be observed by the man on the watch, and take something to
drink at the bar. "If," said I, "the stranger comes down and asks
questions, say you don't know, but will enquire and let him know; if
he offers you anything, accept it, and he will then believe that he can
trust you." No sooner had my engineer entered the public-house
than the stranger came downstairs and asked him: "Do you work at
the bronze-powder factory opposite?" "Yes," was the reply. "Why
I ask you," said the stranger, "is this: I have invented a machine
for making 'hooks and eyes,' and I want some clever engineering firm
to make me these machines; I have been told that you have beautiful
machinery over the way, and I should like to give an order for my
machines to so eminent an engineer; do you know who made all the
machinery at your works?" "I don't know," said the wary Scotchman,
"but I can enquire." "Well," said the stranger," meet me here when
you leave work to-night, and if you can let me know who made your
machinery, I shall reward you handsomely." All this was told me
on my engine-driver's return from the public-house, and I was determined
to have an interview with the stranger. I told my engineer to meet
him as arranged, and simply to tell him that he had ascertained that
the whole of the machinery at the bronze factory was planned by a
Mr. Henry, who resided at No. 4, North Street, New Road, and that
he would probably be there to-morrow at 11 A.M. This was my
brother's address, to which I went before the hour named, telling
my brother's servant that I expected a gentleman to call at 11 o'clock
to ask if Mr. Henry was at home; that she was to say yes, and ask

him into the dining-room, where I would await his arrival. Punctual
to the hour the stranger came. I offered him a chair, and awaited
his communication. "Have I the pleasure," he said, " of seeing Mr.
Henry, the engineer who designed all the machinery at the bronze
factory at St. Pancras?" "Yes," I replied, " I designed the whole of it."
" Ah," said my visitor, " I am so glad thus to make your acquaintance;
for this purpose I have come over from Bavaria, and wish you to
construct a duplicate of it for me." "Well," I replied," this is not
possible, for I have quite given up mechanical engineering, and am so
deeply engaged with some new inventions that I could not even undertake
to furnish you with plans or drawings of the machinery." "But," said
the stranger, "I shall pay you anything you demand in reason; so it
may answer your purpose to lay aside other things for a time." He
pressed me very hard, and I did not know how to get rid of him. I
knew exactly what his object was in watching my premises, and was
satisfied. "Well," he said, " at least you can give me some idea of the
nature of the process, and I shall pay you any fees you like to name."
I replied : "I cannot accept a fee for any information I may give you,
nor would it be fair on my part to furnish you with detailed plans
of the machinery I have constructed for another manufacturer; but
as you have come such a long distance, I may just tell you that to
make cheap bronze powder, you need not go further than making your
alloy in what you call 'long metal'; you will not require any parchment
books to beat in, and you will avoid the use of gold-beater's skins, and
all the expensive labour of beating it into thin leaves. At the Baxter
House factory neither parchment nor gold-beater's skins are ever used; and
you will be surprised to hear that I have no secret to tell you. The
principle on which they work is so simple that a child could understand
it in a moment; you know, of course, what ordinary millstones, used
to grind flour, are like. Well, suppose you take two circular discs,
say, 2 ft. in diameter; divide their surface into eight compartments
by radial lines, and cut small parallel sloping grooves, diagonally arranged
in each compartment : then you have a pair of what may be called 'steel
millstones,' which may be driven by usual wheel-gearing; cut up your
thin sheets of long metal, with a pair of shears, into pieces about 2 in.

square, which a boy can feed into a round hole in the centre of the upper millstone, into which a thick stream of soap and water is constantly running. You cannot fail to understand the principle involved, and you will be not a little astonished to see the result of this simple operation. As far as this information is concerned, you are perfectly welcome to it, and I must now close the interview." My visitor was delighted, and profuse in his compliments and thanks. I have often wondered whether, on his return to Bavaria, he tried to put in practice this impossible mode of making bronze powder; if he did, the disappointment he would experience would be only a fitting punishment for his meanness in trying to bribe those who were in possession of my secret.

Before long my bronze powder was fully recognised in the trade, and found its way into every State in Europe and America; it had, in fact, become the one staple manufacture I had so long and so earnestly sought for, and which I hoped would some day replace and render unnecessary the constantly-recurring small additions to the business I had so laboriously built up. The bronze powder business, however, no longer required my personal attention, and was well managed by those I had chosen as the guardians of a secret, which was long and honourably kept. The large profits derived from it not only furnished me with the means of obtaining all reasonable pleasures and social enjoyments, but, what was even a greater boon in my particular case, they provided the funds demanded by the ceaseless activity of my inventive faculties, without my ever having to call in the assistance of the capitalist to help me through the heavy costs of patenting and experimenting on my too numerous inventions. The importance of this steady supply of the sinews of war may be easily imagined from the fact that I have obtained no less than 110 separate patents, the mere stamp duties and annuities on which have gone far to absorb £10,000, to say nothing of legal fees, and the costly labour of writing long specifications, coupled with the work of making the necessary drawings required to illustrate and define the precise nature of these varied inventions. Only about a dozen of these inventions are referred to in this hasty ramble through fields of thought and labour; the whole, if thoroughly described and gone into

M

on their merits, would utterly weary, and wear out the patience of, my most indulgent reader.

While referring to patents for inventions, I cannot refrain from pointing to this particular invention of bronze powder as an example that may advantageously be borne in mind by those short-sighted persons who object to grants of letters-patent. There can be no doubt of the fact that the security offered by the patent law to persons who expend large sums of money and valuable time in pursuing novel inventions, results in many new and important improvements in our manufactures, which otherwise it would be sheer madness for men to waste their energy and their money in attempting. But in this particular case the conditions were most unfavourable for patenting, owing to the fact that the article produced was only a powder, and could not be identified as having been made by any particular form of mechanism. Therefore it could not be adequately protected by patent; moreover, by my machinery, the cost of production, if only paid for at the ordinary rates of wages, did not exceed one-thirtieth of the selling price of the article. This fact alone offered an irresistible temptation to others to evade the inventor's claims, and so rendered the patent law a most inadequate protection. On the other hand, the great value of a small bulk of the material made it possible to carry on the manufacture in secret, and this method of manufacture was rendered the more feasible by making each different class of machine self-acting, and thereby dispensing entirely with a host of skilled manipulators. It may therefore be fairly considered, so far as this particular article was concerned, that there were, in effect, no patent laws in existence.

Now let us see what the public has had to pay for not being able to give this security to the inventor. To illustrate this point, I may repeat the simple fact that the first order for bronze powder obtained by my traveller was for two pounds of pale-gold, at eighty shillings per pound net, for the Coalbrookdale Iron Company. I may further state that, in consequence of the necessity for strict secrecy, I had made arrangements with three young men (my wife's brothers), to whom salaries were paid far beyond the cost of mere manual labour (of which, indeed, but little was required). My friend Mr. Young desired to occupy the position of sleeping partner only, and not be troubled with any details of the manufacture; so I entered into a contract with him to pay all salaries, find all raw materials, pay rent, engine power, and bring the whole produce of the manufactory into stock, in one-ounce packages, ready for delivery, at a cost, for all qualities, of five shillings and sixpence per pound; after which he and I shared equally all profits of the sale. It is rather a curious coincidence that the one ounce bottles of gold paint were labelled five shillings and sixpence each, off which the retailer was allowed a liberal discount.

Had the invention been patented, it would have become public property in fourteen years from the date of the patent, after which period the public would have been able to buy bronze powder at its present market price, viz., from two shillings and threepence to two shillings and ninepence per pound. But this important secret was kept for about thirty-five years, and the public had to pay excessively high prices for twenty-one years longer than they would have done had the invention become public property in fourteen years, as it would have been if patented. Even this does not represent all the disadvantage resulting from secret manufactures. While every detail of production was a profound secret, there were no improvements made by the outside public in any one of the machines employed during the whole thirty-five years; whereas during the fourteen years, if the invention had been patented and published, there would, in all probability, have been many improved machines invented, and many novel features applied to totally different manufactures.

I have lingered long over this subject of bronze powder, because it is one which has had great influence on my career; it was taken up at a period when my energy and my endurance, and my faith in my own powers, were at their highest; and as I look on all the incidents surrounding it, through the lapse of time and the many changes of the fifty years since it was undertaken, I wonder how I had the courage to attack a subject so complicated and so difficult, and one on which there were no data to assist me. There were not even the details of former failures to hold up the finger of warning, or point out a possible path to pursue, for no one had yet ventured to try and replace the delicate manipulation which experts had made their own, both in Japan and China, where texts or prayers printed with bronze were offered up at the shrine of Confucius two thousand years before I had ever seen a particle of bronze powder.

I cannot conclude this imperfect account of the bronze powder manufacture without a tribute to those on whose scrupulous integrity hung the whole value of this invention from day to day through all those long years. The eldest brother of my wife had previously been connected with the watch manufacture in London, while the next to him in age had not yet commenced his career; and I could offer a position sufficiently remunerative to induce both of them to assist in carrying on the bronze manufacture. The younger brother, Mr. W. D. Allen, had been with me as a pupil for a year or two; finding him a bright, intelligent lad, when he was about to leave school, I prevailed upon his father to let me have charge of him, and impart, as far as I was able, some knowledge of engineering. Thus, living in the same house with me, he grew up more like one of my own sons than a brother-in-law. In due time he also took up his position in the bronze works, and kept my secret with the same silent caution as his elder brothers had done. He also assisted me in my early steel experiments at Baxter House, and, later on, when I determined to build a steel works at Sheffield, the great confidence I felt in his judgment and integrity induced me to offer him a partnership. He became the managing partner of Messrs. Bessemer and Co., of Sheffield, and after fourteen years of the most successful management, I and each of the other partners retired from the business, leaving Mr. Allen in sole possession of the works, which he purchased at a sum mutually agreed upon.

Many of my readers will be more or less acquainted with Mr. W. D. Allen, whose intimate knowledge of every detail of the Bessemer process enabled him to pay large dividends to the present Limited Company, even in bad times. Thus my brother-in-law's position in life was assured; his brother John had died several years previously, and there only remained his brother Richard to carry on the business at Baxter House.

In closing these details of the bronze powder manufacture, I may say that, later on, the handsome royalties paid by my steel licencees rendered the bronze powder business no longer necessary to me as a source of income; and I had then the extreme satisfaction of presenting the works to my brother-in-law, Richard Allen, who had, with so much caution, successfully kept, for more than thirty years, a secret for which, he perfectly well knew, some thousands of pounds would have been given him at any moment.

CHAPTER VI

IMPROVEMENTS IN SUGAR MANUFACTURE

IN the early part of the year 1849, I had formed an intimate acquaintance with a Mr. Cromartie, a Jamaica sugar-planter, and at many of our friendly meetings we had discussed the question of the sugar manufacture as then carried on in the West Indian Islands. The more I heard of the state of this important industry, the more astonished I became on finding out how rude, how unmechanical, and how unscientific were many of the processes then employed, not only in extracting the saccharine juices of the cane, but also in its after-treatment. By a curious coincidence, at this very period the imperfection of the Colonial sugar manufacture had attracted the attention of the Society of Arts, and his Royal Highness Prince Albert had taken a very special interest in this subject, and generously offered a gold medal to be awarded to the person who should, during the ensuing year, effect the greatest improvement in the mode of expressing the saccharine juice of the sugar cane. I was much interested on hearing this, and applied myself to the problem with great zest, for I heard that the contest was to be an unusually sharp one. I was informed that the manufacturers of Colonial sugar machinery looked on it as a question that would decide which firm was in future to do the bulk of the Colonial engineering work, and that powerful vested interests were supposed to be at stake. This rendered it the more necessary that I should make every effort to gain such a knowledge of the subject as would enable me to devise a machine capable of extracting, as completely as possible, the whole of the juice from the cane. I, therefore, in the first place, obtained from Madeira a bundle of sugar canes, and I may say that up to that time I had never seen a cane. Those I had ordered to be sent to London arrived fresh and full of juice, as

I had directed that their ends should be dipped in melted pitch, so as to prevent decay, and the escape of any juice from them.

These canes were from $1\frac{1}{2}$ in. to $1\frac{3}{4}$ in. in diameter, having dividing knots at from 5 in. to 7 in. apart, throughout their length. The cane consists of an outer tubular part of hard fibrous wood, thinly coated with very hard pure silica; the interior of the thin wooden tube is filled with a soft pithy matter, almost like a sponge, saturated with juice, of which the ripe mature cane contains about 88 to 90 per cent. of its

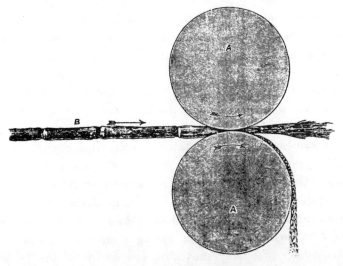

Fig. 17.   Sugar-Cane passing between Rolls

whole weight. I put short lengths of these canes to many tests in different ways, and especially noted their great elasticity; a 6-in. length, suddenly pressed between two flat surfaces, would lie in a complete pool of juice, and if the pressure were quickly released, the flattened elastic tube would again expand and as quickly reabsorb a very large portion of the fluid with which it was in contact. Here, I saw at a glance, was the weak point in the roller-mill, in which the cane quickly enters between a pair of rolls, and is for the moment collapsed. But as it emerges from them it again expands by its elasticity, drawing into the expanding spongy mass a large portion of the juice, which is rapidly flowing in contact with it, over the lower roll of the mill. This will be readily understood by reference to the engraving, Fig. 17, page 87, showing in section a pair of iron rolls A, A, between which a cane B is passing in the direction shown by arrows. It will be observed that at the central part the cane is crushed very thin; but as it emerges, it, in part, recovers its former dimensions, and in doing so absorbs a very large percentage of the juice previously expressed.

These and other observations, carefully made and noted at the time, forced on my mind the conviction that no form of roller-mill could, from the inherent nature of its action, give satisfactory results; and that a slower and longer continued pressure on the cane must be resorted to, if the greater part of this valuable fluid was to be extracted.

By means of the hydraulic press, 86 per cent. of juice could be obtained; but this system was far too slow, and entailed so much labour as to render it impossible to deal with the enormous mass of canes grown on a moderate-sized plantation. Following, however, the general idea of the press, I designed an entirely novel system of extracting juice from canes, the main feature of which was the cutting of the cane into lengths of about 6 in., thus leaving both ends of these short pieces open for the escape of the juice, instead of operating in the usual way upon canes of 4 ft. to 6 ft. in length, having numerous transverse knots or partitions, which effectually prevented any escape of the juice endwise. The two convex surfaces of a pair of rolls of 2 ft. in diameter, pressed on less than 6 in. of cane, at any moment, and if they revolved as slowly as five revolutions per minute, the 6 in. of cane passing between them commenced and finished the period of pressure in just one second. In the cane press about to be described, every one of these open-ended 6 in. lengths would be subjected to intense pressure for a period of two and a-half minutes; in practice, it has been found that the juice was vigorously given out for the first minute, and then gradually declined; finally ceasing to yield one drop more of juice for about half a minute before it was discharged from the open end of the press tube.

In order that this new system of continuous pressure might be fairly tested, I erected a complete press and steam engine combined, at my experimental premises at Baxter House. I also imported a large quantity of canes from Madeira and from Demerara, for the purpose of studying their structure, and making experiments with them, under varying conditions of pressure and time. The quantity of juice which this small apparatus was found capable of expressing exceeded 600 gallons per hour. The juice was much more free from pithy fragments than that which was obtained from the roller-mill, while the quantity of colouring matter and chlorophyl extracted from the knots was much smaller, because in the press these hard knots sank into the softer surrounding parts, while between the rolls they got far more pressure than the softer parts of the cane, because of their greater solidity. But the most important result, which was fully established, was the high percentage of juice obtained.

In our first experiment, made immediately after the arrival of the canes, the quantity of juice obtained exceeded 80 per cent. ; in another experimental trial, when the canes had been four months cut, $73\frac{3}{4}$ per cent. was expressed; and, later on, in a public experiment, when the canes had suffered from drying, $65\frac{1}{2}$ per cent. was expressed. In reference to the far smaller quantity of juice obtained in practice by the old system of rolling-mills, I may quote from the Seventh Report of the Parliamentary Committee on Sugar and Coffee Planting, where, at page 259, will be found a memorandum dated "Colonial Laboratory, Georgetown, 3rd February, 1848," from Dr. John Shier, Agricultural Chemist, who—speaking on Sugar Mills—says:—

From numerous trials on various estates, I am satisfied that the average yield does not exceed 45 per cent. ; the first of all improvements then seems to be to obtain a larger percentage of juice from the cane.

It is a curious fact that throughout this competition no one but myself came forward with any plans to do away with the roller-mill. There were plenty of improvements in this class of machine ; two rollers and three rollers, new gearing, and combined engines and mills. In one case a magnificent mill had been patented. It was a combined engine and mill, weighing no less than forty tons—no light matter to pass over half-made Colonial roads—and it was designed by Messrs. Robinson and Russell, who were large sugar-mill manufacturers in London.

N

The extreme lightness of my cane press formed a strong, and from a Colonial point of view, a most important, contrast to this. The press was put to work, and publicly exhibited to dozens of persons who were owners of sugar plantations in our various sugar-growing Colonies, and great expectations were formed by them. They saw the canes weighed and operated upon, then the squeezed mass again weighed, the reduction in weight clearly showing the quantity or percentage of juice obtained by the press, which was admittedly at least 20 per cent. more than the average produced by the old roller-mills then universally employed. The juice obtained was very rich in quality, in consequence of a considerable evaporation from the canes which had gone on during the three or four months since they were first cut. As a matter of curiosity, I manufactured from the juice obtained about half a hundred-weight of crystallised sugar of very good quality, which I presume was the first sugar ever produced direct from the sugar-cane in London, and was much prized as a matter of interest by some of my friends for that reason.

Without going into the minutiæ of detail, it may be interesting to give a short description of the cane press, which is here illustrated by engravings copied from drawings of the press, as erected at my experimental works, Baxter House.

The first engraving, Fig. 18, on Plate VIII, shows a side elevation of the press, and the steam-engine with which it was combined, on one large bed-plate. The second engraving, Fig. 19, on page 91, shows a vertical section through one of the gun-metal perforated pressing tubes; the interior of these was of rectangular form in cross-section, being 6 in. in height by $3\frac{1}{2}$ in. wide.

In the centre of each of these tubes there was a massive plunger fitting accurately. A square steel bar passed through the two plungers, and also through slots made in the sides of the tubes for that purpose, the outer ends of these bars being rounded and fitted into the ends of two massive connecting-rods, which were actuated by a pair of short-throw cranks formed one on each side of the central crank of the steam-engine. This arrangement is best seen in Fig. 20, page 91, which is a plan of the cane press and engine.

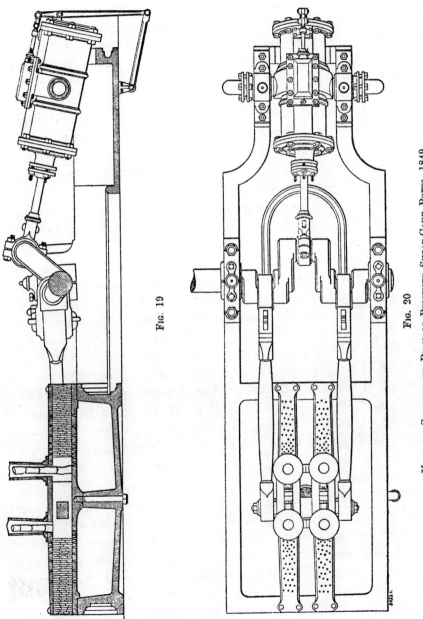

Fig. 19

Fig. 20

Vertical Section and Plan of Bessemer Sugar-Cane Press, 1849

From the upper surface of each of the pressing-tubes, two tall circular hoppers stood vertically, and were attached at their upper ends to a stage or floor on which the canes were delivered, and where two attendants were stationed, whose business it was to continually drop canes into these tubular hoppers. When the several parts of the apparatus were in the position shown in Fig. 19, page 91, the plunger had cut a 6-in. length off the lower ends of the canes in the left-hand hopper, and had pushed them against the compressed mass of canes occupying that end of the pressing tube, the result being that this mass was moved a little way further along, the fluid parts escaping from the numerous perforations in the tube.

While this had been going on the canes in the right-hand hopper had fallen down into the pressing tube, and the return stroke of the plunger would then cut off a 6-in. length from these canes, and force them up against the mass of canes occupying the right-hand end of the press tube, moving the mass of flattened canes a small distance forward, and discharging a portion of them from the open end of the tube. In this way every rotation of the crank cut off portions of the canes in each of the hoppers, and carried them forward, thus keeping the tubes always filled with a mass of compressed canes, which were jammed so tightly in the tubes as to offer an immense resistance to the plunger, governed by the length of the tube. The two cranks which actuated the plungers were at right angles to the crank operated on by the steam power; hence, when the engine was exerting its greatest power, the cranks actuating the plungers were passing their dead points and thus exerted an enormous force on the mass of canes, which moved forward but a very small distance at each stroke.

With the engine running at only 60 strokes per minute, each plunger cut off two 6-in. lengths from each cane in the hoppers; and as there were four hoppers with two canes in each, 4 ft. of cane were operated upon at each revolution, or at 60 strokes per minute only, some 240 ft. of cane were cut and pressed per minute. It was found that the canes thus passing along the tubes were forced out of the open ends of the latter adhering together, and looking like a polished square bar of wood; the juice of the cane passing through the numerous perforations and falling into the square cistern formed beneath them by the massive bed-plate, was conveyed away by a pipe to the evaporating pans.

The committee appointed to judge of the various plans submitted in competition for the gold medal offered by his Royal Highness, Prince Albert, came in force to Baxter House, and witnessed the cane press in operation. Although the committee did not openly express their views to me, I could not doubt that their convictions were entirely in my favour, a natural result of the incontrovertible facts I had placed before them. In due course I received a notice that the prize so much coveted was about to be awarded to me, an entire outsider, wholly unknown to any of the sugar-mill manufacturers of this country.

How often it has occurred to me, and how often have I expressed the opinion that, in this particular competition—as in many other previous cases—I had an immense advantage over many others dealing with the problem under consideration, inasmuch as I had no fixed ideas derived from long-established practice to control and bias my mind, and did not suffer from the too-general belief that whatever is, is right. Hence I could, without check or restraint, look the question steadily in the face, weigh without prejudice, or preconceived notions, all the pros and cons, and strike out fearlessly in an absolutely new direction if thought desirable. Indeed, the first bundle of canes I ever saw had not arrived from Madeira a week before I had settled in my own mind certain fundamental principles, which I believed must govern all attempts to get practically the whole juice from the cane; but of course, there were many circumstances that rendered it necessary to modify first principles, having reference to the cost of the machine, its easy transit across country, freedom from repairs in isolated situations, etc., etc.

In due course I had to attend a meeting at the Society of Arts, where I was much surprised to find the large hall crowded with spectators. At one side of the room was a raised daïs, on which his Royal Highness, Prince Albert, was seated at a small table, and at his side was the Chairman of the Committee of Mechanical Experts, who had reported to the Prince the result of their deliberations. In front of the platform occupied by the Prince Consort there was a long avenue covered with crimson cloth, and skirted on each side by rows of seats, occupied by ladies, who added to their personal charms all that the milliner's art could accomplish to give grace and *éclat* to the occasion. It was, I found, my rôle to brave all the dangers of this double battery of youth and beauty; and, like the good St. Anthony, I had to keep my eyes fixed upon the crimson cloth, for I did not dare to look. If anything could add to the satisfaction of the moment, it was the presence on this occasion of the Chairman of the Committee of Experts, who was about to read his Report, for this gentleman was no other than that talented and well-known engineer, Mr. John Scott Russell, than whom no one in all Great Britain was more able to do justice to the subject reported on. His firm of Robinson and Russell were extensive manufacturers of Colonial Sugar Machinery, but they had refrained from competing on this occasion, thus allowing Mr. Scott Russell to add another to the many proofs of the high code of honour so conspicuous in the whole body of Civil Engineers in this country, by giving publicly unqualified testimony to the merits of what was, in fact, the scheme of a rival manufacturer. The honourable distinction received from such a source, while it was most gratifying to myself, was more than reflected upon the speaker.

Among many other things, Mr. Scott Russell, in addressing the Society and reading his report, said, " the new cane press of Mr. Bessemer has the merit of introducing a principle at once new and of great beauty into the process, while reducing the weight and cumbrousness of the machinery ; much has been done by Mr. Bessemer towards removing the main obstacle to improvements in the working machinery of the Colonies in the Tropics, viz., the difficulty of transport." Mr. Scott Russell further pointed out that : "When these facts of facility of transport, simplicity of foundation, and other advantages come to be considered in reference to cost, it will at once be perceived that notwithstanding the great advantages it offers in respect of quality and quantity of juice, certainty and uniformity of action, and freedom from accident by wear and tear, the cane press, when placed in working condition upon an estate, will have cost less than the most ill-constructed mill and engine to be obtained from the cheapest and most inferior makers."

At the conclusion of Mr. Scott Russell's address there was a round of applause, and this was followed by the rising of his Royal Highness Prince Albert, who complimented me in the kindest manner on the success of my invention—an invention which I had taken such unusual steps to prove, by bringing, as it were, the Colonies to us, and by resting my claims to recognition on actually accomplished facts. His Royal Highness then placed in my hands a beautiful Gold Medal. In briefly expressing my thanks, I said that whatever advantages might in the future result from this invention, they would be entirely due to the encouragement held out by his Royal Highness; and amid the warmest recognition from the assembled spectators, I beat a retreat with the prize I had received.

CHAPTER VII

A HOLIDAY IN GERMANY

I HAD been working pretty hard up to the time of the trials of the cane press, and felt that I was entitled to a little relaxation. One of my German friends, who had ceased to import bronze, was about to visit his native town, and pressed me to join him in a pleasure excursion up the Rhine; my wife preferred taking the children and governess to some quiet English town, and so I set off with my friend, stopping first at Cologne, which, with its quaint old buildings and magnificent Cathedral, afforded us much pleasure for our first week's holiday. After this, we went up the Rhine as far as Düsseldorf, where we arrived on the day of St. Ursula, the patron Saint of Düsseldorf. The streets were all alive with spectators viewing the long religious processions to be seen issuing from the various churches; the large white caps of the lady processionists formed a strong contrast with their simple black dresses; then came numerous bands of children, carrying flowers and various emblems, the clergy heading each procession, and carrying coloured wax candles of several feet in height, all of which was both novel and interesting to an untravelled Englishman like myself, but which has been so often seen by many of my readers that I will not "repeat the oft-told tale." After a short stay here we pursued our journey up the Rhine, passing many well-known points of interest that skirt that beautiful river, and eventually landed at Biebrich, whence we pursued our journey to Frankfort, with which town I was very much pleased. I have still a distinct remembrance of my visit while there to Beth-mann's Museum, to see the celebrated statue of Ariadne gracefully seated on a tiger, the room in which it is shown being provided with crimson curtains, through which a rich glow of light falls on to the cold white marble, producing a unique and charming effect.

From Frankfort we journeyed on to Nüremburg, where we took up our abode at Bayrischer Hof. We determined to see all we could, in a week, of this charming, quaint old town. A few days later, my friend told me he wished to go over to Fürth, some miles distant. This little town is the principal seat of the German bronze manufacture, and my friend, having some connection there, we went together to Fürth, where he called on a manufacturer with whom he had done business in former years. We spent a very pleasant day with this gentleman's family; the weather was delightful, and we were able to sit under the trees in the open square until a late hour in the evening, enjoying not a few glasses of their light beer, and returning at night to Nüremburg, to renew our search for amusement among its quaint old streets and public buildings. On the second day after our visit to Fürth, on our arrival at the Hotel, the landlord told us there was something wrong, and that two police-officers were waiting our return, and had papers for our arrest. We were, of course, greatly astonished, but had no doubt that it was some huge mistake; however, it was not so, and we found that the order was to arrest an Englishman of the name of Bessemer. After a little discussion, the landlord very kindly suggested that we should remain in charge of one of the officers at the Hotel, while he and the other went to the police-office, where he became bail for our appearance before the magistrate at eleven on the following morning; so fortunately we were allowed to pass a quiet night at the Hotel.

Next day, after some little bustle and annoyance, I found myself in court, face to face with my accuser and the magistrate, who fortunately could speak enough of my own language to make himself perfectly well understood. He told me that I was charged with what was a very grave offence in Bavaria, viz.: attempting by bribery to induce a workman in the employment of a bronze manufacturer at Fürth to betray the secrets of his employer, and go over to England to assist in establishing a manufactory on the model of that of his employer. It was added that I had offered the man 2,000 thalers (about £200). This was the main feature of the charge which was read over to me in German, and then in English by the magistrate, who demanded to know what I had to say

o

in my defence. I then explained, at some length, the fact that I had many years previously discovered a system of making bronze powder by machinery, and that, with three attendants, I could manufacture daily as much bronze powder as eighty men could produce by the system then in use at Fürth; that I had lowered the price of the article 30 or 40 per cent; and that the people of Fürth had, no doubt, lost a large part of their trade, a circumstance likely to cause much irritation to the workmen engaged in this manufacture. I said that the idea of my wishing to establish the old mode of manufacture in England, and to learn any secrets connected with it, was simply ridiculous. I further stated that I had come there purely for pleasure and recreation; and the landlord of the Hotel where I was staying would be able to tell them that, in the absence of my German friend, I was wholly unable to ask for a single article of food in the German language. "If you, sir," I said "will ask my accuser what I offered him, and what was said on both sides before finally settling to give him 2,000 thalers for his services, you will readily convince yourself of the absolute falsehood of the charge, which could only have been made in pure spite or envy." A long talk in German between the magistrate and my accuser ended in the magistrate saying that I was dismissed, and found not guilty of the charge laid against me; "but," added the magistrate, "you must leave by post-wagon this afternoon." I expressed my astonishment of this treatment, telling him that I wished to stay in Nüremburg for several more days, and I intimated that I should at once ask the protection of our Minister at Munich. "It is for your own protection that I wish you to go," said the magistrate; "if you stay here you will be stoned." "Surely," said I, "after such an abominable charge has been brought against me, I cannot sheer off in so cowardly a manner, and must look to you for protection during my stay here." "Well, if you wish, you can have the protection of two officers wherever you go." I thanked him, and accepted the escort he had offered. This was rather good fun at first, but it soon began to be very irksome. We were stared at by all the visitors at the hotel. We had to pay for the admission of these men at all the places of amusement we visited, etc.; so we hurried our explorations of this very interesting old town, and on the third morning after my arrest we commenced our return journey. Our guards appear to have had strict orders; they went on the coach with us all the way until we passed the frontier, and found ourselves in Prussia, and not until then did we get rid of their really unnecessary services. I have never found out the facts, but I have always strongly suspected that this charge was got up against me to pay off the little trick on the German spy who wanted to get at the secrets of my manufacture by his pretended invention of a machine for making hooks and eyes. However, " All's well that ends well ;" and I was glad to return home from a very enjoyable holiday, invigorated in health, and quite ready to set to work again on whatever might come first on the *tapis*.

CHAPTER VIII

IMPROVEMENTS IN GLASS MANUFACTURE

RETURNED once more to dear old Baxter House, I came face to face with the *débris* of former mechanical investigations piled up here and there in some of the outbuildings, where quantities of old glass pots, and the ruins of a pair of large furnaces, lay scattered among heaps of wheels and pulleys on long shafts, and fragments of old iron framing. Each single piece of this wild mass brought back to memory the particular part it had played in one of those fierce contests with the mechanical powers, in which it may have come off victoriously, or, through want of foresight of the guiding mind, have been ignominiously beaten, to remain a mute witness to the shortcomings of so many plausible theories. Few men have made more mistakes than I have; perhaps there are few men who have so boldly grappled with absolutely novel problems about which no published data existed to guide and modify the first ideas whence all elaborate mechanical structures naturally spring, just as a plant does from its seed. There were many remains in this old storehouse which reminded me of investigations, interesting enough in themselves, but which I must leave wholly unmentioned if I am ever to arrive in this imperfect history at that part of my life's most energetic labours in which my colleagues of the Iron and Steel Institute are more immediately interested. So I must hasten on, and, in mercy to them, leave unsaid so much that I should have to tell if the limits of my little history, and the kind patience of my readers, permitted me to inflict it on them. The ventilation of mines by my combined steam fan, the centrifugal pumps which formed so interesting an exhibit in the International Exhibition of 1851; the compression of pure bituminous coal rendered plastic by superheated steam, and pressed into rectangular polished blocks by a continuous feeding and continuous discharge from a machine similar to the cane press: these and several other minor inventions must be passed over.

But there is one subject of deep interest that I desire to save from absolute oblivion, since its record may at some future time set some active and ingenious mind to work on the lines briefly indicated, and thus add another triumph to the many lately achieved in the domain of optical science.

For some years previous to the period of which I am writing, I was deeply interested in the question of "burning glasses," such as those of Buffon, Parker, and others; my aim being to construct an instrument of sufficient power to act on several ounces, instead of several grains, of the material, which was to be operated upon in crucibles, into which the focus of the lens was directed. In following up this idea, my attention was naturally turned to the enormous difficulty of producing perfectly homogeneous discs of optical glass of large diameters. Fraunhöfer's magnificent lenses of small size had for many years attracted universal admiration, and learned societies were intent on further investigation of the subject. Thus it was that Faraday commenced an enquiry which only ended in failure.

Fraunhöfer's system of manufacture was at that time a profound secret, and the small discs of glass which he sold at fabulous prices were the envy of all other optical glass makers. Faraday, whose scientific knowledge and attainments pointed him out as the most likely scientist to succeed in this new field of enquiry, was, I doubt not, led absolutely astray by the appearance of Fraunhöfer's small discs; had Faraday never seen one of them, and been left to his own resources, he would most probably have succeeded.

The small discs produced by Fraunhöfer, four or five inches in diameter and from half to three-quarters of an inch in thickness, showed what really appeared to be incontrovertible evidence that they were made in small open flat dishes, of the form shown in Fig. 21, page 102.

These little cakes of glass, *a*, had a flat shining upper surface, evidently the natural, or fire, polish, as it is called, and were rounded at the top edges as shown at *b*, the periphery of the flat cake and its lower surface having the unmistakeable impress of the shallow fireclay dish shown in section at *c*. These apparently irresistible proofs that the glass was made in small quantities, and was very fusible and very fluid, no doubt deceived Faraday, and so misdirected his experiments as to lead to failure; all of which became self-evident when the mode of producing these little cakes was known. Glass made in large pots and at the highest attainable temperature is only semi-fluid, and is found to be of different densities in the upper and lower portions of the mass, owing to the varying specific gravities of its constituents. A partial admixture slowly going on in consequence of unequal expansion by heat in so bad a conductor as glass, and the motion induced by air bubbles slowly rising to the surface, have the effect of introducing veins, or striæ, consisting of streaks of more or less dense portions carried upwards by the rising air bubbles, running throughout the general mass, and entirely spoiling it for optical purposes. Now Fraunhöfer, knowing no means

FIG. 21.   SECTION OF FIRECLAY SAUCER AND GLASS DISC

of preventing the formation of these veins or striæ, proceeded on this simple but laborious mode of counteracting these defects. He made a large potful of glass as perfect as he could by simple fusion ; he allowed it to get cold in the pot; he then sawed the mass horizontally into slices, polished their surfaces, and thus examined their internal structure ; and wherever there was a line or streak of more or less dense glass, the defective part was applied to a glass-grinder's wheel and cut away, not as a deep narrow notch but by a wide shallow indent; the surface was again polished for re-examination, and this process was repeated until no more veins, or striæ, were visible. The mutilated and indented disc of glass, sometimes cut nearly half-way through, was then put into one of the shallow fireclay dishes already described, gently heated at first, and finally made sufficiently soft to sink down and acquire the form and dimensions of the dish, the impress of whose surface it bore, while its upper surface assumed the polished appearance of ordinary molten glass.

What I desired to achieve was the production, at a small cost, of large and massive discs or lenses, which could not be produced by Fraunhöfer's system.   Among the several plans I proposed, I will describe only two, each of which attacked the problem from an entirely different standpoint.   First, I may mention that I made a series of laboratory experiments with viscid transparent fluids, contained in glass vessels of various forms and under varied conditions.   Venice turpentine was first tried, but very viscid castor oil was the nearest to glass in its indications of movement within itself.   Small grains of broken red sealing-wax, by their greater specific gravity, showed well the tendency of the oxide of lead (used in flint glass) to subside; and how, by rotating this vessel with one small fragment of sealing-wax, its movement was restrained within a circle the diameter of which was equal to the subsidence of the particle during a semi-rotation of the vessel containing the oil.   The effect of the gentle rotation or rolling of the vessel was also experimented on in various ways.   A small portion of the viscid oil was poured out, and a very minute quantity of blue powder ground up in it, just enough to give a faint blue colour.   This blue oil was then poured back again into the nearly globular-shaped glass vessel, which must be considered as the glass pot; a little movement of the vessel produced streaks of blue colour like veins in marble, dispersed throughout the general mass of viscid fluid. But by continuing to roll the glass globe slowly for about two or three hours, not the slightest trace of veins or streaks of blue remained visible, while a very slight tint of blue pervaded the whole mass of oil, which was now perfectly homogeneous.   It will be observed that the motion so given to the whole mass did not divide it, as the insertion of a stirrer would have done.   I also demonstrated the fact that stirring from the surface by a rod was wholly impossible without the introduction of air in large quantities.   So extraordinary is this fact that I cannot refrain from putting it on record.   Take a glass jar or vessel, say ten inches deep and two inches in diameter, open at the top and closed at the bottom, as shown in Figs. 22 and 23, on page 104.   Nearly fill it with clear, but viscid, castor oil, carefully removing all traces of air from the fluid by exhaustion under the glass bell of a common air-pump; place the jar on a table, take a polished metal, or, preferably, a glass, rod about the size of a blacklead pencil, and having a smooth, rounded end, wipe it, and very slowly and steadily lower it some six inches into the oil, as shown in Fig. 22; then as slowly and carefully withdraw it, occupying quite a minute in doing so. There will remain no trace that anything has entered the oil. Now place the jar again under the bell of the air-pump, take a few strokes with it, and there will appear a line of ill-defined mist, standing vertically upwards about six inches in height in the centre of the jar; at each stroke of the pump it becomes more visible, and enlarges in diameter. It soon assumes the appearance of innumerable little globes, like the hard roe of a herring, as shown at Fig. 23. A little

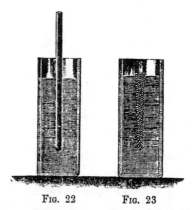

Fig. 22   Fig. 23

Experiment showing Air carried into a Viscid Fluid by a Stirrer

more exhaustion, and these still further expand and rush upwards by the thousand, until at last all the air adhering to, and taken down by, the glass rod has been removed. What you may do, and what you may not do, with molten glass was thus beautifully illustrated by some of these preliminary experiments with viscid fluids. You may move the glass about; you may rotate these viscid fluids in a closed vessel; and you may even pour them, provided the last part of your stream does not fall on the poured-out mass.

To return to actual glass, the subject divides itself into two main systems of procedure, viz., you may make glass by the fusion of pure silica, lime, and potash, or other alkali, with or without the addition of a considerable quantity of oxide of lead, which is used where great density and refractive power are required. Then it becomes more desirable that an intimate mixture of the materials should take place, and throughout the twelve to twenty hours required for fusion, no subsidence of the heavy portions of the mixture, or flotation of the lighter ones, should be suffered to take place, or a homogeneous mass will not be obtained. It is manifestly easy to remove the heavy, sweet particles from the lower part of a cup of tea by one or two gentle movements of a spoon, and so get the whole cup of fluid equally sweet, but we have been warned by our oil experiment that the fluid glass must not be stirred by a rod or we introduce air; and if we wait long enough for the air slowly to find its way again to the surface, we inevitably have an interval in which the difference in specific gravity of the several materials will assert itself, and we get subsidence, lose the homogeneity of the mass, and all our stirring will have been in vain. The outcome of these and other observed conditions was the proposal to employ oscillating, semi-rotating, or continuous slow-rotating melting crucibles, the latter being preferred. The crucibles and the proper kind of furnace for this purpose may be largely varied; one of the simple forms is given in Fig. 24, Plate IX.; its leading features are representative of all the others.

The furnace there shown consists of a cylindrical casing of iron, A, lined with firebrick, B; it is divided into two chambers, the lower one, C, being provided with firebars on which the fuel rests; while the upper chamber, D, is cylindrical in form with a curved roof, having an opening at J formed in a circular piece of moulded firebrick J*, which is removed when putting in the rotating crucible H. Above the opening, J, is a suspended hood, E, which communicates with a tall chimney; the lower part of the chamber, D, is conical in form, having a small central opening, F, and four larger cylindrical openings, G, surrounding it, each of which communicates with the fire chamber, C, and allows the flame to ascend and play up and around the crucible, H. This crucible is conical in form both above and below its largest diameter, and terminates in a raised neck or mouth at H*; the furnace, A, is suspended on axes, occupying the position indicated by the letter M; these axes are fitted to a strong iron ring or hoop, N, which surrounds the furnace, and is itself supported on the axes, P, which rest on iron frames, O.

P

The axes, P, are placed at right angles to the axes, M, so as to allow the crucible to roll or gyrate on its axis, its upper and lower ends moving in a circular path. It will be observed that the crucible, H, rests on one of its sides on the conical floor of the chamber, D, and is kept in position by its lower spherical end, M, moving in the cylindrical opening, E. Now if the furnace be moved quietly on its axis, the crucible gravitating to the lowest inclined side of the conical surface on which it rests, will roll round on its own axis so long as the furnace is kept in motion. This motion of the furnace may be easily effected by means of a short-throw half-crank on a vertical axis passing upward through the floor in a line through the centre of the furnace, the crank-pin having a spherical end fitting into the cylindrical socket projecting downwards from the underside of the ashpit. The motion of the furnace should be very slow, so as to give about one revolution of the crucible in five or ten minutes, and thus allow a constant movement of the whole of the material to take place without dividing or breaking the continuity of the mass, preventing any subsidence of the heavier particles, and securing the perfect homogeneity of the whole. When the fusion and mixing is judged to be complete, the crucible can be pushed with a rod into an upright position, and, by drawing the fire, cooled as rapidly as possible by the current of air flowing through the furnace.

Homogenous optical glass, free from those long "wreaths" or lines of varying density, so common in ordinary glass, was also proposed to be made in the following manner. A large potful of glass of the required composition must be allowed to get cold, and then broken up, the central portions only being selected for use. These pieces are to be crushed, all the glass being reduced to absolute powder, and separated by sifting; all pieces exceeding the size of a grain of rice should be rejected. The very small and nearly equally-sized fragments that remain are then to be carefully washed in distilled water and put into a lenticular-shaped crucible, the exterior surface of which should be glazed, so as to render it impervious and air-tight. The crucible having been put into a suitable furnace and gradually heated, a small platinum pipe communicating with the upper part of the crucible is also connected with an exhaust pump, so as to remove every particle of air from the crucible and from between the granules of glass while these still retain their granular condition. As soon as the glass becomes fluid, it forms a homogenous mass, the law of diffusion equalising any minute differences in composition of continuous grains, while wholly avoiding those long "wreaths" or streaks so fatal in large masses of glass. On the strength of these crude notions a number of various-shaped clay crucibles were ordered to be made, with a view to carry on an elaborate series of experiments on the lines indicated; but as these crucibles required at least three months to dry, I had ample time to pursue some other interesting investigations relative to the production of glass for ordinary commercial purposes.

In going over a glass-works some years previously, I had noticed what I, at the moment, thought was a great oversight in the mode of proceeding. The materials employed, viz., sand, lime, and soda in ascertained quantities, were laid in heaps upon the paved floor of the glasshouse, and a labourer proceeded to shovel them into one large heap, turning over the powdered materials, and mixing them together; a certain quantity of oxide of manganese was added during the general mixing operation, for the purpose of neutralising the green colour given to glass by the small amount of oxide of iron contained in the sand. The materials were then thrown into the large glass pots, which were already red-hot inside the furnace. What appeared to me to be wanting in this rough-and-ready operation was a far more intimate blending of these dry materials. A grain of sand lying by itself is infusible at the highest temperature attainable in a glass pot, and the same may be said of a small lump of lime; but both are soluble in alkali, if it be within their reach. These dry powders do not make excursions in a glass pot and look about for each other, and if they lie separated the time required for the whole to pass into a state of solution will greatly depend on their mutual contact. In such matters I always reason by analogy, and look for confirmation of my views to other manufactures or processes with which I may happen to have become more or less acquainted. I may here remark that I have always adopted a different reading of the old proverb "A little knowledge is a dangerous thing"; this may indeed be true, if your knowledge is equally small on all subjects; but I have found a little knowledge on a great many different things of infinite service to me. From my early youth I had a strong desire to know something of any and all the varied manufactures to which I have been able to gain access, and I have always felt a sort of annoyance whenever any subject connected with manufacture was mooted of which I knew absolutely nothing. The result of this feeling, acting for a great many years on a powerful memory, has been that I have really come to know this dangerous little of a very great many industrial processes. I have been led into this long digression because I meant to illustrate my observations on the extreme slowness of the fusion of glass by an analogy in the manufacture of gunpowder. I have shown how impossible it is for the dry powdered materials employed in the formation of glass to chemically react upon each other when they are lying far apart. Now, if we take the three substances—charcoal, nitre, and sulphur, of which gunpowder is composed, and break them into small fragments, then shake them loosely together, and put a pound or two of this mixture on a stone floor and apply a match, the nitre will fizzle a little briskly; the sulphur will burn fitfully or go out, and the charcoal will last several minutes before it is consumed. If, instead of this crude and imperfect mixture, we take the trouble to grind these matters under edge-stones into a fine paste with water, and then dry and granulate it, we have still the precise chemical elements to deal with as we set fire to on the stone floor; but they now exist in such close and intimate contact as to instantly act upon each other, and a ton or two of these otherwise slow-burning materials will be converted into gas in a fraction of a second. The inference I drew from this analogy was simple enough, viz.: grind together the materials required to form glass, and when the heat of the furnace arrives at the point at which decomposition takes place, the whole will pass into the state of fluid glass much more quickly, and will yield a more truly homogenous glass than is obtained in the usual manner.

I was at this time engaged in constructing a large reverberatory furnace for the fusion of glass on the open hearth, and I may forestall what I have to say respecting this mode of founding glass, by stating that when I employed a mixture of raw material merely shovelled into the bath as practised in ordinary glass-making, it took from ten to twelve hours to fuse half a ton of sand and lime in my new furnace; but when I took precisely the same quantity and quality of materials which had been reduced to a uniform powder, as fine as flour, by grinding the mixed materials under edge stones, my glass, instead of requiring ten or twelve hours for fusion, became beautifully fluid in four and a-half to five hours. When I first tried this fine ground material in my furnace, I patiently watched the whole process hour after hour; the inert mass of dry white powder lay quietly under the rushing current of flame passing over it, without showing any symptom of fusion. At last I sought relief for my over-fatigued eyes by half an hour's turn up and down the yard; and on my return into the glass-house, I was astonished to hear a curious sound issuing from the furnace, closely resembling the noise given out by a frying-pan when cooking fish; on the application of my eye to the peep-hole of the furnace, I saw that the level of the glass had risen an inch or two, and that a rapid boiling was going on, caused by the disengagement of gas resulting from the rapid reaction of soda on the silicic acid. I scarcely need say how greatly I was pleased at witnessing in a first experiment so important a result, and so distinct an example of the value of a little of this so-called "dangerous knowledge."

Up to this period the fusion of glass in large crucibles was universal, and the reverberatory furnace which I had erected at Baxter House for this purpose was the first in which glass was made on an open hearth, and the parent of all those bottle furnaces in which the fusion of glass is carried on in open tanks. It was here also that the hollow box roof was first used in reverberatory furnaces—a form of roof afterwards employed by me so economically in the reverberatory melting furnaces used in the early days of the Bessemer steel manufacture. The immense economy, in time, consumption of fuel, and cost of large melting-pots, resulting from the fusion of glass on the open hearth of a reverberatory furnace was accompanied by one great disadvantage, viz., the tendency of molten matter to fall from the roof of the furnace into the bath, and thus spoil the glass. It was found that whenever the underside of the furnace roof was exposed to an excessively high temperature, the alkaline vapours from the bath beneath caused a fusion of the brickwork, and tears, with their long tails, would fall slowly from above and discolour the glass in the bath beneath. It was mainly to counteract this injurious action that I invented the thin box roof which entirely cured this defect, while the durability of the furnace arch was at least four to one as compared to the ordinary solid form. How well I still remember the trouble and anxiety these tears from the roof caused me, and how I watched through the eye-holes of the furnace the effects of the alkaline vapours on the hollow box roof when it was first under trial. I looked ceaselessly into the fierce glare of the furnace, with but a piece of thick glass between my eye and the bright molten mass, only eighteen or twenty inches distant. When watching by the hour at a time to see if a single tear was formed on the roof, the eye accommodated itself to the intense light, and all within that glorious mass of incandescent matter could be seen in its minutest details. I remember one peculiar circumstance that stood out from all the rest ; while one of the hollow firebricks of the roof was in a condition of plastic clay, the brickmaker had taken hold of it, and a hollow caused by his thumb was beautifully delineated on the underside of this particular brick ; it happened to be opposite the eye-hole, and was an excellent mark whereby any change in the state of the roof could, from time to time, be observed. This must have been as far back as 1847, but that thumb-mark is as indelibly impressed on my memory as it was on the plastic clay. How many hours in succession I have watched that mark through the fierce heat and blinding light of the incandescent furnace I cannot now take upon myself to say, but my whole heart and mind were so absorbed in the investigation that I never gave a thought to the fearful risk I ran of destroying my sight. Now, when I recall these facts vividly to memory, I can realise the folly I was guilty of, and can, in all humility, thank Heaven that I am not at this moment a blind old man.

It is now just forty-nine years since I succeeded in fusing the materials used in the manufacture of glass on the open hearth of a reverberatory furnace, in about one-third the time and with one-third the fuel required for its fusion in the large and expensive glass pots then in use. But there was still one great desideratum : the glass fused in pots was usually blown into long round-ended cylinders or muffs, the ends of which had to be opened while the glass was still hot and plastic— an operation requiring great skill and dexterity on the part of the glass-blower. These open-ended cylinders, when cold, were slit from end to end by a diamond, and again heated until sufficiently soft to be spread out flat on the smooth stone bed of a furnace specially constructed for that purpose; after which they had to be ground and polished, if made into what is known as patent plate.

Now, what I proposed to do, instead of this slow, laborious, and expensive series of operations, was simply to allow the semi-fluid molten glass to escape by an opening extending along the whole length of the bath, and about $1\frac{1}{2}$ in. in width, and to flow gently between a pair of cold iron rollers, so as to determine its breadth and thickness at a single operation. I aimed at converting the whole contents of the furnace into one continuous sheet of glass in ten or fifteen minutes, wholly without skilled manipulation of any kind, or the employment of the other furnaces, which are necessary for opening and spreading the blown cylinders before referred to. It will be obvious that the continuous sheet as it passed from the rolls might be cut into any desired lengths, and thus very much larger sheet glass could be made than it was possible to obtain by blowing it into cylinders.

Having thus foreshadowed the design I had in view, I will briefly explain the nature of the apparatus which I erected at Baxter House to test the practicability of the scheme; and for this purpose I give the engravings, Figs. 25 and 26, Plate X, by way of illustration. Fig. 25 is a cross-section taken through the centre of the bath of the reverberatory furnace, looking towards the fire-bridge A, over which the flame passes. This flame is deflected downward on to the molten glass B, occupying the hearth of the furnace C, which is a sort of rectangular tank, having all along one side a slot or opening D, against which an iron bar E is fixed, so as to close the slot and prevent the escape of the semi-fluid glass. The arched roof, F, of the furnace is formed of hollow boxes of firebrick, each box having a round opening in each of four of its sides, while the upper side is quite open and the lower one closed, and forms the underside of the furnace roof.

In the front of the furnace is a cast-iron door frame D, lined with firebrick. It extends the whole length of the tank in which the glass is melted, and it is removed from its position when necessary, by slings from a jib-crane attached to hooks H, which project from each end of the frame. The flame, after passing over the glass materials in the bath, travels downwards to an underground flue connected to a tall chimney-shaft. There is also a narrow passage I, running downward into the same underground flue, and extending upwards, as shown at I*, so as to admit a current of flame, as indicated by arrows in front of the opening D, and round the curved underlip c* of the cistern c, in order to keep all the front part of the cistern in a highly-heated state.

In front of the furnace a rolling machine is fitted to a suitable slide, so that it may be removed from the furnace a short distance, as shown in Fig. 25, which is a side elevation of the apparatus. It consists of a pair of smooth cast-iron hollow rollers M and N, into which a current of cold water is allowed to flow through the pipes P and Q, and from which the water escapes by similar pipes at the other end of the rollers. A telescopic pipe R slides in and out of a fixed pipe s, and thus keeps up an uninterrupted communication with the water supply. The rollers are brought nearer together, or further apart, by means of screws T in the usual manner, and thus regulate the thickness of the sheet of glass.

Fig. 26 represents in section the furnace, from which the large fire-door has been removed; the rolling machine has also been moved along its slide, until its lower roller N is in almost close contact with the lip c* of the melting cistern; this movement is effected by turning the handle U, which actuates the wheel V, and the rack W, and moves the whole rolling apparatus into position. When this has been done, the rollers are thrown into gear with a shaft, not shown in the drawings, and are caused to revolve at the desired speed. As soon as the machine has been thrown into gear the iron bar E is withdrawn, when a slowly-moving, white-hot, semi-fluid mass creeps out of the long slot, and coming into contact with the lower revolving roll N, is moved by it into the space between the rolls, and is compressed into a thin continuous sheet from an eighth to a quarter of an inch in thickness, as desired ; a projecting V-shaped rib on the upper roller M, will cut the glass into lengths equal to its circumference. The sheet of glass thus severed from the general mass will rapidly slide down the smooth curved surface X of the machine, and deposit itself on the flat stone bed at the foot of the incline, from which it may be transferred into a suitable annealing oven. It will be understood that, as soon as the bath is empty, the rolling machine will be run back on its slide to the position shown in Fig. 25. The bar E and the door G will then be replaced, the bath charged with a fresh supply of raw material, and the process be repeated as soon as the glass is in the proper condition.

From this general description of the process, and the simple mechanism employed, it will be seen that a large quantity of glass could be produced with a very small plant. Thus, suppose that the glass materials are melted in five hours and that the time of casting is, say, fifteen minutes, a cast would easily be made every six hours, or four times per day. A bath only 4 ft. by 3 ft. in area, and 12 in. deep, when making strong horticultural glass $\frac{1}{10}$th in. thick, would yield theoretically 5,760 ft. per day (say 5,000), equal to, at least, 400 blown cylinders 4 ft. long by 1 ft. in diameter.

I was quietly pursuing my experiments with the apparatus described, when I was unexpectedly called upon by an eminent glass-manufacturer. He said that he had heard that I was doing something novel in the production of sheet glass, and if my patent was secured, he should much like to know what was the nature of the invention. I told him my patent was secure, and that I should be happy to give him a general outline of the scheme. He was greatly interested, and the shadows of the evening had imperceptibly fallen upon us in my little private room before my visitor rose to depart. He was very desirous to see the experimental apparatus ; and knowing that my guest, Mr. Lucas Chance, was at the head of the largest glass-works in the kingdom, and worthy of all confidence, I acquiesced in his strongly-expressed desire, and said if he would call again the next day at noon, I would have a charge of glass ready to roll into sheet in his presence.

The following morning, all was got in readiness for a cast. Mr. Chance

Q

critically examined the rolling apparatus, and looked into the furnace from time to time, just as a man would who thoroughly knew what he was about; and when I said we had better now get to work, there were myself, Mr. W. D. Allen, my eldest son Henry, a carpenter, and my engine-driver present in the small room in which the furnace and machine had been erected. As soon as the bar retaining the charge was removed, and the tenacious semi-fluid glass touched the lower roll, the thick round edge of the slowly-moving mass became engaged in the narrow space, where the second roll took hold of it, and the bright continuous sheet descended the inclined surface, darkening as it cooled slightly. I had intentionally omitted the cutter in the roll so as to make a continuous sheet; this had to be pulled away, for my little room was not half long enough to accommodate it. The heat suddenly thrown off from so large a white-hot surface threatened our garments if we stood too near, and unfortunately some oily cotton waste took fire, causing a momentary panic. Mr. Chance called out, "Cease the operation; cease the operation!" We were all in a perspiration, and the long adhesive sheet of glass, 70 ft. long by 2½ ft. wide, was gathered up before the door. The heat was very great, and throwing the rolls out of gear, we all beat a hasty retreat. However, as far as the rapid formation of thin sheet-glass was concerned, there could be no doubt whatever, and I and my visitor sat down quietly to cool ourselves, and think over what had taken place. Notwithstanding the mistake of not putting in the cutter, and making the glass into small sheets, I had the satisfaction of knowing that I had just made a sheet of glass more than three times the length of the longest piece that had ever been produced, and that Mr. Chance had seen, for the first time in his life, a continuous sheet of glass flowing from a machine, wholly without any skilled manipulation. "Well," said Mr. Chance, "you have gone a good way, but you have much further to go yet before you touch the real point—the commercial point. Now it has struck me that we have so many appliances, and so many skilled employés in all departments, that perfecting such a novel process must be more easy and less expensive to us than it can be to you. After all, should it not become so perfect as to be a commercial success, or should some other way be found of effecting the same result, you might have all your labour in vain; but I freely admit that you have done enough to constitute an actual value as the invention now stands. Just think it over, and determine whether you will sell me your invention as it stands and make at once a profit on what you have done, or whether you will spend more labour and money with the chance of much greater remuneration, if you succeed commercially, and no one else supersedes you? I am going down to Birmingham this evening by the 9 P.M. train. Dine with me at seven o'clock at the Euston Hotel, and tell me, yes or no, whether you are disposed to sell your invention in its present state." With this he took his leave, and I had still three hours to reflect over a most unlooked-for proposition. It was very exciting, and I talked the matter over with Mrs. Bessemer, and the general consensus of opinion was : "Realise, by all means, if you can get an adequate amount, but don't give it away." This decision, however, was a long way off a positive fixed sum, to which a "yes" or "no" was to be uttered by both sides. Time slipped on, and when the clock struck seven I found myself in a snug private room at the Euston Hotel. We had a nice little dinner, and, when the table was cleared, my host said : "Well, have you decided?" I said that I had thought the matter over, and felt that any sum must now be a sacrifice, but that I was prepared to sell on one condition, viz., there was to be no discussion of price. I would name what I had fixed on ; would he give me a simple yes or no ? To this he agreed, and then I said, "The sum I have fixed on is six thousand pounds." "Well," said Mr. Chance, "I will give you that sum for your patent; I shall not go down to Birmingham to-night, and to-morrow at 11 A.M., if you will call at Messrs. Hooper and Co.'s, my solicitors, in Sackville Street, we can settle the whole matter there and then." We met as arranged, a short agreement was drawn up, Mr. Chance handed me a cheque for £1,000 with a short-dated bill for £5,000, and we parted very good friends, mutually pleased with our bargain. As a fitting tribute to the memory of Mr. Chance's able solicitor, Mr. Hooper, I may mention that I found him so shrewd and careful of his client, and so just withal, that I from that day gave his firm all my legal business as far as patents were concerned.

At this period I was deeply interested, from a scientific point of view, in the plate-glass manufacture. I was, and ever shall be, a great admirer of plate glass, which I hold to be one of the most beautiful and most marvellous productions of all our varied manufactures; and I must confess that, at the present day, I am disgusted with that idiotic fashion which rejects this splendid production for the small lead panes of a greenish bubbly glass, which, with difficulty is now made bad enough to imitate the early and most imperfect state of the glass manufacture; and which the bad taste—or rather the absence of taste—of the present generation admires and "tries to live up to."

It need not, therefore, be a matter of surprise that I felt a strong desire to cheapen and facilitate the production of plate glass, a manufacture which, in my enthusiasm, I attacked at all points, beginning with the preparation, sorting, cleansing and blending of the raw materials employed, followed by the novel device of a circular reverberatory furnace, in which the founding pots were arranged in a large circular chamber surmounted by a flatly-curved dome. There were also similar furnaces designed for refining the glass, having a crane revolving with the reverberatory dome of the furnace. The crane was, in fact, a veritable automaton, that would remove the one small cover, which, as the dome revolved, gave access in turn to a dozen large glass pots placed in a circle. The three arms or grips of the crane descended vertically into the furnace, and brought up the huge crucible, and when emptied, replaced it in three or four minutes, within half an inch of the exact spot whence it had been lifted. The casting table and all the annealing ovens were arranged in a circle, all accessible from a circular railway laid down in the great casting hall. I may also mention that every detail of the grinding and polishing machinery had undergone an entire change, rendering these operations more rapid and more accurate.

I feel, however, that I dare not trouble my readers by entering into further details. Suffice it to say that a revolution in the appliances and mode of working pervaded the whole manufacture, to properly describe which would fill an illustrated volume. Various portions of the scheme were practically tested; I built a circular furnace for six large pots and erected the automatic crane before referred to, which, like a living thing, dived for a minute or two into the raging heat, and brought forth noiselessly the pot of molten glass (as easily as the human hand could take a tumbler of water off the table), and returned it empty to the same spot.

I was well satisfied with the whole scheme, and wished a few friends to join me in the erection of a plate-glass works in London. My partner, Mr. Robert Longsdon, with his usual architectural skill and good taste, designed the necessary buildings for a complete works, embodying all the novel modes of conducting each department of the manufacture. On Plate XI., in Figs. 27 and 28, illustrations, showing an elevation and section of his design are given, just to save the whole project from oblivion. I have, at this moment, no sort of doubt that, had I convinced others of one-fourth of the improvements embodied in this new scheme, there would have been no difficulty in finding privately a few friends who would have joined their capital to mine, and the works would have been started. But it is difficult to impress one's ideas on others, and I desired to have the personal and entire conviction of its value on the part of all those I asked to join me in the enterprise; without this I was resolved not to move further in the matter, and failing to obtain it, the whole scheme was abandoned.

There is one point in connection with patented inventions upon which I have always felt strongly. I have maintained that the public derive a great advantage by useful inventions being patented, because the invention so secured is valuable property, and the owner is necessarily desirous of turning that property to the greatest advantage; he either himself manufactures the patented article, or he grants licenses to others to do so. In either case the public reap the advantage of being able to purchase a better or a cheaper article than was before known to them, due to the inventor's perseverance in forcing his property upon the market. But if a novel article or manufacture is simply proposed by a writer, and published in the technical press or in newspapers, as a rule (almost without a single exception) no manufacturer will go to the trouble and expense of trying to work out the proposed invention. He says to himself: "I shall not risk the expense necessary to develop this new idea, for it may entirely fail; or even if I succeed, its development will cost me much more than it will cost other manufacturers, who will immediately avail themselves of it if I succeed; no, let some one else try it;" and so the invention is lost to the world in consequence of having been given away. This loss to the public is equally the case with patents that are not taken up; and one of the simplest and most effective inventions which I ever made may be here cited as an example, as it formed part of the novel system of plate-glass manufacture just referred to. When a sheet of plate glass some 10 ft. or 12 ft. long and 6 ft. or 7 ft. wide has been ground perfectly flat on both sides, it is still dull and grey, and has to be polished. For this purpose it is usual to fix it firmly on to a large stone polishing table, so that the powerful alternate pushing and pulling of the polishing rubbers over its surface may not break or displace the sheet. To do this a large quantity of plaster-of-Paris is mixed with water, and spread as quickly as possible over the surface of a stone table much larger than a billiard table. Then the sheet of glass is dexterously laid upon the semi-fluid plaster, and carefully bedded by expert workmen so as to be well supported at all parts of its extensive surface; the superfluous plaster lying beyond the edges of the plate of glass is then scraped away. The polishing machine must remain idle until the plaster is sufficiently firm and hard to retain the glass safely in place. Care must also be taken to thoroughly remove all smears of plaster around the edges and any splashes on the surface, for the plaster is always more or less gritty, and one or two particles of sharp grit will play havoc with the polished surface, scratching it terribly. Let us suppose that one side of the great glass sheet has been polished. It is then necessary to unbed it, and this requires much skill. A man at each corner inserts a thin blade of steel and gently prises the sheet up; he must not spring it much, or the corner will snap off, and considerably diminish the size of the sheet when squared up. With much risk and trouble, the plate of glass is eventually released, and lifted off the stone bed; then the workmen proceed to chip off the hard plaster which firmly adheres to the table. This makes a great mess all round the polishing machine by the flying about of chips of plaster. The stone table having been chipped all over, and scraped quite clean, a fresh lot of plaster is again mixed up, dexterously spread, and the sheet of glass, with its unpolished surface uppermost, is again bedded on the table, and all superfluous surrounding plaster carefully cleared away.   Again the powerful polishing machine remains inactive, until the sheet of glass is firmly stuck to the bed, and, after polishing, the same dangerous process of springing the glass loose from the table has to be repeated.   After its removal, the bed has to be chipped all over, and the hard coating of the plaster-of-Paris removed, for the reception of another plate.

Such is the laborious, dirty, and risky process to which every sheet of plate glass is subjected in the ordinary course of its manufacture.   Now let us see what was the simple mode which I patented of holding down a sheet of plate glass securely during the polishing process.   I employed (see Figs. 29 to 32, page 120) a cast-iron ribbed plate of the size of the polishing stone table, on the upper side of which a large slab of slate (such as is used for billiard tables) was supported and bedded on the ribs of the iron plate.   This surface was then ground flat, in the same manner as plate glass, the space beneath the slate and between the iron ribs forming a shallow box.   A number of round holes of about a quarter of an inch in diameter were made through the slab of slate all over its surface, at a distance of 4 in. or 5 in. apart, so that air could enter the iron box freely from all parts of its surface; a pipe of 1 in. in diameter led to a steam jet or other exhauster, so that air could be withdrawn from the box or let into it by a small hand-tap, as desired.   This, then, was the whole apparatus which constituted the invention; it was extremely inexpensive, and once made, almost indestructible.

This device took the place of the stone table under the ordinary polishing machine.   Its operation may be described as follows:—The sheet of glass to be polished is gently slid upon the table, and covers all the small holes in the slate bed; and if then the small tap which connects the underside of the slate table with the steam jet or other exhaust apparatus is turned on, a partial vacuum is formed beneath the sheet of glass, and it becomes in an instant immovably fixed and adherent to the slate bed on which it rests.   There is no plaster employed, and consequently none of the mess or labour attending its mixture and chipping off; there is no delay in the use of the polishing machine while the plaster

HENRY BESSEMER

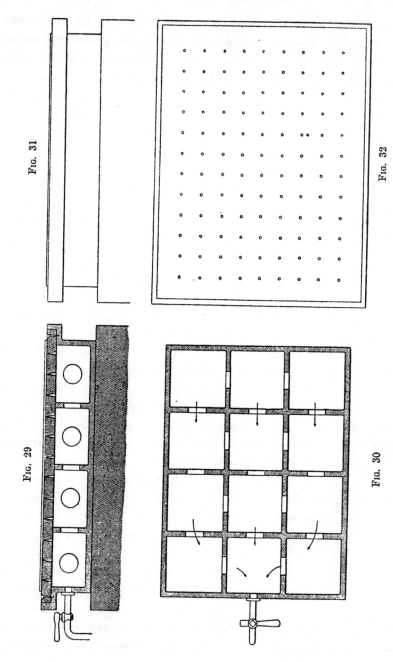

FIG. 31

FIG. 32

FIG. 29

FIG. 30

FIGS. 29 TO 32. BESSEMER'S PNEUMATIC POLISHING TABLE FOR PLATE GLASS

is becoming hard, or when it is being cleared away. The plate of glass is an absolute fixture in less than a quarter of a minute after the exhaust is turned on; and it is as rapidly released by reversing the tap and readmitting the air to the box. The plate is then turned over, and the tap being opened to the exhaust, it instantly becomes re-fixed to the slate surface; there is no cost of plaster, there is no labour, and no risk of snapping off the corner of the plate to release it. Take a moderate-sized plate of 6 ft. by 10 ft. (60 square feet), and take a very low exhaust, say, 2 lb. per square inch, equal to 288 lb. per foot or 17,280 lb. of atmospheric pressure, holding it immovably fixed. Every schoolboy who has seen how powerfully the glass bell of an air-pump is held in place by atmospheric pressure, must understand this simple, effective, cleanly, inexpensive, safe and rapid way of holding down a plate of glass.

This invention, which formed one item of the many improvements in the plate-glass manufacture which I did not carry out, has been available for the free and unrestricted use of the public for nearly fifty years, and yet no plate-glass works in this, or any other, country has taken advantage of it. The simple fact is that an invention must be nursed and tended as a mother nurses her baby, or it inevitably perishes. Nor is this almost incredible indifference to their interest the result of the invention being unknown to the public; for I exhibited a polishing table so constructed, among many other things, in the International Exhibition of 1851, where it became one of the most attractive of my exhibits. I well remember that on one occasion I was requested to be present two hours prior to the opening of the Exhibition to the public, and had the honour of showing and explaining the device to Her Most Gracious Majesty, who was on that occasion accompanied by His Royal Highness the Prince of Wales and other distinguished persons.

The slate table was about 4 ft. by 3 ft., and I used a plate of polished glass a little less than a square yard, weighing about 36 lb. The firm way in which it was held was most easily demonstrated by placing one side of the plate about four or five inches on the edge of the slate bed, and allowing the remainder to project Not only did

R

the atmospheric pressure sustain the plate overhanging in this way, but no one could lift it up or force it down. I was also able to illustrate a fact but little known, viz., that a plate of perfectly flat ground glass lying in absolute contact with a true plane surface cannot be smashed by a blow from a wooden mallet with a curved face. I have struck my yard-square sheet of glass at the Exhibition, when held down by atmospheric pressure, dozens of times before the public, as hard as I could strike it with a wooden mallet, and never broke a single sheet in doing so. If it lay hollow and not in absolute contact with the table, a child could fracture it in a dozen places, but when in contact all over its surface, no amount of force less than that at which glass crushes to powder will crack a properly supported sheet. From what I have said I think I have shown that, however self-evident an invention may be, or however advantageous it might be to a manufacturer, if it is public property he will not touch it.

I have already so far trespassed on the patience of my readers in reference to the manufacture of glass that I must bring these remarks to a close. But there is just one little point that I may be excused for mentioning; it has reference to the silvering of glass, which every-one knows was effected by the amalgamation with mercury of a large sheet of thin tinfoil, the amalgam adhering to the surface and remaining on the side next the glass, a beautifully-polished and highly-reflecting surface. But it had a bluish or leaden hue that was most unfavourable to the fair sex, and spoiled the best complexion. I thought much over this defect, and at last succeeded in greatly improving the whiteness of the reflection. This I effected by the use of pure silver powder. The sheet of tinfoil was employed as before, and amalgamated with mercury, the greater part of which was drained off the surface of the foil, and then pure silver in the form of an impalpable powder (known as silver-bronze powder), was freely dusted all over the amalgamated surface. The fine silver particles became rapidly amalgamated or dissolved by the mercury, and when the sheet of glass was slid on and pressure applied, an amalgam of pure silver coated the glass, greatly improving the brilliancy and colour of the mirror. This method seemed likely to have a great future, but before it got into use, a process suggested by Liebig some years before was developed and applied in a practical form by Professor Henry Draper.   By this method of working, which he used for the silvering of glass mirrors for reflecting telescopes, Professor Draper entirely dispensed with the tinfoil and mercury process, and deposited pure silver direct on to the glass from its solution.   This was a far more perfect mode than my own of putting pure silver on to the glass, and quite put an end to my process.   At the present day all glass mirrors are silvered by one or the other of several modified forms of Leibig's admirable invention.

CHAPTER IX

THE EXHIBITION OF 1851

ABOUT this period everyone was interested in the forthcoming International Exhibition of 1851. I had applied for space to exhibit the process of separating molasses from crystallized sugar by my combined steam and centrifugal apparatus; this formed a very attractive display. The crystallised sugar, with its adhesive coating of brown treacle, was spun round in the wire cage at a speed of 1,800 revolutions per minute; and on throwing a bowlful of cold water into the machine, in thirty seconds the dark sticky mass was like a snowdrift, with its sparkling crystals compactly spread round the revolving basket. Crowds of people would stand round the machine, and seemed never tired of witnessing its operations.

I took a deep interest in the development of the International Exhibition, and as an exhibitor I used to pass long mornings in the building very frequently, prior to its public opening in May, 1851. On one of these occasions I chanced to meet my esteemed old friend, the late Mr. Bryan Donkin, F.R.S., and he went with me to see how my exhibits were being fixed up. Seeing my centrifugal machine, he said: " Why do you not show that old scheme of yours for raising large volumes of water by centrifugal force?" "Oh," I replied, "I had almost forgotten it." He said, "Everybody is fond of looking at a cascade, and a large body of water such as you can lift would make one of the most interesting exhibits in the mechanical department." Thus encouraged, I next day sat down to my drawing-board, and schemed a combined-engine and centrifugal pump, which I afterwards exhibited. There was very little time to make all new patterns and large loam castings; indeed, it seemed almost impossible to do so. But I posted my drawings to Messrs. George Forester and Co., Engineers, of Liverpool, who had previously executed some important orders for machinery for me. My instructions were : "If you can make the combined steam engine and centrifugal pump in thirty-two days from the receipt of this, and undertake to deliver it at the Exhibition building on the thirty-third day, set to work at once and make it ; but if you cannot undertake to do so, do not touch it at all, as I must hold you responsible. I know it is a most arduous task, but if you execute so important an order in so short a period it will do you much credit, and I will put on the side of the machine a conspicuous brass plate, giving the full address of your firm as makers, with the date of order and date of delivery engraved thereon." The result was that the whole apparatus was admirably finished and delivered one day before the prescribed limit. Some months later Messrs. Forester and Co. executed for me two combined engines and pumps, each of which, when set temporarily at work in Toxteth Park, Liverpool, was found to discharge 109 tons of water per minute at a height of 7 ft. above the source of supply. These pumps were afterwards erected for the drainage of some sugar estates in Demerara, which lay 5 ft. below high-water mark in the tidal river into which they were drained, and each of them lifted a small rivulet 10 ft. wide by 18 in. deep, flowing at a speed of three miles an hour.

To my no small surprise, I found that I did not stand alone in the Exhibition as the inventor of centrifugal pumps, for there were two others, one by Mr. Appold, and another by Messrs. Gwynne, from the United States ; each of these was doubtless a separate and distinct invention.

Notwithstanding my frequent visits to the Exhibition to superintend the erection of my exhibits, the place remained as fresh and as full of interest as though I had never been inside those magic walls of glass. How vividly still my mind retains the impression of the opening day ; what a glorious May morning, the crowning day of expectation to so many thousands ! We were warned that unless we started from home very early, we should never reach the building by 11 A.M. I lived then on the road to Highgate, only two miles off. My wife and my eldest sister left home with me in a brougham at 8 A.M. Even at that early hour the streets were thronged ; all London was astir, and as we slowly neared the Park the streets were densely crowded; everyone in holiday attire, all looking joyous and brimming over with eager expectations. Very soon our quiet trot had dwindled to a walking pace, and as we entered the Park by the Marble Arch, our progress ceased in an absolute stop, followed by a little move preparatory to another long stop. We had got into the Park by 9 A.M., and there were yet two hours, but we had begun even then to fear we might not reach the building in time. There was no intentional obstruction, and the police did all they could under impossible conditions; we were hemmed in on all sides by carriages and pedestrians, and were almost immovable. An hour's intermittent motion had brought us from the Marble Arch to the Piccadilly entrance, from which rolled another avalanche of almost hopelessly struggling humanity. Yet all was good-humour and high expectation; tickets were flourished from innumerable carriage windows, and fair ladies in their sweetest and most persuasive tones, asked aid of the police, who were powerless to help them. Another half-hour from the Piccadilly entrance brought us in full sight of the fairy palace, which sparkled in the sun, but was as yet a few hundred yards distant. But we were in an almost solid mass of carriages, horses, policemen, and pedestrians. A look at my watch showed me that there was no hope for us if we kept our seats in the brougham. We were within fifty yards of the building, and we agreed to get out and chance struggling up to the door on foot. Hundreds of ladies in their satin shoes descended from their carriages to the gravel, and with their beautiful dresess pulled tightly round them, trusted to their feet. In charge of my two ladies I showed my tickets and got, at last, passed on to the doors, which we entered ten minutes before the appointed time, and just three hours after we had started. We hurried to our places, and could now breathe more freely; the air was full of perfume from the sweet flowers that filled all vacant places, and added a lustre to the gorgeous scene. When the formal processions had gone round the building, there came the one great treat of the opening day, never to be forgotten by those who heard it. The sister of my old friend Alfred Novello, Miss Clara Novello, sang the National Anthem, and by a supreme effort, her full melodious voice filled the whole space with a glorious volume of sound that could not fail to inspire the deepest feeling of loyalty. And as her voice rose and fell to the cadences of the beautiful Anthem, the thousands of faces of those present showed at a glance how all were moved by feelings of deep emotion and loyalty to Her Most Gracious Majesty, and to Prince Albert, whose cherished dream of the International Exhibition was thus so happily realised.

Returning again to the quiet daily routine, life at that period found me pretty regular in my attendance at the office in Queen Street Place, where I often spent a few hours with some client, who had sought advice in reference to an invention, possibly more or less crude and impracticable, or, it may be, of great value if only a little more mechanical knowledge had been expended on its details. Such investigations were sometimes very interesting; and I well remember several inventions which were brought before me at that time, and which have since taken their place among the important mechanical improvements of the present century; while many others that were essentially bad and wholly impracticable were fought for by their luckless inventors with a tenacity worthy of a better cause. It was just this class of inventors that one could not convince of the false notions under which they laboured; if a man knew so little of mechanical laws as to suppose that by some tricky arrangement of levers which he had devised he could make the descent of 20 lb. lift up 40 lb. to the same height, it took a vast deal of labour to convince him of his error; and he paid consultation fees with the inward belief still clinging to him that, somehow or other he was right, only he could not make me see things in the proper light. But generally I found it possible to bring home to the most prejudiced minds such unpleasant facts, and I have in many cases received the most frank and friendly acknowledgments from men who would have spent hundreds of pounds in search of the impossible had not an hour's discussion shown the fallacy of their convictions.

During the years 1852 and 1853, I was very busy with inventions of my own, for I find that in those two years I took out no less than twelve patents, that is, on an average, one every two months. These being mechanical inventions relating to manufactures, there arose, in each case, much studying of details, and many original drawings had

to be made in addition to the specifications to be written and claims
to be settled. Some of these were followed up with results that were
highly satisfactory ; but it was my misfortune that inventions sprung
up in my mind without being sought, and as soon as a new idea
presented itself there was no peace for me until the first crude notions
were shaped and moulded into a tangible form, and this again criticised
and improved upon. Then came experimental research, or in many
cases the invention was patented as a mere theoretical deduction
because it had to make room for the next : whereas each invention,
to be made a commercial success, required to be carefully and forcibly
brought under the notice of the particular trade to which it referred.

In regard to one of these patents of 1853 I will just say a word
or two, as a mere record of a first proposal to stop a railway train
by the simultaneous application of a brake on every carriage wheel
of the train. I fully appreciated the advantages of this simultaneous
action on every wheel, because by such means a train of twenty or
thirty carriages could be stopped just as quickly and as easily as a
single carriage, since each vehicle was subjected to the same retarding
action ; but it was necessary that this should be effected, as far as possible,
without any complicated mechanical arrangements likely to get out
of order in practice. My invention, which I call the hydrostatic brake,
was one of extreme simplicity. Under ordinary circumstances the application of hydraulic power means packed pistons, water-tight stuffing-boxes,
inlet and outlet valves, etc., all of which mechanical appliances were,
in my plan, entirely dispensed with, and a rectangular cell of vulcanite
rubber was used for transmission of the pressure.

The arrangement was as follows :—A rectangular iron box was held
by bolts passing through flanges at each end, by means of which it was
secured to the underside of the carriage frame. The interior of the box
was 7½ in. long by 4 in. broad and 8 in. deep, and there was fitted inside
it a block of wood, not unlike one of the blocks used in street paving,
and having a curved lower surface fitting the tyre of the carriage wheel.
This block projected downwards from the mouth of the box, and left a
space of 1½ in. between it and the upper closed side. Into this space was
fitted a hollow rectangular chamber or box of vulcanised rubber, capable

of expanding and contracting; it was attached to the wood-block on its lower surface and to the box on its upper side. There was a small pipe which connected the chamber with a continuous pipe leading to the locomotive, where the driver could turn water pressure on or off instantly whenever necessary. Each of the rubber chambers contracted by external atmospheric pressure if connected to the exhaust, and lifted the wood-block from off the wheel; but the instant that pressure was applied, each of the chambers expanded, and pressing on its wood-block forced it in contact with the wheel and retarded the motion. Thus if the wood block were $7\frac{1}{2}$ in. by 4 in., it presented a surface of 30 square inches, and every 10 lb. to the inch pressure on its surface was equal to 300 lb. on the wheel. The main leading pipe was always charged with water, which is non-elastic, and was permanently in communication with each of the chambers. If half a pint of water was exhausted from each chamber, the block was raised more than half an inch from the wheel, and relieved the pressure. Modifications of this simple brake have been made; but it came before its time, and was not accepted by the railway companies. I am pleased to have lived long enough to see continuous brakes universally adopted, for by them vast numbers of persons have been saved from injury or death.

CHAPTER X

EARLY GUNNERY EXPERIMENTS

AT the time when the Crimean War broke out, the attention of many persons was directed to the state of our armaments, and I, like others, fully shared the interest which was excited. The question of elongated projectiles had been previously considered, but we were quite unprepared at that time with rifled ordnance. In thinking over this subject, it occurred to me that it would be possible to give rotation to a projectile, when fired from a smooth-bore gun, by allowing a portion of the powder gas to escape through longitudinal passages formed in the interior, or on the outer surface, of the projectile. If such passages terminated in the direction of a tangent to the circumference of the projectile, the tangential emission of powder gas (under enormous pressure) would act as in a turbine, and produce a rapid rotatory motion of the projectile. It may at first sight appear that such a method would be attended with great loss of power, but it must be remembered that in any system of rifled ordnance enormous energy is required to revolve a heavy projectile, to say nothing of the power lost by the friction of the studs in the rifled grooves.

It was under the impression that my invention would enable all existing smooth-bore guns to be at once utilised for discharging elongated shells and solid projectiles, and would at the same time solve a problem of great national importance, that I applied for and obtained a patent on the 24th November, 1854. It will be evident that this system of giving rotation to elongated projectiles might, in some cases, have rendered it desirable to hoop, or otherwise strengthen, existing guns, or to construct new guns of greater strength than those then in general use. But the main question was: Can rotation be given efficiently without the manifold disadvantages of rifling? As a matter of fact, I submitted my plans to the War Office, and, after some considerable delay, I was informed that the invention was not of a nature to be used, or experimented upon, by the War Department. Our War Department had at that time no artillery that could throw an elongated projectile; yet with that ever-ready tendency of our military authorities to pooh-pooh every proposition of the civil engineer, my scheme was set aside, and so simple and inexpensive an experiment as the manufacture of half a dozen cast-iron elongated projectiles was refused. Nothing more than this was required, as they had plenty of cast-iron guns in

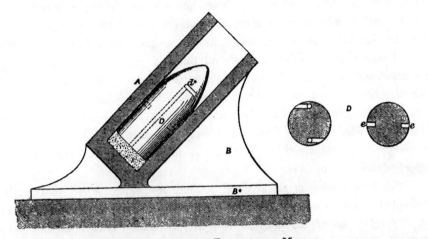

FIG. 33.   SECTION OF EXPERIMENTAL MORTAR

store, and all other needful appliances. I, however, was determined to ascertain for myself whether I was right or wrong in my belief that rotation could be effected simply by the emission of a portion of the gases in the manner described, and for this purpose I made a simple cast-iron gun of 5½-in. bore, and of short length. As I had no butt to fire into, I thought it best to use the gun as a mortar, and discharge it into the air at an angle of 45 deg. of elevation; by using small charges I ensured the projectiles falling in my own grounds near Highgate, where I was then living The gun was a simple bored cylinder, cast all in one piece with the framing, which, with its wide base-plate B,* served for a carriage, as shown in the section in Fig. 33. The projectiles weighed

60 lb. each, and were turned and truly fitted to the bore of the gun; the form of projectile employed is given at D, which is an elevation showing by dots one of its longitudinal passages with the tangential aperture at *d*.* With this simple apparatus I commenced my trials, using extremely small charges of powder, which I gradually increased until the projectile reached an estimated altitude of 200 ft., and fell to earth well within my own grounds. In order to clearly see that the projectile revolved during flight, I bored on its opposite sides two holes, ¾ in. in diameter and 2 in. in depth, in a radial direction, as shown in section at *e e*. These holes were tightly rammed with damp meal powder, and on firing the gun (as I used no wad) the powder became ignited and fizzed away like a squib. Standing beside the gun I saw the shot soaring, with its flat end presented to me, and by its rapid rotation the two squibs formed a sort of revolving Catherine-wheel, which burned until after the shot had fallen to earth, thus proving beyond all controversy that the projectile both rotated and went end-on during its whole flight. But still I was no nearer my object, and might for ever have remained as I was, but for an accidental circumstance which I will relate.

Some few months after these preliminary experiments were made, I happened to be one of a house party, staying with Lord James Hay at the residence of his married daughter, Madame Gudin, in the Rue Balzac, Paris. Our host gave a farewell dinner to General Hamlin, and a number of other French officers, who were going to the Crimea. Among the guests present on that occasion was Prince Napoleon, to whom I was introduced by my host as the inventor of a new system of firing elongated projectiles from smooth-bore guns. I happened to have with me a tiny pocket model, made in mahogany, of one of these new projectiles, which, in order that it might be more easily understood, had the passages for the escape of gas formed in its exterior surface, instead of in the interior, as will be seen from the annexed engraving, Fig. 34, representing in full size this little model projectile, which I made more than forty years ago, and which is still in my possession. Its action was very prettily shown in this way : an upright glass tube of 1½ in. internal diameter (accurately fitting the shot) had its lower end stopped up so that it resembled the barrel of a gun.   If the model shot were put into the upper end of this glass tube, when held in a vertical position, it could not sink down to the bottom without displacing the air contained in the tube.   This air necessarily found an escape through the external passages; and by the force induced by the escape of air in the direction of a tangent to the circumference, a slow and steady rotation of the little mahogany projectile was observed, as it gradually sank down to the lower end of the glass tube.   Prince Napoleon was very much pleased with the idea, and said that he was sure that his

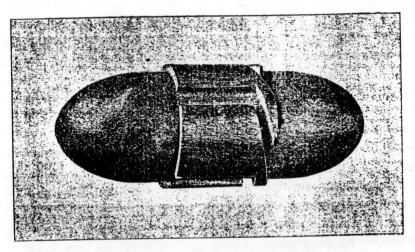

Fig. 34.   Model of Bessemer's Revolving Shot

cousin, the Emperor, would take great interest in my invention, and that he would get an appointment for me to show it to him.   A few days later, I received a note from Colonel Belleville requesting my attendance on the following morning at the Tuileries, where I had a most interesting interview with the Emperor, who gave me *carte blanche* to go to Vincennes, and there order to be made everything that was necessary to fairly test my invention.   I, however, found that it was much more difficult to get what I wanted made at Vincennes than it would have been at my own works in London, where other matters required my attention.   I consequently sought another interview with the Emperor, when I explained this fact to him, and asked permission to make the projectiles in London, and bring them over. No objection was raised to this proposal, and as I was about to take my leave the Emperor said : " In this case you will be put to some expense ; I will have that seen to."

A few days after my return to London, I received a letter from the Duc de Bassano, enclosing an autograph letter from the Emperor, addressed to Messrs. Baring Bros., Bankers, London, giving me credit for "costs of manufacturing projectiles," but without naming any sum. Whatever private instructions there may have been given as a limit to the amount of credit, none were visible to me ; and I could not help forcibly contrasting this delicate and generous treatment by the Emperor with the curt refusal of our own military authorities to give my invention a trial at home.

In a few weeks, the projectiles necessary for the experiments were all made under my own eye, and packed ready for transport. There were 24-lb. and 30-lb. elongated shots of 4.75 in. in diameter, fitting the 12-pounder smooth-bore French guns, gauges for which had been sent to me in order to ensure accuracy in size. I had been provided with a special permit to pass the Customs House at Calais, notwithstanding which, my passport was rigidly examined, and I was looked at and questioned by all sorts of officials before I was allowed to proceed on my journey.

I, as specially directed, went straight on to Vincennes on the following morning, and was met by Commandant Minié (the inventor of the rifle which bears that name), who had received instructions to superintend the experiments and report thereon to the Emperor.

In the large open plain known as the Polygon, at Vincennes, a series of thin wooden targets had been set up, one behind the other at about 100 metres apart. My projectiles were fired point-blank at these targets, and generally passed through five or six of them before reaching the ground, making round holes in each, and showing that all the shots went end-on. In order to enable us to measure the amount of rotation of the shots, I had given them a thin hard coating of black Japan varnish, which was partly scratched off and scored in lines when passing through the thin planks. There were a few inches of snow on the ground at the time these experiments were made, and we could observe the projectiles ricocheting away to the left as a result of their continued rotation after striking the ground, and sending up the snow in little jets, thus indicating where they were to be found. On recovery, the spiral lines scored on the japanned surface in its passage through the target gave every facility for ascertaining the angle, and consequently the amount of rotation. It was thus ascertained that from one and a-half to two and a-quarter rotations had taken place in the length of the gun, or a greater amount of twist than was usually given at Woolwich to projectiles of that calibre. Evidence was thus afforded that the dogmatic way in which the invention had been ignored by our military authorities was in no way justified. Whatever the real merits or demerits of my invention may have been, it was at least shown that, at a time when we had no established rifled system, this early attempt at a solution of the difficulty had sufficient merit to render it worthy of a trial.

By the time the experiments were concluded the winter sun had almost disappeared, and both weary and cold, the several officers, who took part in the day's trials, and myself walked back to the grim old fortress of Vincennes, and after threading our way along the cold stone passages, we found ourselves in the officers' quarters. A bright blazing fire of logs on the low hearth looked so inviting that we all instinctively gathered round it, and under the happy influence of a steaming cup of good mulled claret, there was much noisy talking and gesticulation. During one of our more quiet intervals, Commandant Minié remarked that it was quite true that the shot revolved with sufficient rapidity, and went point forward through the targets; and that, he said, was very satisfactory as far as it went. But he entirely mistrusted their present guns, and he did not consider it safe in practice to fire a 30-lb. shot from a 12-pounder cast-iron gun. The real question, he said, was; Could any guns be made to stand such heavy projectiles? This simple observation was the spark which has kindled one of the greatest industrial revolutions that the present century has to record, for it instantly forced on my attention the real difficulty of the situation, viz. : How were we to make a gun that would be strong enough to throw with safety these heavy elongated projectiles ? I well remember how, on my lonely journey back to Paris that cold December night, I inwardly resolved, if possible, to complete the work so satisfactorily begun, by producing a superior description of cast-iron that would stand the heavy strains which the increased weight of the projectiles rendered necessary. At that moment I had no idea whatever in which way I could attack this new and important problem, but the mere fact that there was something to discover, something of great importance to achieve, was sufficient to spur me on. It was indeed to me like the first cry of the hounds in the hunting field, or the last uncertain miles of the chase to the eager sportsman. It was a clear run that I had before me—a fortune and a name to win—and only so much time and labour lost if I failed in the attempt. When, a few days later, I personally reported to the Emperor the results of the trials at Vincennes, I told His Majesty that I had made up my mind to study the whole question of metals suitable for the construction of guns, a proposal which he encouraged by many kind expressions, and a desire that I should communicate to him the result of my labours.

My knowledge of iron metallurgy was at that time very limited, and consisted only of such facts as an engineer must necessarily observe in the foundry or smith's shop; but this was in one sense an advantage to me, for I had nothing to unlearn. My mind was open and free to receive any new impressions, without having to struggle against the bias which a lifelong practice of routine operations cannot fail more or less to create.

A little reflection, assisted by a good deal of practical knowledge of the properties of copper and its several alloys, made me reject all these from the first, and look to the metal iron, or some of its combinations, as the only material suitable for heavy ordnance. At that time nearly all our guns were simply unwrought masses of cast iron, and it was consequently to the improvement of cast iron that I first directed my attention.

The experiments at Vincennes took place on or about the 22nd

December, 1854, and before the close of that year I found myself once more at Baxter House, busy with plans for the production of an improved metal for the manufacture of guns, which improvement in the quality of the iron I proposed to effect by the fusion of steel in a bath of molten pig-iron in a reverberatory furnace. I soon determined on the form of furnace, and applied for a patent for my " Improvements in the Manufacture of Iron and Steel," which was dated as early as January 10th, 1855—that is, within three weeks after the experiments in the Polygon at Vincennes.

CHAPTER XI

THE GENESIS OF THE BESSEMER · PROCESS

IT will, perhaps, assist the non-technical reader to understand what follows if I explain, in a few words, the forms in which iron and steel existed at the time when I commenced the experiments which resulted in the creation of the Bessemer process. At that date there was no steel suitable for structural purposes. Ships, bridges, railway rails, tyres and axles were constructed of wrought iron, while the use of steel was confined to cutlery, tools, springs, and the smaller parts of machinery. This steel was manufactured by heating bars of Swedish wrought iron for a period of some six weeks in contact with charcoal, during which period a part of the carbon was transferred to the iron. The bars were then broken into small pieces, and melted in crucibles holding not more than 60 lb. each. The process was long and costly, and the maximum size of ingot which could be produced was determined by the number of crucibles a given works could deal with simultaneously. Such steel when rolled into bars was sold at £50 to £60 a ton. The wrought iron bars from which the steel was made were manufactured from pig-iron, as was all wrought iron, by the process known as "puddling." Naturally, such a process was costly; puddling demands great strength and endurance on the part of the workmen, combined with much skill.

Practically, all objects in iron, except such as were simply castings, were at that time made from wrought iron manufactured by puddling. The object I set before myself was to produce a metal having characteristics comparable with those of wrought iron or steel, and yet capable of being run into a mould or ingot in a fluid condition. I was aware that Fairbairn and others had sought to improve cast iron by the fusion of some malleable scrap, along with the pig iron, in the cupola furnace. This fusion of scrap-iron, intermixed with a mass of coke, was found to convert the malleable iron into white cast iron, which was at the same time much contaminated with sulphur. Therefore, to a great extent, this system had failed in its object. In my experiments I avoided the difficulties inseparable from Fairbairn's method, by employing a reverberatory furnace in which the pig-iron was fused. Into the bath so formed I put broken-up bars of blister-steel, made from Swedish or other charcoal - iron, its fusion taking place without its being further carburised by contact with the solid fuel, or contaminated by the absorption of sulphur. The high temperature necessary for the fusion of a large proportion of steel in the bath was obtained by constructing the fire-grate much wider than the bath, by contracting the width of the furnace considerably at the bridge, and also by continuing to taper slightly the furnace all the way from the fore-bridge to the downcast flue, which was connected with a tall chimney-shaft. Many alterations and modifications of this furnace were made from time to time, but it was found that the large volume of flame sweeping over the open hearth of the furnace was mixed with a considerable quantity of combustible gas. To consume this gas a hollow fire-bridge was employed, having numerous perforations made in the clay lumps of which it was composed, and so arranged as to allow jets of hot atmospheric air to mingle with these combustible gases, and produce an intense heat close down to the surface of the bath. It was also found that this admission of hot air all along the back of the fire-bridge produced a decarbonising action on the bath; hence the state of carburation of the metal might be altered by regulating the admission of air. This passage of air through the hollow fire-bridge served also to keep down the temperature of the latter and render it more durable.

Some of the samples of metal which I produced were, when annealed, of an extremely fine grain, and of great strength. At this stage of my experiments I cast a small model gun, which in the lathe gave shavings slightly curled, and closely resembling the turnings from a steel ingot; the metal, when polished, also looked white and close-grained like steel. I was so well pleased with this little model gun that I took it over to Paris, obtained an audience with, and showed it to, the Emperor, who had encouraged this attempt to improve the iron employed in founding heavy ordnance. His Majesty, who had desired me to report progress, accepted this experimental gun, remarking that some day it might have an historical interest. It was in recognition of this circumstance that His Majesty, later on, intimated, through Colonel Belleville, his desire to confer on me the decoration of the Legion of Honour, provided I could obtain permission to wear it, a privilege which our Ambassador twice refused. His Majesty also sanctioned the erection of my furnace at the Government Cannon Foundry at Ruelle, near Angoulême, to which place I went with proper introductions for the purpose of arranging all the necessary details. I also sent over from England several thousand special fire-bricks, etc., for the erection of the furnaces.

But, on resuming my further researches, after my return to London, an incident occurred which suddenly put a stop to the intended works at the Ruelle gun - foundry, and in fact altered all my future plans and investigations.

The furnace, as then arranged, is shown in vertical section in Fig. 35, and in horizontal section, on the line passing above the fire-bridge, in Fig. 36, Plate XII., the bath being empty and showing the tapping-hole, and the way in which the furnace narrows at the fire-bridge. Fig. 37, on the same Plate, is also a horizontal section, taken on a line passing through the openings in the perforated hollow fire-bridge, and clearly shows how the jets of air were directed so as to produce an intense ignition of the combustible gases, mingled with, and passing over with, the large volume of flame from the overcharged fire-grate.

The small scale on which this experimental furnace was built (a capacity of 3 cwt. only) was much against my obtaining the high temperature necessary to melt a large proportion of steel in a pig-iron bath. I was, of course, fully aware that a furnace of sufficient capacity to cast a 5-ton or a 10-ton gun would acquire a much higher temperature than was possible in my small furnace. I knew also that forced draught, obtained by closing in the ashpit and forcing air into it, would still further increase the temperature. That this forced draught was in my mind at the time is shown by the fact that I took out a patent for the manufacture of cast steel, dated October 17th, 1855 ; that is, about two months after the casting of the model gun, in which specification I fully described the forcing of air by a fan into the closed ashpits of the furnaces employed in the manufacture of cast steel. It has since often occurred to me that, with these additional resources still untried, I did not act wisely in so suddenly abandoning these open-hearth experiments, in favour of an entirely different system, suggested to my mind by the incident to be presently referred to. But with my impulsive nature, and intense desire to follow- up every new problem that presented itself, I at once threw myself unreservedly into this new study, which seemed to open the way to the rapid production of bars, rails, and plates of malleable metal direct from the blast-furnace.

Before dismissing this subject, it may be interesting, even at this distant period, to speculate on what would have been the natural outcome of my open-hearth furnace experiments, had I not been so suddenly diverted from their further pursuit.

Such a furnace, with forced draught and a capacity of 10 tons, would undoubtedly have melted malleable iron or steel in a bath of pig iron, and have decarburised the latter to the desired extent; for I had, in fact, already fused steel, in a bath of pig iron, on the open hearth of this small reverberatory furnace ; and as far back as January, 1855, I had claimed in my patent, " *The fusion of steel in a bath of melted pig or cast iron in a reverberatory furnace, as herein described.*"

This was about ten years prior to the patent taken out by M. Emile Martin, and now generally known as the " Siemens-Martin process." This latter patent was obtained in England in the name of A. Brooman, the patent agent of Emile Martin, and is dated August 18th, 1865, or more than ten years after my patent of January 10th, 1855. M. Emile Martin in his patent says : " *The manufacture is effected upon the principle of fusion of iron or natural steel in a bath of cast iron, maintained at a white heat in a reverberatory furnace, such as Siemens gas furnace.*"

I, however, desire to say that I make no claim to the prior invention of the Siemens-Martin process, nor do I assume that my patent of 1855 furnished any information which either of these gentlemen had availed themselves of, although my patent for melting steel in a bath of cast iron on the hearth of a reverberatory furnace had been granted, and the specification published, some nine years prior to M. Martin's application for his patent. But seeing how many years I was in advance of M. Martin, I feel perfectly justified in saying that the fusion of steel in a bath of pig iron on the open hearth of a reverberatory furnace, which I had patented and accomplished ten years prior to the Siemens-Martin patent, was, to use a favourite expression of Mr. Gladstone, "approaching within measurable distance" of that successful process known as the open-hearth manufacture of mild steel.

On my return from the Ruelle gun-foundry I resumed my experiments with the open-hearth furnace, when the remarkable incident, mentioned above, occurred in this way. Some pieces of pig iron on one side of the bath attracted my attention by remaining unmelted in the great heat of the furnace, and I turned on a little more air through the fire-bridge with the intention of increasing the combustion. On again opening the furnace door, after an interval of half an hour, these two pieces of pig still remained unfused. I then took an iron bar, with the intention of pushing them into the bath, when I discovered that they were merely thin shells of decarburised iron, as represented at A, Fig. 37, Plate XII., showing that atmospheric air alone was capable of wholly decarburising grey pig iron, and converting it into malleable iron without puddling or any other manipulation. Thus a new direction was given to my thoughts, and after due deliberation I became convinced that if air could be brought into contact with a sufficiently extensive surface of molten crude iron, it would rapidly convert it into malleable iron. This, like all new problems, had a special interest for me, and I became impatient to test it by a laboratory experiment. Without loss of time I had some fire-clay crucibles made with dome-shaped perforated covers, and also with some fire-clay blow-pipes, which I joined on to a 3 ft. length of 1 in. gas pipe, the opposite end of which was attached by a piece of rubber tubing to a fixed blast-pipe. This elastic connection permitted of the blow-pipe being easily introduced into and withdrawn from the crucible, as shown at Fig. 38, Plate XIII., which represents a vertical section of an air furnace containing a crucible that, in this case, forms the "converter." About 10 lb. of molten grey

pig iron half filled the crucible, and thirty minutes' blowing was found
to convert 10 lb. of grey pig into soft malleable iron.   Here at least
one great fact was demonstrated, viz., the absolute decarburisation
of molten crude iron without any manipulation, *but not without fuel*, for
had not a very high temperature been kept up in the air furnace all the
the time this quiet blowing for thirty minutes was going on, it would
have resulted in the solidification of the metal in the crucible long before
complete decarburisation had been effected.   Hence arose the all-important
question : can sufficient internal heat be produced by -the introduction
of atmospheric air to retain the fluidity of the metal until it is wholly
decarburised in a vessel not externally heated?   This I determined to try
without delay, and I fitted up a larger blast-cylinder in connection with
a 20 horse-power engine which I had daily at work.   I also erected
an ordinary founder's cupola, capable of melting half a ton of pig iron.
Then came the question of the best form and size for the experimental
"converter."   I had very little data to guide me in this, as the crucible
converter was hidden from view in the furnace during the blow.   I found,
however, that slag was produced during the process, and escaped through
the holes to the lid.   Owing to this, I determined on constructing a very
simple form of cylindrical converter, about 4 ft. in height in the interior,
which was sufficiently tall and capacious, as I believed, to prevent anything
but a few sparks and heated gases from escaping through a central hole
made in the flat top of the vessel for that purpose, as shown in the vertical
section at Fig. 39, Plate XIII.   The converter had six horizontal
tuyères arranged around the lower part of it ; these were connected by six
adjustable branch pipes, deriving their supply of air from an annular
rectangular chamber, extending around the converter, as shown.

All being thus arranged, and a blast of 10 or 15 lb. pressure turned
on, about 7 cwt. of molten pig iron was run into the hopper provided
on one side of the converter for that purpose.   All went on quietly
for about ten minutes ; sparks such as are commonly seen when tapping
a cupola, accompanied by hot gases, ascended through the opening on
the top of the converter, just as I supposed would be the case.   But
soon after a rapid change took place ; in fact, the silicon had been
quietly consumed, and the oxygen, next uniting with the carbon, sent

up an ever-increasing stream of sparks and a voluminous white flame. Then followed a succession of mild explosions, throwing molten slags and splashes of metal high up into the air, the apparatus becoming a veritable volcano in a state of active eruption. No one could approach the converter to turn off the blast, and some low, flat, zinc-covered roofs, close at hand were in danger of being set on fire by the shower of red-hot matter falling on them. All this was a revelation to me, as I had in no way anticipated such violent results. However, in ten minutes more the eruption had ceased, the flame died down, and the process was complete. On tapping the converter into a shallow pan or ladle, and forming the metal into an ingot, it was found to be wholly decarburised malleable iron.

Such were the conditions under which the first charge of pig iron was converted in a vessel neither internally nor externally heated by fire.

I, however, desired to convert a second charge of pig iron which had been put into the cupola; and in order to prevent this dangerous projection upwards of sparks and molten slags, a temporary expedient was resorted to, which, however, failed in its object.

I procured one of those circular, chequered cast-iron plates so much used in the London pavements to allow coals to be put into the cellars below the pavement. This plate, which was about a foot in diameter, was suspended by a chain at a distance of about 18 in. above the central opening in the top of the converter, as shown in Fig. 39, Plate XIII.

This, as a mere temporary device, was deemed sufficient to allow the conversion of another 7 cwt. charge to be effected, without any danger of setting fire to the premises. The converting operation went on quietly as before, but when the eruption commenced, I saw the suspended plate get rapidly red-hot, and in a few minutes more it melted and fell away, leaving the chain dangling over the opening, and allowing the slags and splashes of metal to shoot upwards as before. Thus it happened that the first converter that I constructed was at once condemned as commercially impracticable, owing to this vertical eruption of cinder, and for this reason only.

All attempts to lessen the violence of the process by the reduction of the number of tuyères, or by lessening their diameter, or by diminishing the pressure of the blast, only resulted in a reduction of the necessary temperature, and in preventing the conversion of the molten pig into malleable iron. In one case the trial of a diminished area of tuyère openings resulted in nearly the whole charge of metal, after more than an hour's blowing, being converted into a solid mass of brittle white iron, similar to ordinary refiner's plate metal. Indeed, I may say the result of all my early investigations proved to me, beyond the possibility of a doubt, a fact which has since been confirmed in every Bessemer steel works throughout Europe and America, viz. : that rapidity of action, ending in a violent eruption, is an absolutely necessary condition of success. Not only must the converted metal acquire an enormously high temperature, so that it may not be chilled when pouring it out of the converter, or when a relatively large quantity of much cooler metal be added to deoxidise it, but it must not chill and form a shell in the ladle during the comparatively long time required for casting the ingots. Hence, to carry out the Bessemer process successfully, a temperature must be obtained very considerably above the mere melting temperature of malleable iron ; and in order to secure this it is necessary to drive powerful streams of air into the metal, so as to divide it into innumerable tiny globules diffused throughout the whole body of iron under treatment which, for the time being, may be likened to a fluid sponge with the active combustion of carbon with oxygen going on in every one of its myriads of ever-changing cavities.

It has been found that the union of carbon and oxygen takes place so rapidly at this high temperature as to produce a series of mild explosions. In the large converters in common use, a space some 8 ft. or 10 ft. in height above the normal level of the metal is provided, in which this violent action expends itself unseen, and is only partially recognised by a small quantity of slags leaping out of the mouth of the converter.

With these facts before us, it must be self evident that all attempts to produce malleable iron in a plain cylindrical vessel that has no top to it, and in which the metal normally rises to within 6 in. of the open

U

mouth, must utterly fail from two causes: first, because heat would fly off so freely that the temperature of molten malleable iron could never be reached; and secondly, because nearly all the metal contained in such a shallow, open-topped vessel would have leaped out of it, and have been scattered in all directions on the occurrence of the explosive eruption, without which no charge of molten pig iron has, or can be, converted into fluid malleable iron by a blast of air.

I had no sooner condemned my first cylindrical converter than I commenced to remedy its defects. The most obvious and ready way of doing this would have been simply to make an opening on one side of it near the top, and thus allow the escape of the ejected matter to take place horizontally, directing it against a wall, or allowing it to fall into a pit. But I desired to prevent this discharge of metal splashes as much as possible. Hence I determined on constructing a new converter with an upper chamber, having an arched roof and a conical sloping floor. This converter is represented in Figs. 40 and 41, on Plate XIII., the last-named view being a horizontal section through the tuyères. When a converter is so constructed, the ejected fluid, that would otherwise pass vertically upwards into the air, is thrown against the arched roof, and any metal that may be emitted falls again on the sloping floor of the upper chamber, and returns to the lower one. The flame and a portion of the slags find their way out of the two square lateral openings provided for that purpose. This upper chamber also served as a receptacle for heating up any metal intended to recarburise, or alloy with, the steel in course of being converted. The sectional plan, Fig. 41, shows six well-burned fire-clay or plumbago tuyère pipes fitted to openings left in the lining for that purpose. Their outer ends were made conical to facilitate the ramming in of loam around them, which effectually held them in position, and at the same time admitted of their easy removal when worn out; a jointed piece of iron tube, with a catch to hold it in place, conveyed the blast to each tuyère.

Another view, Fig. 42, Plate XIV., of this converter, taken at right angles to Fig. 40, shows on one side the hopper by which the molten iron was run into it by a movable spout direct from the cupola. This view also shows the tapping-hole open, and the spout which conducted the converted metal into a movable shallow pan or receiver, supported by a long handle (not shown). A fire-brick plug attached to a long handle was fitted to a fire-brick ring or opening in the bottom of the pan, and prevented any *débris* from the tapping-hole being carried into the mould. As this apparatus was intended to exhibit the process, it was essential that an easy way should be provided for getting away the ingots and quickly repeating the operation. This casting apparatus, constructed precisely as represented in Fig. 42, was erected at my Bronze Manufactory in London, about two months prior to my reading the "Cheltenham" paper, in August, 1856, to which I shall refer later. The mould was 10 in. square, and about 3 ft. in length inside; it was made in two pieces planed quite parallel, and then permanently bolted together. The base was a massive square flange, resting on four dwarf columns, which stood on the square upper flange of an hydraulic cylinder; bolts passed through these dwarf columns, and through the square flanges, thus uniting the ingot mould and hydraulic cylinder. To the latter a ram or plunger was fitted, having a movable square head, which accurately fitted the mould, and formed a movable bottom to it. Both the ram and the external surface of the mould were kept cool by a water-jacket, provided with supply and waste pipes. Matters being thus arranged, the converted metal was allowed to fall in a vertical stream from the receiver on to the head of the ram. The receiver was then removed, and as soon as the steel was solidified, water under pressure was turned on to the hydraulic cylinder, when a beautiful ingot, 10 in. square, and weighing about 7 cwt., steadily rose and stood on end ready for removal, the head of the ram rising one or two inches above the top of the mould. There are, no doubt, many persons still living who witnessed this combined converting and casting apparatus in successful operation.

Two 10-in. square ingots, made with this apparatus, were sent to the Dowlais Iron Works in Wales, and, without hammering, were rolled into two flat-footed rails on the 6th September, 1856; that is, twenty-four days after the reading of the "Cheltenham" paper. They were rolled under the personal superintendence of Mr. Edward Williams, Past President of the Iron and Steel Institute. Two pieces of these rails are still kept at the Institute in a large glass case containing many other examples of the early working of my process in London and in Sheffield.

Before concluding this brief sketch of the earliest forms of apparatus designed by me to facilitate or improve the process, I must revert to the difficulties inseparable from a fixed converter. In this form of apparatus much heat is dissipated by the blowing which takes place during the running in of the metal, and by the continuation of the blast after the metal is converted, and during the whole time of its discharge, which is a period of uncertain length. There is also the difficulty of stopping the process if anything goes wrong with the blast engine, or if a tuyère gives way. I searched diligently for a remedy for these and other grave defects, which at that time appeared impossible to remove, until the happy idea occurred to me of mounting the converter on axes, so as to be able to keep the tuyères above the metal until the charge of molten iron was run in, thus permitting the blowing of the whole charge to be commenced at one and the same time, and admitting also of the cessation of blowing during the discharge. This movement of the converter permitted a stoppage of the process to take place at any time for the removal of a damaged tuyère if necessary, and afforded great facilities for working.

The special form of the movable converter was also a matter of great importance, and there were several requirements to provide for. First, in order to make the heavy lining secure when turned upside down, a more or less arched shape in all directions was necessary. A long oval form seemed best adapted to the purpose, as it allowed some eight or nine feet in height for the metal to throw itself about in without leaving the converter. Then the large mouth or outlet pointing to one side was desirable, so that the sparks could be discharged away from the casting pit. After much study, I arrived at the form shown at A, Fig. 43, Plate XV., which is an external elevation ; B is a vertical section showing the position in which the vessel is retained during the running-in of the metal ; c shows it during the blow, and D the position it assumes when the converted metal is poured into a loamed-up casting ladle. This ladle is shown at E and F : it is provided with a discharge valve at the bottom, so that it can be moved from mould to mould by closing the valve during such movement, and then permit a vertical stream to descend into the mould perfectly free from any mixture of slags. The advantage of this mode of filling the moulds will be understood when it is borne in mind that they are necessarily narrow upright vessels. It is well known that a stream of molten metal, poured from the lip of a ladle, will describe a parabolic curve in its descent, tending to strike the further side of the mould before reaching' the bottom. The surface of the cast-iron mould so struck is instantly melted by the incandescent stream of steel, and the ingot and the mould thus become united, causing great inconvenience. Nor is it easy, in pouring the steel from the lip of the open ladle, to prevent some of the fluid slag floating on its surface from flowing over with the steel and spoiling the ingot. All of these difficulties are avoided by the ladle fitted with a bottom valve discharging a vertical stream down the centre of the mould, the quantity and flow being regulated with great facility by the hand-lever on the side of the ladle. At G and H, Fig. 43, are shown the bottom of the converter and the form of tuyères.

Many other mechanical contrivances were necessary to perfect the process, such, for instance, as my patent blast engine, with its noiseless self-acting valves; the hydraulic crane carrying the pouring ladle over every mould in the semi-circular casting pit, and designed to rise and fall in accordance with the movement of the converter when filling the ladle for casting; the direct-acting ingot cranes, which clear the pit and refill it with another set of moulds rapidly, and with very little manual labour; the elevated "valve-stand," from which safe position a single workman can overlook the whole converting apparatus, and control all their movements, govern the blast, and work the hydraulic cranes, etc.

The mode of transmitting semi-rotating motion to the converter was another important problem which I had to solve. I was of opinion that ordinary shafting and straps were inapplicable to this fiery monster. Five or ten tons of fluid metal had to be lifted in one direction, this load diminishing until the fluid running to the opposite end of the converter tended to reverse the driving gear. If anything went wrong, or slipped, the converter might swing itself round and discharge the incandescent metal on to the floor or among the workpeople. These considerations led me to adopt the hydraulic apparatus now universally employed for governing the motions of the converter: for, with this simple and reliable means, a lad at a safe distance can start or stop it instantly, can alter its speed and motion, and control the pouring of a 10-ton charge with ease and certainty.

The first movable converter was erected at my steel works at Sheffield, and was moved by hand-gearing, because at that early date I had not

FIG. 47   INGOT CRANE; BESSEMER PLANT AT SHEFFIELD

invented the hydraulic apparatus just described. This early converting plant did good work at Sheffield, and was constructed precisely as represented in Fig. 44, Plate XVI., which shows also the first modification of the hydraulic casting crane, and its ladle with valve, afterwards elaborated by me and rendered suitable for casting heavy charges of steel. The development of this earliest form of plant is shown in Figs. 45 and 46, Plates XVII. and XVIII., and Fig. 47, annexed. The early experiments at Baxter House were so far successful, as to justify myself and some of my friends in entering into partnership, and erecting in the town of Sheffield, a steel works which still remains in active operation under the style of " Henry Bessemer and Company, Limited." These works were established both for commercial purposes, and also to serve as a pioneer works or school, where the process was for several years exhibited to any iron or steel manufacturers who desired to take a license to work under my patents. All of these were allowed, either personally or by their managers, to see their own iron converted prior to their taking a licence.

CHAPTER XII

THE BESSEMER PROCESS

I WELL remember how anxiously I awaited the blowing of the first 7-cwt. charge of pig iron. I had engaged an ironfounder's furnace-attendant to manage the cupola and the melting of the charge. When his metal was nearly all melted, he came to me, and said hurriedly: "Where be going to put the metal, maister?" I said: "I want you to run it by a gutter into that little furnace," pointing to the converter, "from which you have just raked out all the fuel, and then I shall blow cold air through it to make it hot." The man looked at me in a way in which surprise and pity for my ignorance seemed curiously blended, as he said: "It will soon be all of a lump." Notwithstanding this prediction, the metal was run in, and I awaited with much impatience the result. The first element attacked by the atmospheric oxygen is the silicon, generally present in pig iron to the extent of $1\frac{1}{2}$ to 2 per cent.; it is the white metallic substance of which flint is the acid silicate. Its combustion furnishes a great deal of heat; but it is very undemonstrative, a few sparks and hot gases only indicating the fact that something is going quietly on. But after an interval of ten or twelve minutes, when the carbon contained in grey pig iron to the extent of about 3 per cent. is seized on by the oxygen, a voluminous white flame is produced, which rushes out of the openings provided for its escape from the upper chamber, and brilliantly illuminates the whole space around. This chamber proved a perfect cure for the rush of slags and metal from the upper central opening of the first converter. I watched with some anxiety for the expected cessation of the flame as the carbon gradually burnt out. It took place almost suddenly, and thus indicated the entire decarburisation of the metal. The furnace was then tapped, when out rushed a limpid stream of incandescent malleable iron, almost too brilliant for the eye to rest upon ; it was allowed to flow vertically into the parallel undivided ingot mould. Then came the' question, would the ingot shrink enough, and the cold iron mould expand enough, to allow the ingot to be pushed out ? An interval of eight or ten minutes was allowed, and then, on the application of hydraulic force to the ram, the ingot rose entirely out of the mould, and stood there ready for removal.

This is all very simple now that it has been accomplished, and many of my readers may, from their intimate knowledge of this subject, have felt impatient at its mere recital. But it is, nevertheless, impossible for me to convey to them any adequate idea of what were my feelings when I saw this incandescent mass rise slowly from the mould : the first large prism of cast malleable iron that the eye of man had ever rested on. This was no mere laboratory experiment. In one compact mass we had as much metal as could be produced by two puddlers and their two assistants, working arduously for hours with an expenditure of much fuel. We had obtained a pure, homogeneous 10-in. ingot as the result of thirty minutes' blowing, wholly unaccompanied by skilled labour or the employment of fuel; while the outcome of the puddlers' labour would have been ten or a dozen impure, shapeless puddle-balls, saturated with scoria and other impurities, and withal so feebly coherent, as to be utterly incapable of being rendered, by any known means, as cohesive as the metal that had risen from the mould. No wonder, then, that I gazed with delight on the first-born of the many thousands of the square ingots that now come into existence every day. Indeed, at the date I am writing (1897), the world's present production of Bessemer steel, if cast into ingots 10 in. square and 30 in. in length, weighing 7 cwt. each, would make over 90,000 such ingots in every working day of the year.

I had now incontrovertible evidence of the all-important fact that molten pig iron could, without the employment of any combustible matter, except that which it contained, be raised in the space of half an hour to a temperature previously unknown in the manufacturing arts, while it was simultaneously deprived of its carbon and silicon, wholly without skilled manipulation. What all this meant, what a perfect revolution it threatened in every iron-making district in the world, was

x

fully grasped by the mind as I gazed motionless on that glowing ingot, the mere contemplation of which almost overwhelmed me for the time, notwithstanding that I had for weeks looked forward to that moment with a full knowledge that it meant an immense success, or a crushing failure of all my hopes and aspirations. I soon, however, felt a strong desire to test the quality of the metal, but I had no appliances to hammer or roll such a formidable mass; indeed, we had no means at hand even to move it. But I saw that there was one proof possible to which I could subject the ingot where it stood, and calling for an ordinary carpenter's axe, I dealt it three severe blows on the sharp angle of the

Fig. 48. Malleable Iron Ingot

prism. The cutting edge of the axe penetrated far into the soft metal, bulging the piece forward but not separating it, as shown in the sketch, Fig. 48. Had it been cast iron those angle-pieces would have been scattered all over the place in red-hot fragments, but their standing firm and undetachable assured me that the metal was malleable.

Notwithstanding the strong views I entertained of the value of my invention, I desired to obtain the unbiassed opinion of some eminent engineer, who might possibly take a very different view from my own. I did not wish to live in a fool's paradise, and was most anxious to know how my ideas would be received by others. I knew Mr. George Rennie very well by reputation, and I invited him to a private view of the process, as carried on in the upright converter. He kindly consented to give me his opinion, came to Baxter House and saw the process, with the result that he took a very deep interest in it. While discussing the subject, after the blow, he said : " This is such an important invention that you ought not to keep the secret another day." " Well," I said, " it is not yet quite a commercial success, and I think I had better perfect it before allowing it to be seen." " Oh," he said, " all the little details requisite will come naturally to the ironmaster ; your great principle is an unquestioned success ; no fuel, no manipulation, no puddle-balls, no piling and welding ; huge masses of any shape made in a few minutes." This truly great engineer was fairly taken by surprise, and his enthusiasm was as great and as genuine as it could have been had he himself been the inventor. All at once he said : " The British Association meets next week at Cheltenham, and I advise you strongly to read a paper on that occasion. I am President this year of the Mechanical Section. I wish I had known of this invention earlier. All our papers are now arranged for the meeting, and yours would be at the bottom of the long list, and it might simply be taken as read and would not be heard at all. But so important is this new process to all engineers that, if you will write a paper, I will take upon myself the responsibility of putting it first on the list." I could not withstand so handsome an offer from so distinguished a source. I told him that I much doubted my ability to write a paper in any way worthy of being read before the British Association, as I had never written or read a paper before any learned society. " Do not fear that," he said. " If you will only put on paper just such a clear and simple account of your process as you have given verbally to me, you will have nothing to fear." Soon after this he took his departure, with many words of encouragement, and I was left face to face with a task that I had not bargained for. I, however, at once set to work, and, having completed my paper in a few days, I left London on Tuesday, the 12th August, 1856, for Cheltenham.

On the following morning, while finishing my breakfast at the hotel, I was sitting next to Mr. Clay, the manager of the Mersey Forge, at Liverpool, to whom I was well known, when a gentleman who turned out to be Mr. Budd, a well-known Welsh ironmaker, came up to the breakfast-table, and, seating himself opposite my friend, said to him; "Clay, I want you to come with me into one of the Sections this morning, for we shall have some good fun." The reply was: "I am sorry that I am specially engaged this morning, or I would have done so with pleasure." "Oh, you must come, Clay," said Mr. Budd. "Do you know, that there is actually a fellow come down from London to read a paper on the manufacture of malleable iron without fuel? Ha, ha, ha!" "Oh," said Mr. Clay, "that's just where this gentleman and I are going." "Come along, then," said Mr. Budd, and we all rose from the table and proceeded towards the rooms occupied by the Mechanical Section. It was getting rather late, the room was well filled, and I, dropping the arm of my friend, ascended the raised platform and was cordially received by the President. Soon after, when the general bustle had subsided, Mr. George Rennie stood up, and in a few appropriate words explained that, at the eleventh hour, he had become acquainted with the fact that a most important discovery had been made in the manufacture of iron and steel, and he had considered it desirable that a paper describing the invention should be read at that meeting. As the papers for that section had already been arranged, he had ventured on a step which he hoped would be excused by all those gentlemen who had favoured them by preparing papers for that occasion. He considered that the paper about to be read was too important to be put at the tail end of the list, and, as the only alternative, he had ventured to put it at the head. He had great pleasure in introducing to the meeting the inventor, Mr. Henry Bessemer, who would now read his Paper on "The Manufacture of Iron Without Fuel."

The audience received me very kindly, and I had the honour of reading my paper, of which a *verbatim* copy is here given.

The manufacture of iron in this country has attained such an important position that any improvement in this branch of our national industry cannot fail to be a source of general interest, and will, I trust, be sufficient excuse for the present brief, and, I fear, imperfect paper. I may mention that for the last two years my attention has been almost exclusively directed to the manufacture of malleable iron and steel, in which, however, I had made but little progress until within the last eight or nine months. The constant pulling down and rebuilding of furnaces, and the toil of daily experiments with large charges of iron, had already begun to exhaust my stock of patience; but the numerous observations I had made during this very unpromising period all tended to confirm an entirely new view of the subject which, at that time, forced itself upon my attention, viz., that I could produce a much more intense heat without any furnace or fuel than could be obtained by either of the modifications I had used, and consequently that I should not only avoid the injurious action of mineral fuel on the iron under operation, but I should at the same time avoid also the expense of fuel. Some preliminary trials were made on from 10 lb. to 20 lb. of iron, and although the process was fraught with considerable difficulty, it exhibited such unmistakable signs of success as to induce me at once to put up an apparatus capable of converting about 7 cwt. of crude pig iron into malleable iron in thirty minutes. With such masses of metal to operate on, the difficulties which beset the small laboratory experiments of 10 lb. entirely disappeared. On this new field of inquiry I set out with the assumption that crude iron contains about 5 per cent. of carbon; that carbon cannot exist at a white heat in the presence of oxygen without uniting therewith and producing combustion; that such combustion would proceed with a rapidity dependent on the amount of surface of carbon exposed; and, lastly, that the temperature which the metal would acquire would be also dependent on the rapidity with which the oxygen and carbon were made to combine; and consequently that it was only necessary to bring together the oxygen and carbon in such a manner that a vast surface should be exposed to their mutual action, in order to produce a temperature hitherto unattainable in our largest furnaces. With a view of testing practically this theory, I constructed a cylindrical vessel 3 ft. in diameter, and 5 ft. in height, somewhat like an ordinary cupola furnace. The interior of this vessel is lined with firebricks, and at about 2 in. from the bottom of it, I insert five tuyère pipes, the nozzles of which are formed of well-burned fireclay, the orifice of each tuyère being about ⅜ in. in diameter; they are so put into the brick lining (from the outer side) as to admit of their removal and renewal in a few minutes when they are worn out. At one side of the vessel, about half-way up from the bottom, there is a hole made for running in the crude metal, and on the opposite side there is a tap-hole stopped with loam, by means of which the iron is run out at the end of the process. In practice this converting vessel may be made of any convenient size, but I prefer that it should not hold less than one, or more than five, tons of fluid iron at each charge. The vessel should be placed so near to the discharge hole of the blast furnace as to allow the iron to flow along a gutter into it; a small blast cylinder will be required capable of compressing air to about 8 lb. or 10 lb. to the square inch. A communication having been made between it and the tuyères before named, the converting vessel will be in a condition to commence work; it will, however, on the occasion of its being used after re-lining with firebricks, be necessary to make a fire in the interior with a few bucketfuls of coke, so as to dry the brickwork and heat up the vessel for the first operation, after which the fire is to be all carefully raked out at the tapping hole, which is again to be made good with loam. The vessel will then be in readiness to commence work, and may be so continued without any use of fuel until the brick lining in the course of time becomes worn away and a new lining is required. I have before mentoned that the tuyères are situated close to the bottom of the vessel;

the fluid metal will therefore rise some 18 in. or 2 ft. above them. It is therefore necessary, in order to prevent the metal from entering the tuyère holes, to turn on the blast before allowing the fluid crude iron to run into the vessel from the blast furnace. This having been done, and the fluid iron run in, a rapid boiling-up of the metal will be heard going on within the vessel, the metal being tossed violently about and dashed from side to side, shaking the vessel by the force with which it moves. From the throat of the converting vessel flame will then immediately issue, accompanied by a few bright sparks. This state of things will continue for about fifteen or twenty minutes, during which time the oxygen in the atmospheric air combines with the carbon contained in the iron, producing carbonic acid gas and at the same time evolving a powerful heat. Now as this heat is generated in the interior of, and is diffused in innumerable fiery bubbles throughout, the whole fluid mass, the metal absorbs the greater part of it, and its temperature becomes immensely increased, and by the expiration of the fifteen or twenty minutes before-named, that part of the carbon which appears mechanically mixed and diffused through the crude iron has been entirely consumed. The temperature, however, is so high that the chemically-combined carbon now begins to separate from the metal, as is at once indicated by an immense increase in the volume of flame rushing out of the throat of the vessel. The metal in the vessel now rises several inches above its natural level, and a light frothy slag makes its appearance, and is thrown out in large foam-like masses. This violent eruption of cinder generally lasts about five or six minutes, when all further appearance of it ceases, a steady and powerful flame replacing the shower of sparks and cinder which always accompanies the boil. The rapid union of carbon and oxygen, which thus takes place, adds still further to the temperature of the metal, while the diminished quantity of carbon present allows a part of the oxygen to combine with the iron, which undergoes combustion and is converted into an oxide. At the excessive temperature that the metal has now acquired, the oxide as soon as formed undergoes fusion, and forms a powerful solvent of those earthy bases that are associated with the iron. The violent ebullition which is going on mixes most intimately the scoria and the metal, every part of which is thus brought in contact with the fluid oxide, which will thus wash and cleanse the metal most thoroughly from the silica and other earthy bases which are combined with the crude iron, while the sulphur and other volatile matters which cling so tenaciously to iron at ordinary temperatures, are driven off, the sulphur combining with the oxygen and forming sulphurous acid gas. The loss of weight of crude iron during its conversion into an ingot of malleable iron was found on a mean of four experiments to be 12½ per cent., to which will have to be added the loss of metal in finishing rolls. This will make the entire loss probably not less than 18 per cent., instead of about 28 per cent., which is the loss on the present system. A large portion of this metal is, however, recoverable by treating with carbonaceous gases the rich oxides thrown out of the furnace by the boil. These slags are found to contain innumerable small grains of metallic iron, which are mechanically held in suspension in the slags, and may be easily recovered. I have before mentioned that after the boil has taken place a steady and powerful flame succeeds, which continues without any change for about ten minutes, when it rapidly falls off. As soon as this diminution of flame is apparent the workman will know that the process is completed, and that the crude iron has been converted into pure malleable iron, which he will form into ingots of any suitable size and shape, by simply opening the tap-hole of the converting vessel and allowing the fluid malleable iron to flow into the iron ingot-moulds placed there to receive it. The masses of iron thus formed will be perfectly free from any admixture of cinder, oxide, or other extraneous matters, and will be far more pure, and in a more forward state of manufacture, than a pile formed of ordinary puddle-bars. And thus it will be seen, that by a single process requiring no manipulation or particular skill, and with only one workman, from three to five tons of crude iron pass into the condition of several piles of malleable iron in from thirty to thirty-five minutes, with the expenditure of about one-third part the blast now used in a finery furnace with an equal charge of iron, and with the consumption of no other fuel than is contained in the crude iron. To those who are best acquainted with the nature of fluid iron, it may be a matter of surprise that a blast of cold air forced into melted crude iron is capable of raising its temperature to such a degree as to retain it in a perfect state of fluidity after it has lost all its carbon, and is in the condition of malleable iron, which in the highest heat of our forges only becomes softened into a pasty mass. But such is the excessive temperature that I am enabled to arrive at with a properly-shaped converting vessel and a judicious distribution of the blast, that I am enabled not only to retain the fluidity of the metal, but to create so much surplus heat as to re-melt the crop-ends, ingot-runners, and other scrap that is made throughout the process, and thus bring them without labour or fuel into ingots of a quality equal to the rest of the charge of new metal. For this purpose a small arched chamber is formed immediately over the throat of the converting vessel, somewhat like the tunnel-head of the blast furnace. This chamber has two or more openings on the side of it, and its floor is made to slope downwards to the throat. As soon as a charge of fluid malleable iron has been drawn off from the converting vessel the workmen will take the scrap intended to be worked into the next charge, and proceed to introduce the several pieces into the small chamber, piling them up around the opening of the throat. When this is done, he will run in his charge of crude metal, and again commence the process. By the time the boil commences, the bar-ends and other scrap will have acquired a white heat, and by the time it is over most of them will have been melted and run down into the charge. Any pieces, however, that remain may then be pushed in by the workman, and by the time the process is completed they will all be melted, and ultimately combined with the rest of the charge; so that all scrap iron, whether cast or malleable, may thus be used up without any loss or expense. As an example of the power that iron has of generating heat in this process, I may mention a circumstance that occurred to me during my experiments. I was trying how small a set of tuyères could be used; but the size chosen proved to be too small, and after blowing into the metal for one hour and three-quarters, I could not get up heat enough with them to bring on the boil. The experiment was, therefore, discontinued, during which time two-thirds of the metal solidified, and the rest was run off. A larger set of tuyère pipes were then put in, and a fresh charge of fluid iron run into the vessel, which had the effect of entirely remelting the former charge, and when the whole was tapped out it exhibited, as usual, that intense and dazzling brightness peculiar to the electric light.

    To persons conversant with the manufacture of iron it will be at once apparent that the ingots of malleable metal which I have described will have no hard or steely parts, such as are found in puddled iron, requiring a great amount of rolling to blend them with the general mass; nor will such ingots require an excess of rolling to expel cinder from the interior of the mass, since none can exist in the ingot, which is pure and perfectly homogeneous throughout, and hence requires only as much rolling as is necessary for the development of fibre. It, therefore, follows that, instead of forming a merchant bar or rail by the union of a number of separate pieces welded together, it will be far more simple, and less expensive, to make several bars or rails from a single ingot. Doubtless this would have been done long ago, had not the whole process been limited by the size of the ball which the puddler could make.

The facility which the new process affords of making large masses will enable the manufacturer to produce bars that, on the old mode of working, it was impossible to obtain; while, at the same time, it admits of the use of more powerful machinery, whereby a great deal of labour will be saved, and the process be greatly expedited. I merely mention this fact in passing, as it is not my intention at the present moment to enter upon any details of the improvements I have made in this department of the manufacture, because the patents which I have obtained for them are not yet specified. Before, however, dismissing this branch of the subject, I wish to call the attention of the meeting to some of the peculiarities which distinguish cast steel from all other forms of iron: namely, the perfect homogeneous character of the metal, the entire absence of sand-cracks or flaws, and its greater cohesive force and elasticity, as compared with the blister steel from which it is made—qualities which it derives solely from its fusion and formation into ingots, all of which properties malleable iron acquires in like manner by its fusion and formation into ingots in the new process. Nor must it be forgotten that no amount of rolling will give to blister steel (although formed of rolled bars) the same homogeneous character that cast steel acquires by a mere extension of the ingot to some ten or twelve times its original length.

One of the most important facts connected with the new system of manufacturing malleable iron is, that all iron so produced will be of that quality known as charcoal iron: not that any charcoal is used in its manufacture, but because the whole of the processes following the smelting of it are conducted entirely without contact with, or the use of, any mineral fuel; the iron resulting therefrom will, in consequence, be perfectly free from those injurious properties which that description of fuel never fails to impart to iron that is brought under its influence. At the same time, this system of manufacturing malleable iron offers extraordinary facility for making large shafts, cranks, and other heavy masses; it will be obvious that any weight of metal that can be founded in ordinary cast iron by the means at present at our disposal may also be founded in molten malleable iron, and be wrought into the forms and shapes required, provided that we increase the size and power of our machinery to the extent necessary to deal with such large masses of metal. A few minutes' reflection will show the great anomaly presented by the scale on which the consecutive processes of iron-making are at present carried on. The little furnaces originally used for smelting have assumed colossal proportions, and are made to operate on 200 or 300 tons of materials at a time, giving out 10 tons of fluid metal at a single run. The manufacturer has thus gone on increasing the size of his smelting furnaces, and adapting to their use the blast apparatus of the requisite proportions, and has by this means lessened the cost of production in every way; his large furnaces require a great deal less labour to produce a given weight of iron than would have been required to produce it with a dozen furnaces; and in like manner he diminishes his cost of fuel, blast, and repairs, while he insures a uniformity in the result that never could have been arrived at by the use of a multiplicity of small furnaces.

While the manufacturer has shown himself fully alive to these advantages, he has still been under the necessity of leaving the succeeding operations to be carried out on a scale wholly at variance with the principles he has found so advantageous in the smelting department. It is true that hitherto no better method was known than the puddling process, in which from 400 lb. to 500 lb. of iron is all that can be operated upon at a time; and even this small quantity is divided into homœopathic doses of some 70 lb. or 80 lb., each of which is moulded and fashioned by human labour, carefully watched and tended in the furnaces, and removed therefrom one at a time to be carefully manipulated and squeezed into form. When we consider the vast extent of the manufacture and the gigantic scale on which the early stages of the process is conducted, it is astonishing that no effort should have been made to raise the after-processes somewhat nearer to a level commensurate with the preceding ones, and thus rescue the trade from the trammels which have so long surrounded it.

Before concluding these remarks, I beg to call your attention to an important fact connected with the new process, which affords peculiar facilities for the manufacture of cast steel. At that stage of the process immediately following the boil, the whole of the crude iron has passed into the condition of cast steel of ordinary quality; by the continuation of the process the steel so produced gradually loses its small remaining portion of carbon, and passes successively from hard to soft steel, and from soft steel to steely iron, and eventually to very soft iron; hence, at a certain period of the process, any quality of metal may be obtained. There is one in particular, which, by way of distinction, I call semi-steel, being in hardness about midway between ordinary cast steel and soft malleable iron. This metal possesses the advantage of much greater tensile strength than soft iron. It is also more elastic, and does not readily take a permanent set; while it is much harder, and is not worn or indented so easily as soft iron, at the same time it is not so brittle or hard to work as ordinary cast steel. These qualities render it eminently well adapted to purposes where lightness and strength are specially required, or where there is much wear, as in the case of railway bars, which, from their softness and lamellar texture, soon become destroyed. The cost of semi-steel will be a fraction less than iron, because the loss of metal that takes place by oxidation in the converting vessel is about $2\frac{1}{2}$ per cent. less than it is with iron; but, as it is a little more difficult to roll, its cost per ton may fairly be considered to be the same as iron. But, as its tensile strength is some 30 or 40 per cent. greater than bar iron, it follows that for most purposes a much less weight of metal may be used, so that, taken in that way, the semi-steel will form a much cheaper metal than any with which we are at present acquainted.

In conclusion, allow me to observe that the facts which I have had the honour to bring before the meeting have not been elicited from mere laboratory experiments, but have been the result of working on a scale nearly twice as great as is pursued in our largest iron works: the experimental apparatus doing 7 cwt. in thirty minutes, while the ordinary puddling furnace makes only $4\frac{1}{2}$ cwt. in two hours, which is made into six separate balls, while the ingots or blooms are smooth, even prisms 10 in. square by 30 in. in length, weighing about equal to ten ordinary puddle-balls.

During the reading of the paper, I made a chalk sketch of the converter on the blackboard, and answered several questions put by

Y

members present; at its conclusion, an enthusiastic vote of thanks was accorded me.

On the table in front of the raised platform I had exhibited a few samples hastily got together for the occasion; one of them was a flat iron bar, about 3½ in. wide by ¾ in. in thickness, which had been rolled direct from a cast ingot at the Royal Arsenal at Woolwich, then under the superintendence of Colonel Eardley Wilmot. Another, but smaller, bar of iron had been rolled, cut up and piled, and again rolled into a long bar of small section. One of the ends cut off from this bar, showing the overlapping of some parts of the pile, has fortunately been preserved, and is now in the glass-case of old specimens which I presented some years ago to the Iron and Steel Institute. I also exhibited a large mass of fractured decarburised iron of silvery whiteness, and some broken ingots of malleable iron, etc.

The first person to rise after the reading of the paper was the late Mr. James Nasmyth, who occupied a seat near me on the platform. He held up between his thumb and finger a small fragment of wholly decarburised iron, and enthusiastically exclaimed, " Gentlemen, this is a true British nugget." Then in glowing terms he referred to the novelty of the process, the rapid conversion into malleable iron of the molten iron as it came direct from the blast furnace, the power the process afforded of dealing with immense masses, the absence of all skilled labour, and the non‑employment of fuel. All this, he said, pointed to results so vast and so commercially important, that it was impossible to grasp the full effect it must have both on the iron and engineering interests of this and of every other country. This paper had come upon him quite unexpectedly, and the true instinct of the engineer and man of science rose above all other considerations. He forgot how his own personal interests might be affected by it, and in his enthusiasm he said: " I am not going in any way to claim priority of thought or action, but I cannot forget that a few years ago I patented, in the puddling process, the use of steam, which was blown through the bar or 'rabble' with which the puddling operations are carried on. This might be called a first step on the same road; but Mr. Bessemer has gone miles beyond it, and I do not hesitate to say that I may go home from this meeting and tear up my now useless patent." Mr. Nasmyth resumed his seat amid a storm of cheers. Surely all who heard that noble speech, however much they might have honoured Mr. Nasmyth as an improver of the puddling process, must have honoured him infinitely more for thus throwing over his own production, and fearlessly advocating an invention that so utterly destroyed the value of his own.

I must not forget to mention that Mr. Budd—who may be well excused for the feeling of ridicule inspired by the extraordinary title of my paper—was the next to rise at the meeting. He said he had listened with deep interest to the important details of this invention, and if Mr. Bessemer desired an opportunity of commercially testing it, he should be most happy to afford him every possible facility. His ironworks were entirely at Mr. Bessemer's disposal, and if he liked to avail himself of this offer, it should not cost him a penny. This generous proposal made ample amends for the little joke at the breakfast-table, and was received with hearty cheers; after some further discussion, and the reading of some other papers, the meeting broke up. As I was about to leave, *The Times* reporter was introduced to me, and he told me that he had not paid sufficient attention to the first part of my paper, as the ironmasters present seemed to treat it rather as a good joke than as a reality, and, taking his cue from them, he had not made so full a report as he desired. But the enthusiastic way in which the latter part of my paper was received on all sides, made him desirous of giving a much fuller report than he had done. He further said: " If you will be kind enough to lend me your paper, I will promise you that every word of it shall appear in *The Times* to-morrow." I was much pleased with his proposal, and at once handed him my paper, which duly appeared *in extenso* on the following morning as promised, and from *The Times* report of August 14th, 1856, the copy just given is reproduced. It is impossible to gauge with any degree of accuracy the effect, social or political, of the hundreds of articles that, from time to time, have appeared in that influential and widely-circulated journal, but when we view the publication of this particular paper from a national point of view, it simply defies any estimate of the magnitude of the interests involved.

And yet this high appreciation of my invention by Mr. George Rennie, and the announcement of it to the whole world through the columns of *The Times*, was like a two-edged sword; for, while on the one hand it was the direct cause of bringing to my aid the sinews of war, and assisted me in fighting the great battle of vested interests arrayed against me, on the other hand it had a fearful disadvantage, which might have wrecked all. In listening to the kind words of Mr. George Rennie, I too readily allowed myself to bring my invention under public notice. I should not have done so until all the details of the process had been worked out, and I had made it a great commercial (and not merely a scientific) fact. My premature disclosure brought down upon me a wild pack of hungry wolves, fighting with me, and with each other, for a share of what was to be made by this new discovery. To these eager adventurers, the conversion of five tons of crude molten iron into cast steel, in a few minutes, was the realisation of the fabled philosopher's stone, that transmuted lead into gold. It was not a question with these people of improving my process, but of an endeavour to imitate it, or to do something similar by some dodge or other that was not covered by my patent. If they could simply surround me and hem me in with possible or impossible claims, I must surely, they thought, pay them to get out of my way. The agent of one of these so-called inventors told me to my face that he had a little bit of land in the middle of my road, and that there was not room for me to pass on either side, and that I dared not run over him. Many examples might be adduced of the wild schemes propounded in this mad race to appropriate the principle of my invention. One inventor, instead of forcing air upward through the metal, proposed to suck it out of the vessel by directly pumping out the fire and showers of sparks, instead of driving clean, cold, atmospheric air into it, as I had claimed in my patent. Another would force down air upon the surface with such great pressure as to penetrate the metal from the top instead of letting the air pass naturally upwards. Another would allow the molten iron to flow down steps, and blow on it as it fell from step to step. Another claimed to spread the metal in a thin sheet and blow on to it, but not into it, as I did. Another so-called inventor proposed to let the molten iron fall down a deep well in the form of a shower, and collect it at the bottom as malleable iron, not thinking that his process would simply make iron shot. Another claimed the exclusive use in my process of that kind of pig iron that had been most commonly used in Styria for the last hundred years for making steel, the ore of which was known as " stahl stein," or steel ore ; nor was I to use manganese either as a metal, an oxide, or a carburet, although that metal was in daily use in all the hundreds of steel pots in Sheffield.

I had used the word " pig-iron " from which, after various processes, all iron and steel then in use was made ; had I used the more scientific term, " carbonate of iron," instead of the accepted trade term, " pig " or crude iron from the blast furnace, I should have been safe from one scheme intended to circumvent me by a play on words. According to this plan, malleable scrap iron was put into a tall cupola furnace, and during its descent absorbed so much carbon as to issue therefrom as a white cast-iron. It was claimed that this was not pig-iron or crude molten iron, as mentioned in my patent, as it was assumed that white iron so made, with two per cent. of carbon, might be blown into steel by my process without my being able to prevent it. These, and all other discreditable attempts to make use of a colourable imitation of my patent, utterly and ignominiously failed.

Within a few days of the publication of my Cheltenham paper, many eminent engineers and ironmasters from various parts of the kingdom did me the honour to come up to London, and see the process carried out at my bronze factory at St. Pancras. Many and strange were the opinions expressed on these occasions, and many questions were asked as to the terms on which I proposed to allow the trade to use the process. At that time the steel manufacturer took no interest in the question, and it was left to the ironmaster to secure the huge advantage of the new discovery. I and my partner, Mr. Longsdon, had thought the subject well over, and we came to the conclusion that it would be wise not to have the whole trade opposed to us, but to give a special interest to one ironmaster in each district, so that his working would prove an example to other iron works, and his special interest would induce him at any future time to help to support my patents, and not join in an adverse movement of the trade. But, at first sight, it did not appear easy to do this without parting with a share of the patents, and thus depriving ourselves of the absolute control of them. At last we fixed a royalty of ten shillings per ton for making malleable or wrought iron. To the first applicant for a licence in each district, we would give a great and permanent advantage over all others, and allow him to take a license to make a given number of tons per annum at a royalty of one farthing per ton during the whole term of the patents, he purchasing this right by paying at once a ten shilling royalty on the annual quantity agreed upon. He would then have a strong interest in the maintenance of the patents, and we should have the advantage of cash in hand with which to fight our battles, if attacked. These terms having been definitely fixed, were communicated to the trade, and we continued to show the process to all who wished to see it.

On August 27th — fourteen days after the publication of my Cheltenham paper in *The Times* — we were visited in the afternoon by Mr. H. A. Bruce (afterwards Lord Aberdare) and Mr. George Clark, trustees of the great Dowlais Iron Works. We said that we were sorry that the experiments were over for the day, but we should be happy to show them on the morrow. "Oh," said these gentlemen, "We do not care about seeing the process, for our chemist (Mr. E. Riley), on reading your paper in *The Times*, extemporised a converting furnace in one of the sheds, had the blast conveyed from our blast-furnace engines, and tried the experiment; the object of our visit is to treat for a license. We want to make 70,000 tons of malleable iron per annum." They were a good deal disconcerted on hearing our terms, and after much discussion it was arranged that we should dine with them that evening at the Tavistock Hotel, and further talk the matter over. This discussion resulted in their agreement to pay us £10,000 for a license under which they should be at liberty to make 20,000 tons of malleable iron per annum, at a royalty of one farthing per ton, during the whole duration of the patent. A memorandum to this effect was drawn up and signed as soon as dinner was over; and, when all was thus settled to our mutual satisfaction, our first licensees returned to Dowlais. It was exceedingly satisfactory to us that these gentlemen should have spontaneously made their own experiments in private, and satisfied themselves of the practicability of the process by the aid of their own chemist and workmen; and, on the strength of the results so obtained, should have come up in haste to London to secure a license for their works, lest the right should pass into other hands. This circumstance gave us great assurance of the practicability of the invention which, everyone knew, had at that time never been commercially carried out at any iron works. Hence the purchase of a licence to work the new process was simply a mercantile speculation in which the purchaser, who paid £10,000 down, stood to save, during twelve years, £120,000, less £125 paid in farthings. The inventor, on the other hand, had the advantage of ready cash to cover the risks he himself had run in expending two years of labour, in bearing the costs of constructing apparatus, taking out patents, and making expensive experiments at a time when the whole scheme was purely ideal, and the risks were much larger to him than they were to those who now speculated on his success.

This sale of licenses for the whole term of the patents made the licensees firm supporters of the patents, while the advantage given to one manufacturer in each of the great iron districts was not calculated to injure the trade, as the owner of the privilege would put the extra profits in his pocket, instead of throwing away his advantage by underselling his neighbours. For instance, the Dowlais Iron Company were making 70,000 tons of rolled iron annually, and would have to pay a full royalty on 50,000 tons, thus reducing their advantage to less than three shillings per ton on their annual production of iron, a sum too small to permit of their underselling the rest of the trade. This was, then, the scheme by which I proposed to force my invention into commercial use, in face of the gigantic vested interests arrayed against it.

Soon after the departure of the Dowlais licensees, two gentlemen from Scotland had a close run as to who should arrive first, and so claim the advantage of being the pioneer for Scotland. This claim was eventually settled in favour of Mr. Smith Dixon, of the Govan Iron Works, Glasgow, who paid £10,000 for a license to make 20,000 tons of iron annually at a royalty of one farthing per ton. This was followed by a license to the Butterley Iron Company, in Derbyshire, to make 10,000 tons annually on the same terms. A license was also granted to make 4000 tons annually to a tin-plate manufacturer in Wales, at one farthing per ton, on payment of one year's royalty of £2000, thus making sales of royalties to the amount of £27,000 in less than one month from the announcement of my invention in *The Times*. Up to this period, and long after it, the only persons interested were the ironmasters, the question not making the smallest impression in the steel trade. Sheffield wrapped itself in absolute security, and believed that it could afford to laugh at the absurd notion of making five tons of cast steel from pig-iron in twenty or thirty minutes, when by its own system fourteen or fifteen days and nights were required to obtain a 40-lb. or 50-lb. crucible of cast steel from pig-iron. So the Yorkshire town was allowed to stand aside while the more enter-prising ironmaster gave the invention a trial, as far as bar-iron making was concerned. At this period the ironmaster would never have dreamed of changing his trade to that of a cast-steel manufacturer, had such a thing been proposed to him.

Among the many persons who called on me from time to time, and made proposals for a license, none was so energetic and thorough-going as Mr. Thos. Brown, of the Ebbw Vale Ironworks. He brought with him an eminent consulting engineer, Mr. Charles May, and with a good deal of quiet tact, beat about the bush, trying to gauge my ideas on the value of my patents. He expatiated on the advantages of turning an invention to immediate account, and being not only well paid, but much overpaid, for all costs and labour expended in perfecting the invention, which, when purchased for cash, might be upset in law without any loss to the inventor, who had been wise enough to realise when he had the opportunity. This was the whole gist and meaning of a rather long introductory speech, and I distinctly remember the reply which I made at the time, and which I have often since repeated. I said: " Mr. Brown, the expense and labour that I may have had over this invention is no measure of its value. If you and I were walking arm-in-arm along the street, and I saw something glittering in the gutter, and if the mere fact of my being the first to discover it gave me a legal claim to its possession, and all the labour and trouble taken by me were simply to lift it out of the gutter with my thumb and finger, and if this little glittering thing on examination turned out to be the Koh-i-noor, then the Koh-i-noor being legally my personal property, I should want a million sterling for it, if that happened to be its ascertained commercial value, notwithstanding the fact of its having come so easily into my possession." I thus quietly gave Mr. Brown to understand that I was in no hurry to sell my birthright for a mess of pottage. Mr. Brown then adopted another method, and attempted to dazzle me at once, so as not to spoil the effect of a grand offer by letting it slide out piece-meal. "Well," he said, "the real object of my visit is to make you an offer to purchase all your patent rights in Great Britain for your iron and steel inventions; and I will tell you at once how far I am prepared to go, and I can go no farther. I am prepared to give you £50,000 cash for them." I said: "Mr. Brown, I cannot but feel that this is a very handsome offer indeed, for an invention that has not yet passed from the scientific to the commercial stage, and it is conclusive evidence of the high appreciation of its value by a practical ironmaster, and manager of a great Welsh iron-works. But, Sir, if my invention successfully passes from the scientific to the commercial stage, as I doubt not it will do, it must inevitably revolutionise the iron industry of the whole world; and even the very handsome sum you offer is not a tithe of its actual value. No, Sir, I cannot accept your very liberal offer; it is a large sum to risk, and whatever risk there is, it is I who should run it. I have had dozens of proofs—none of which you have seen—proofs that make me certain of the ultimate result, and I am content to see the invention through all its trials and vicissitudes, and stand or fall by the result."

Mr. Brown was evidently taken aback by my steady refusal to accept a sum which he no doubt felt, and very reasonably so, would certainly tempt me. Indeed, I presume he brought Mr. Charles May simply to witness the bargain he felt sure of making, the written terms of which were most probably in his coat pocket. Intense disappointment and anger quite got the better of him, and for the moment he could not realise the fact of my refusal; he hesitated, muttered something inaudible,

z

took up his hat, and left me very abruptly, saying in an irritated tone, as he passed out of the room, " I'll make you see the matter differently yet!" and slammed the door after him. We shall see, in a future Chapter, what were the steps taken by Mr. Brown to attain this end, and how far he succeeded.

In the meantime, small, upright, fixed converting vessels had been erected at the iron works of Messrs. Galloway at Manchester, at Dowlais in Wales, at Butterley in Derbyshire, and also at the Govan Iron Works at Glasgow, and in each case the results of the trials were most disastrous. The ordinary pig iron used for bar-iron making was found to contain so much phosphorus as to render it wholly unfit for making iron by my process. This startling fact came on me suddenly, like a bolt from the blue; its effect was absolutely overwhelming. The transition from what appeared to be a crowning success to one of utter failure well-nigh paralysed all my energies. Day by day fresh reports of failures arrived; the cry was taken up in the press; every paper had its letters from correspondents, and its leaders, denouncing the whole scheme as the dream of a wild enthusiast, such as no sensible man could for a moment have entertained. I well remember one paper, after rating me in pretty strong terms, spoke of my invention as " a brilliant meteor that had flitted across the metallurgical horizon for a short space, only to die out in a train of sparks, and then vanish into total darkness."

I was present at some of these trials, and saw the utter failure that resulted with the quality of metal operated upon. It is a curious, and scarcely credible, fact that not one of the ironmasters who had previously felt such abundant confidence in the success of the process as to back their opinions with large sums of money, took any trouble whatever, or offered any practical or scientific help, towards getting over this unlooked-for difficulty. They all stood by, mere passive and inert observers of the fact, not one of them lifting up a finger, or stretching out a hand, to save the wreck. For my own part, stunned as I was for the moment by the first blow, I never lost faith, or gave up the belief that all would yet be well. I had too deep an insight into the principle on which the whole theory was based to doubt of its correctness. By the mere accident of living in London, I had access only to the pig iron used

by London ironfounders. I had sent to a founder who had occasionally made me iron castings, and requested him to send me a few tons of pig iron for experiments. He sent me the grey Blaenavon iron which he was then using in his business, and I accepted it simply as pig iron, without ever suspecting that pig iron from other sources was so different, and would give such contrary results.

There was also another most important factor which accounted for my partial success in those early days, and which was unobserved and unknown until a much later period, viz., in all these early experiments in London, I lined the converter with clay or firebrick, and not with a silicious material such as ganister or sand. When the small converting vessels were erected for trial by my licensees, they were lined with silicious materials which prevented the elimination of any phosphorus from the iron, as was demonstrated later by Thomas and Gilchrist's well-known dephosphorising process. It was, however, no use for me to argue the matter in the Press; all that I could say would be mere talk, and I felt that action was necessary, and not words. I therefore determined to justify myself by the only possible means left to me. After a full and deliberate consideration of the whole case, I resolved to continue my researches until I had made my process a commercial, as well as a scientific, success. I was in possession of a large sum of money, which those ironmasters who believed in my invention had deliberately invested in the speculation, acting just as I myself had done, when I had gone to great expense in carrying out my experiments in hope of reaping a large profit. But I was not content to balance matters thus, and cry "quits." At the same time there were duties which I owed to myself and my family. I had spent two years of valuable professional time, much hard labour, and a great deal of money, over this invention, and a proportion of the proceeds belonged, in all fairness, to my family. Having thought thoroughly over the risks and the powerful opposition I had to fight against, I came to the conclusion that it was my duty to settle the sum of £10,000 on my wife under trustees, so that I could not be absolutely ruined in the further pursuit of my invention, or by litigation in the defence of my patent rights. After this investment I had still left £12,000 to spend in perfecting my process, if found necessary. My partner, Mr. Longsdon, who had implicit faith in me, intimated his resolve to go heart and soul with me in bearing his share of the cost. Although not strictly in the chronological order of events, it may here be briefly stated that these licenses to make malleable iron by my process, for which £27,000 had been paid, and which turned out unfortunately to be of no commercial value, in consequence of being superseded by my steel process, were nevertheless re-purchased by Messrs. Bessemer and Longsdon for the sum of £32,500, or £5500 more than they were sold for to those gentlemen who had ventured to speculate on the success of my invention.

At this period it became essential for me to know exactly what were the constituents of pig iron in all its commercial varieties, and what were the precise proportions in which these substances usually existed. In order to gain this all-important knowledge, we engaged the services of Dr. Henry, a well known professor of chemistry, to make complete and careful analyses of the iron and other materials used in all our future experiments, as well as of the results obtained in the converter. The very numerous investigations of this gentleman were supplemented by the able assistance of Mr. Edward Riley and Dr. Percy, and much information was also gathered from the publications and previous researches of Mr. Robert Hunt, of the Record Office of the School of Mines.

In this way, continued investigations, accompanied by experimental trials in the converter, were always adding to our store of facts, but unfortunately they seemed to bring us scarcely a step nearer to the end we had in view. British pig-iron abounded with this fatal enemy, phosphorus, and I could not dislodge it. Apparatus was put up for the production of pure hydrogen gas, which was passed through the metal ; as also were carbonic oxide, carburetted hydrogen, etc. Metallic oxides and alkaline salts, and many other fluxes, were tried with little or no beneficial results, and the metal was treated in various other ways. It is needless to follow the continuous string of heartbreaking failures and disappointments, which were very costly and very laborious. Eventually, I began to feel that the problem must be attacked from an entirely new position, viz., the production of pig-iron without phosphorus, a subject which I now took in hand.   In the meantime I became very anxious to know how far my converting process would be successful if we succeeded in making, or obtaining, some pig-iron that was wholly or practically free from phosphorus and sulphur; and I determined to set this one vital question at rest for ever by obtaining from Sweden some pure charcoal pig-irons from which such excellent steel was made in Sheffield.

The very large scale on which my experimental trials were at this time carried out involved a considerable outlay in various ways, but there was no slackening of exertion, no cessation of the severe mental and bodily labour.   A long and weary year was consumed in experiments, and but little real progress was made towards the removal of the difficulty; many new paths were struck out, but they led to no practical results.   Several weeks were sometimes necessary to make and fit up the apparatus required to test a new theory, and it too often happened that the first hour's trial of the new scheme dashed all the high expectations that had been formed, and we had again to retrace our steps.   Thus, week after week went on amid a constant succession of newly-formed hopes and crushing defeats, varied with occasional evidences of improvement.   I, however, worked steadily on. Six months more of anxious toil glided away, and things were in very much the same state, except that many thousands of pounds had been uselessly expended, and I was much worn by hard work and mental anxiety.   The large fortune that had seemed almost within my grasp was now far off; my name as an engineer and inventor had suffered much by the defeat of my plans.   Those who had most feared the change with which my invention had threatened their long-vested interests felt perfectly reassured, and could now safely sneer at my unavailing efforts; and, what was far worse, my best friends tried, first by gentle hints, and then by stronger arguments, to make me desist from a pursuit that all the world had proclaimed to be utterly futile. It was, indeed, a hard struggle; I had well-nigh learned to distrust myself, and was fain at times to surrender my own convictions to the mere opinion of others.   Those most near and dear to me grieved over my obstinate persistence.   But what could I do?   I had had the most irrefragable evidence of the absolute truth and soundness of the principle upon which my invention was based, and with this knowledge I could not persuade myself to fling away the promise of fame and wealth and lose entirely the results of years of labour and mental anxiety, and at the same time time confess myself beaten and defeated. Happily for me, the end was nigh.

The pure pig-iron, which I had ordered from Sweden, arrived at last, and no time was lost in converting it into pure, soft, malleable iron, and also into steel of various degrees of hardness. It was thus incontestably proved that with non-phosphoric pig-iron my converting process was a perfect success; and that with pig-iron that had cost me only £7 per ton, delivered in London, we could, and did, produce cast steel commercially worth £50 to £60 per ton, by simply forcing atmospheric air through it for the space of fifteen to twenty minutes, wholly without the use of manganese or spiegeleisen.

Thus was the so-called fallacious dream of the enthusiast realised to its fullest extent, and it was now my turn to triumph over those who had so confidently predicted my failure. I could see in my mind's eye the great iron industry of the world crumbling away under the irresistible force of the facts so recently elicited. In that one result the sentence had gone forth, and not all the knowledge accumulated during the last one hundred and fifty years by the thousands whose ingenuity and skill had helped to build up the mighty fabric of the British iron trade—no, nor the millions that had been invested in carrying out the existing system of manufacture—could reverse that one great fact, or stop the current that was destined to sweep away the old system of manufacturing wrought iron, and to establish homogeneous steel as the material to be in future employed in the construction of our ships and our guns, our viaducts and our bridges, our railroads and our locomotive engines, and the thousand-and-one things for which iron had hitherto been employed.

And yet, with all this newly-developed power, I was paralysed for the moment in the face of the stolid incredulity of all practical iron and steel manufacturers—an incredulity which stood like the wall of a fortress, barring my way to the fruits of the victory I had already won.

I announced the fact of my complete success to the world, and held in my hands the most undeniable proofs of the truth of my assertion, but no one would now believe it. They remembered, but too well, the great expectations that were excited two years previously by the first announcement of my invention at Cheltenham, and were not again to be disturbed by the cry of "Wolf!" Thus it happened that, after the hard battle I had fought for so many years, I found myself as far as ever from the fruits of my labour, for not a single ironmaster or steel manufacturer in Great Britain could be induced to adopt the process.

Anxious to possess still further practical proofs of the value of my invention, I made, at my experimental works at St. Pancras, a few hundredweights of steel ingots of all the special qualities required in an engineer's workshop. This steel we took to Sheffield, where it was tilted, by an experienced steel-maker, into bars of precisely the same external appearance as the ordinary steel of commerce. Either I, or my partner, Mr. Longsdon, was present the whole time occupied in the operation, and as each bar was finished we stamped it, while still hot, with a special punch which we kept in our pockets for the purpose, thus rendering the accidental or intentional change of a bar impossible. These bars we took to the works of my friends, Messrs. Galloway, the well-known engineers, of Manchester, where they were given out to the workmen and employed by them for all the purposes for which steel had previously been used in their extensive business. So identical in all essential qualities was this steel with that usually employed that, during two months' trial of it, the workmen had not the slightest idea or suspicion that they were using steel made by a new process. They were accustomed to use steel of the best quality, costing £60 per ton, and they had no doubt whatever that they were still doing so.

None of the large steel manufacturers at Sheffield would adopt my process, even under the very favourable conditions which I offered as regards licenses, viz., £2 per ton. Each one required an absolute monopoly of my invention if he touched it at all. This I fully made up my mind to resist, by adopting the only means open to me—namely, the establishment of a steel works of my own in the midst of the great steel industry of Sheffield. My purpose was not to work my process as a monopoly, but simply to force the trade to adopt it by underselling them in their own market, which the extremely low cost of production would enable me to do, while still retaining a very high rate of profit on all that was produced. My partner, Mr. Longsdon, and my brother-in-law, Mr. William Allen, to whom I mentioned this project, were quite willing to join me in it as a purely manufacturing speculation, apart from any interest in my patents, which, however, the firm were allowed to use free of royalty, in consideration of their permitting the works to be inspected and the process fully explained to all intending licensees.

It will be remembered that Messrs. Galloway, of Manchester, were the first persons who took a license to manufacture malleable iron by my converting process, having purchased the sole right to manufacture it in Manchester and ten miles round, prior to the reading of my paper at Cheltenham. One of the original upright fixed converters had been erected at their engineering works, and having, like all the rest, failed to produce satisfactory results with ordinary phosphoric pig-iron, it had been at once abandoned. But when the proofs of our success in steel making, two years later, were afforded to Messrs. Galloway by the actual use in their own workshop of steel tools of all sorts made by us in London, it was mutually agreed that they should rescind their original license for Manchester and join us as equal partners in the Sheffield works, which I and Mr. Longsdon had determined to erect, with Mr. William Allen as the resident managing partner.

Mr. Longsdon, with his intimate knowledge of architecture, soon designed our model works—a neat white brick range of buildings with sandstone dressings, and a tall chimney as the usual landmark. Thus were established the first Bessemer Steel Works, and in less than twelve months from its commencement, we had built a dozen melting furnaces and erected the steam and tilt hammers, blast furnaces, and converting apparatus, suitable for carrying on the new manufacture. This we commenced by bringing steel into the market at £10 to £15 per ton below the quotations of other manufacturers. In thus opposing the old-established steel trade in its very midst, we ran the risk of " rattening," or a bottle of gunpowder in the furnace flues, by which the workmen of Sheffield

had earned for themselves an unenviable notoriety, and we had reason to consider ourselves fortunate that we escaped. We were doubtless indebted for this immunity to the entire and absolute disbelief, both of masters and men, in our power to compete with them. It was this obstinate refusal to see and judge for themselves which lost the manufacturers of Sheffield their great monopoly of the steel trade ; for, although the steel makers refused to see, it was abundantly clear to the ironmasters that profits could be realised by working the new process ; hence it was speedily adopted in all the great iron districts of the country.

Some idea may be formed of the importance of the manufacture, and of how much the people of Sheffield lost by their prejudice and incredulity, when I state the simple fact that, on the expiration of the fourteen years' term of partnership of our Sheffield firm, the works, which had been greatly increased from time to time, entirely out of revenue, were sold by private contract for exactly twenty-four times the amount of the whole subscribed capital of the firm, notwithstanding that we had divided in profits during the partnership a sum equal to fifty-seven times the gross capital. So that, by the mere commercial working of the process, apart from the patent, each of the five partners retired from the Sheffield works after fourteen years, having made eighty-one times the amount of his subscribed capital, or an average of nearly cent. per cent. every two months—a result probably unprecedented in the annals of commerce.

Remembering the keen interest which the Emperor of the French had taken in my early experiments with rifled projectiles, I naturally made him acquainted with the success I had achieved ; while, at the same time, I also kept our own Government fully informed. At that period the Foundry and Ordnance Department at Woolwich was ably presided over by Colonel Eardley Wilmot, R.A., who had taken the deepest interest in the progress of my invention from its earliest date.

CHAPTER XIII

BESSEMER STEEL AND COLONEL EARDLEY WILMOT

DURING the time that the works at Sheffield were being erected, I was very busy endeavouring to discover all the non-phosphoric iron ores in this country, and, after many analyses, the chief were found to be the hematites of Lancashire and Cumberland, and the Forest of Dean, and some spathose ores at Weardale and at Dartmoor. The hematite pig irons were, however, fatally contaminated with phosphorus, although some of these rich ores were absolutely free from this deleterious element. I found, on repeated analyses, that the mines of the Workington Iron Company yielded a very pure ore, but that their pig-iron contained much phosphorus. Here, at least, I had a field to work upon; and I wrote to the Secretary of the Company, asking him to name a day when I could go down and meet the Directors. An early date was fixed, and, at our interview I told the Directors that I, and many others, would become large buyers if they could make a pig-iron as free from phosphorus as the hematite ore was before smelting. I further said, that if they had no secrets, and would show me everything they were doing, I did not despair of finding out the source of contamination, and of pointing out a way of producing pure pig-iron that would command a ready sale wherever my process was carried on. The Board expressed their willingness to afford me every facility, and sent for their furnace-manager, who was instructed to take me over the works, answer all my questions, and furnish me with samples for analysis of all the raw materials they employed. I went round with him, and collected small samples for analysis of the coke obtained from different sources, the limestones from all the pits they worked, and samples of the hard and soft hematite ore from each of their different mines. The limestones contained but few shells, and I was quite at a loss to imagine where the phosphorus came from. As we were returning to the offices near one of the railway sidings, we came upon a large heap of slags and cinder. "What is that?" I asked the manager. "Oh, that is what we flux the furnace with," he said. "Yes, but what is it?" "It is a furnace slag, rich in iron," he replied. "We send into Staffordshire lots of our fine ore for fettling the puddling furnaces, and after they have done with it they send it back to us; in fact, we could not get a fluid cinder in our blast furnaces without it." "All right!" I said, "the cat is out of the bag now, and the mystery is all over." And so I found that the Staffordshire iron-master, after purifying his phosphoric iron in the puddling furnace, and transferring its impurities to the hematite ore, sent the ore back again to Cumberland, and succeeded in spoiling the purest iron ore which this country possessed.

I was in high spirits at this discovery, for I now felt certain that we should soon have thousands of tons of British iron suitable for the production of steel by my process.

Before leaving the works, I arranged to take all these samples of raw material to London, and get my own chemist to make a careful analysis. Then, choosing the fittest materials in each case, furnace charges could be formulated by our chemist, Professor Henry, of course omitting the phosphoric slag, and substituting for it the dark shale of the coal measures, so as to give a sufficiently fluid cinder. These theoretical furnace charges were afterwards sent to the Workington Company, with the following offer on our part, viz., the company were to use these charges for at least twelve hours after they believed that all the old materials had passed out of the blast furnace, so as to be quite sure that the old impure matters had been entirely got rid of, and then they were to run me 100 tons of this new pig-iron, which I undertook to purchase, whatever its quality might be. They were instructed to make a large letter B on the mould pattern used for casting, so as to distinguish this pig from all others. This plan of marking was duly carried out, and I got my 100 tons of "Bessemer Pig," the first that ever was made. This brand of iron is, up to the present day, quoted in all price lists, and in all the iron markets of the world, and has placed at our disposal millions of tons of high-class iron, such as had never before been produced in this country.

The new steel works of Henry Bessemer and Company, at Sheffield, had been erected some months, and the first converter mounted on axes was put to work in 1858. At first our attention was chiefly directed to the manufacture of high-class tool steel, for which our quotation was £42 per ton, as against £50 or £60 by other makers. All this tool steel was made from Swedish charcoal pig-iron, costing only about £3 per ton more than English brands. The excellence of the steel so made is best proved by the fact that during the two years that this branch of the steel trade was carried on by us at Sheffield, we supplied such firms as Sir Joseph Whitworth, Messrs. Beyer, Peacock and Company, Messrs. Sharp, Stewart and Company, Sir William Fairbairn and Company, Messrs. Hicks, of Bolton, Messrs. Platt Bros., of Oldham, etc. A moment's consideration will show that such firms as those mentioned would never have continued to use this steel if it had been in the slightest degree inferior to the best steel made by the old process. By way of commercial proof, let us suppose that our price was £14 per ton below that of the trade. This would save precisely five farthings on the cost of a tool weighing 1 lb. Now if such a tool during its whole life occupied a workman (whose wages were sevenpence an hour) only twelve minutes more in extra sharpening on the grindstone, the advantage of £14 per ton would have been wholly lost. Is it, I would ask, probable that the eminent engineering firms quoted would have continued to use this Bessemer tool steel if the smallest shade of inferiority had manifested itself? Our tool steel was also used at the Arsenal, Woolwich, at the time when Colonel Eardley Wilmot, R.A., was Super-intendent of the Royal Gun Factories, prior to the advent of Sir William Armstrong, and in confirmation of this fact I may quote the following passage from the *Proceedings* of the Institution of Civil Engineers, according to which, on May 24th, 1859, Colonel Wilmot, in the course of a speech in reference to Bessemer Iron and Steel, said :—

As regards the steel, he had been using it for turning the outsides of heavy guns . . . . . cutting off large shavings several inches in length, and he has found none other superior to it, although much more costly.

Indeed, Colonel Wilmot exhibited to the meeting a box full of exceptionally large and heavy shavings taken off by this steel, in the ordinary course of turning in the lathe.

We had now a large converting vessel erected at Sheffield, and commenced operations on an extended scale. We were very anxious to see how one of these large ingots would behave under the steam hammer, but a delay had unfortunately taken place in the erection of our own large hammer. In my impatience to see the result, I waited only until the first heavy ingot ever cast at the works had cooled down sufficiently to prevent it setting fire to the truck on which it was carried, before I sent it by rail to Messrs. Galloway, at Manchester, who had a large steam hammer in daily use. I followed by train, and saw the ingot formed into a gun of the old-fashioned type. This gun is, in many respects, a unique specimen of pure iron, and is now in the possession of the Iron and Steel Institute. The ingot was made of Swedish charcoal pig, costing £6 10s. per ton delivered in Sheffield ; it was converted into pure soft iron, and no spiegeleisen or manganese in any form was employed in its production. This was not only the first large ingot made at our works at Sheffield, but it was the first piece of ordnance ever made in one piece of malleable iron, without weld or joint. It is no less remarkable for its extreme purity. The metal of this gun had originally been most carefully analysed, and many years later, during a discussion at one of the Iron and Steel Institute meetings in 1879, mention was made of its purity, a statement that was received with incredulity. It was said that it was so near absolute iron that there must have been some mistake in the analysis ; whereupon it was proposed by the President to have it again analysed.

Mr. Edward Riley, the well-known analyst of iron and steel, was entrusted with this interesting investigation, and for this purpose the gun was removed from the offices of the Institute to my laboratory and workshops at Denmark Hill, and there put into the lathe. Shavings off the muzzle of the gun were received on a sheet of clean white paper held by Mr. Riley under the cutting tool, and were afterwards taken by him to his own laboratory in Finsbury Square for careful analysis. This occurred in the early part of March, 1879. A copy of the analysis, which fully confirmed that originally made, is given below.

Laboratory and Assay Offices,
2, City Road (14A, Finsbury Square),
London, E.C.
March 22nd, 1879.

EDWD. RILEY.

DEAR SIR,

Herewith I beg to forward you the results of my analysis of the sample of steel turned from a small steel gun in my presence on Monday last.

The sample gave :

Carbon	.014	
Silicium	.004	
Sulphur	.052	
Phosphorus	.047	.046
Iron	99.893	99.787
Manganese	nil.	
Copper	minute trace	

100.010

Believe me to remain,
Yours very faithfully,
EDWD. RILEY, F.C.S.
Metallurgist, Analytical and Consulting Chemist.

Henry Bessemer, Esq.

It is very generally known that of all the Swedish bar-irons, hoop L Dannemora bar-iron is the purest brand to be met with in commerce. It is these iron bars, which sell for £30 per ton and upwards, that are used wholly, or in part, in making the highest class of crucible steel produced in Sheffield. As an example of its purity, Dr. Percy, in his well-known work on Metallurgy, gives the analysis of what he justly calls "this world-renowned iron," and in order that there should be no possible mistake on this point, I print below a portion of page 736 of the volume devoted to iron and steel.

SWEDEN.

An examination of the specimens of Bessemer steel from Sweden in the Exposition shows us that the metal there produced is of a far superior character to that made in England, and naturally leads to inquiry as to the cause of the difference, and whether we may hope to attain the same success in the United States. First we observe coils of wire of all sizes, down to the very finest, such as No. 47, or even smaller. This they have not been able regularly to produce in England. In the next place we notice a good display of fine cutlery, and the writer is informed by a competent authority that this metal answers so well for this purpose that it is now used almost to the exclusion of any other. This statement is corroborated by the fact that in the miscellaneous classes of the Swedish department, where cutlery occurs not as an exhibition of steel, but merely as a display of workmanship by other parties in the same manner as other articles of merchandise, cases of razors are exhibited with the mark of the kind of steel of which they are made stamped or etched upon them as usual, and these are all "Bessemer," but from a variety of different works, viz., Högbo, Carsdal, Osterby, and Söderfors. The ore used in Sweden for producing iron for the Bessemer process is exclusively magnetic, and of a very pure quality. An analysis of a mixture of those used for the iron employed at the Fagersta works before roasting gives the following composition :—

Carb. acid	8.00
Silicium	17.35
Alumina	0.95
Lime	6.50
Magnesia	4.35
Protoxide of manganese	3.35
Magnetic oxide	32.15
Peroxide of iron	27.40
	100.05
Phosphoric acid	03

All the pig made from this mixture of ores the exhibitors state will give a steel without the use of spiegeleisen, which is not at all red short.

The analysis of gray iron from the same works, used for the Bessemer process, is given as follows :—

Carbon combined	1.012
Graphite	3.527
Silicium	0.854
Manganese	1.919
Phosphorus	0.031
Sulphur	0.010

From these examples, 2, 3, and 4 of hoop L bar-iron, we have for No. 2, pure iron 99.863 per cent.; for No. 3, pure iron 99.220; and for No. 4, pure iron 98.605; giving a mean of 99.220 of pure iron in these three samples of hoop L.

Now, by Mr. Riley's analysis, we have only two testings of the Bessemer malleable iron gun, the first giving 99.893 per cent. of pure iron, and the second one 99.787, a mean of 99.840 per cent. of pure iron, or 00.611 more than Dannemora bar.

Since the Dannemora iron mines achieved their deservedly high reputation, many new mines had come into operation in Sweden, at which pig-iron only was made, and it was the products of these mines that I had analysed for my special use, and thus discovered that some of them were producing pig-iron of extreme purity. Thus I was enabled to make malleable iron or steel of the highest quality from Swedish pig, costing, delivered in Sheffield, £6 10s. to £7 per ton, and yielding, from my converter, ingots of cast steel of great purity at a cost of less than £10 per ton, fully equal to that made from Swedish bar costing £30 per ton, such bar being only the raw material for the old crucible process of making steel.

From a consideration of these facts, it will be readily understood how we could produce cheap high-class tool steel, while for general uses we had obtained native pig-iron—"Bessemer pig"—smelted with coke, admirably adapted for the production of steel for all structural purposes, for which it was in every way superior to the highest brands of iron previously known in this country.

I had no sooner arrived at these results on a commercial scale than I again put myself in communication with Colonel Eardley Wilmot, the Superintendent of the Royal Gun Factory at Woolwich Arsenal, for I had never lost sight of the original object of my research—a metal suitable for the construction of ordnance. It was, in fact, this idea that had led, step by step, to the discovery of my process. I was the more pleased to communicate these facts without delay to the authorities at Woolwich, because, in the person of Colonel Eardley Wilmot, I found a zealous officer, who took the deepest interest in any improved materials or processes that could be advantageously employed in the founding or construction of ordnance. He, fortunately, had no pet schemes of his own to promote, and was neither a patentee nor a private manufacturer; he was, in fact, an officer whose sole aim and ambition was to arrive at the highest perfection and development of the department over which he so ably presided, wholly without reference to the sources from which such improvements were derived.

It was now many months since I had reported myself at Woolwich, but on my communicating the fact that we were commercially successful in producing both pure and malleable iron in masses, and steel of any degree of carburisation that might be desired, at a price far below that of the best bar iron, and in masses of almost any assignable weight, the information immediately riveted Colonel Wilmot's attention. His old hopes of having a superior metal for guns seemed suddenly to revive, and he became deeply interested in all that I had to communicate. After a very protracted discussion, I left with a promise to send him several different qualities of our steel for analysis, testing for tensile strength, etc.

These investigations at Woolwich lasted over a period of several months, during which time I frequently called to see Colonel Wilmot, and sometimes to see Professor (afterwards Sir Frederick) Abel,* who was at the head of the chemical laboratory, where a great number of analyses were, from time to time, made and communicated to me. Many interesting tests were also made by drawing down a portion of an ingot first, to two-tenths in additional length, and then to four-tenths, and so on. Some portions were elongated to five times their original length, each piece being tested to show the true amount of increased strength given to it by additional forging and elongation of the bar. In fact, Colonel Wilmot left no stone unturned to arrive at the actual facts of the case, and a full knowledge of the strength and properties of the new material. Some of the tests above mentioned have been lost, but I have still twenty-nine well-authenticated records showing the extreme tenacity and toughness of the metal. On one occasion I happened to remark to Colonel Wilmot that such was the extraordinary ductility of our cast malleable iron and mild cast steel, that I had no doubt a thick gun-tube might be collapsed, and hammered up quite flat, under the steam-hammer, whilst perfectly cold, without showing any tendency to crack or burst open. Colonel Wilmot observed that, notwithstanding the numerous proofs he had had of its marvellous tenacity, he thought that no material could possibly undergo such a severe ordeal without fracture. " Well," I said, " it will be an interesting experiment, even if it fails, and I will put it to the test if you wish it." I accordingly had an

* Died, September 6th, 1902.

B B

ingot of mild steel, and one of wholly decarburised iron, forged until
they were extended to about double their original length. Two portions
of each were cut off, turned, and bored in the lathe, and then beautifully
finished both inside and out, the length and diameter of each cylinder
being 6 in. and the thickness of metal $\frac{3}{4}$ in. These pieces of gun-tube were
bored to $4\frac{1}{2}$ in., in diameter—a size suitable for a 40-pounder gun. I
personally took these four tubes down to Woolwich, and was present with
Colonel Wilmot when they were placed in succession (while cold) under
the large steam hammer, and crushed flat, each tube being quite closed
up. In no case was there the slightest indication of either tearing or
rupture at any part of their surfaces. Colonel Wilmot was greatly
astonished, and so was the experienced foreman of the hammer shop
who conducted the experiment, and who expressed his admiration with
a forcible adjective, which I need not repeat. I gave one steel and one
pure iron cylinder to Colonel Wilmot, and retained the other two, which
were exhibited in the International Exhibition of 1862.

After personally inspecting the crushing of the two pure iron cylinders
and the two mild steel ones, Colonel Wilmot was so convinced of the
immense importance to the State of Bessemer mild steel as a material
for guns, that he said he would no longer delay taking active steps
for its manufacture at Woolwich. On his asking me if he might go
over our Sheffield Works, and see for himself how everything was
done, I at once assented. A day was fixed, and Colonel Wilmot and
I went down together to Sheffield, where he passed the greater part
of the following day in making himself fully acquainted with all the
details of what was in reality a very simple process, and with which
he expressed himself perfectly satisfied. I cannot omit to mention a
very curious and somewhat significant fact, which more than justified
Colonel Wilmot in the strong opinion he had formed of the value
and practicability of the process. The well-known and extensive steel
works of Sir John Brown and Co. are only separated by a wall from
the Bessemer Steel Works at Sheffield, but neither Sir J. Brown, nor
any of his people, had taken the smallest apparent interest in what
we were doing, and, indeed, like the rest of the good people at Sheffield,
had a profound disbelief in the production of steel direct from pig-iron

by any conceivable process. Now Colonel Wilmot, during this visit to Sheffield, had occasion to see Sir John Brown on other business, and, so ardent a convert had he become, that he succeeded in persuading Sir John Brown and his partner Mr. Ellis, to go with him next door and see the Bessemer process in operation. They came, and had but a short time to wait before the cupola furnace was tapped, and a charge of molten pig-iron was run by a spout directly into the empty converter. They seemed much interested in watching the great change which took place in the flame and sparks emitted as the process proceeded; but when the eruption of cinder, and the accompanying huge body of flame, were seen to issue from the converter, they were greatly astonished. In about twenty minutes the flame had dropped, the mouth of the huge vessel was gradually lowered, and a torrent of incandescent metal was poured into the casting ladle. Up to this moment they merely expressed surprise at the volume of flame, the brightness of the light, and the entire novelty of the process. But no sooner did they see the incandescent stream issue from the mouth of the converter, than their practised eyes in an instant recognised it to be fluid steel, and they themselves were "converted," never to fall back again into a state of unbelief. They stayed to witness the casting operation, and accepted one of the hot ingots for testing at their own works, the result being that Sir John Brown and Company became the first licensees in Sheffield under my steel patents.

The moral to be drawn from these facts is simply this;—that the state of the manufacture was at that period such, that after once witnessing the process and testing the material at their own works, these eminently practical steel-makers resolved, at the risk of entirely revolutionising their old-established business, to put up plant and become Bessemer Steel manufacturers. Now, I would ask any impartial person if this fact did not justify, and more than justify, Colonel Wilmot in the conclusion to which he had arrived independently — that this cheap and rapid production of steel ought at once to be utilised in the manufacture of guns for the British Government.*

* On the occasion of my reading a Paper at the Institution of Civil Engineers, on The Manufacture of Iron and Steel, in May, 1859, Colonel Wilmot said, in reference to a

After my return to London, I waited on Colonel Wilmot by appointment, went with him to inspect the gun-foundry at the Arsenal, and chose a suitable spot for the erection of the Bessemer Steel plant. It was finally arranged by us to remove one of the three large reverberatory furnaces that had been used to melt pig-iron for casting guns, and in its place put up a pair of converters, utilising the other two furnaces for melting the Bessemer pig. I took accurate measure of the foundry and its contents, so as to enable me, at my own offices, to arrange all the details of a converting plant to be erected in the old gun-foundry and to make an estimate of the cost.

When this was done, I handed to Colonel Wilmot an approximate estimate of £6,000, for erecting a steam-engine, boilers, and converting plant of sufficient size to produce 100 tons of gun steel per day, and I guaranteed that the cost of the steel poured into their own moulds should not exceed £6 10s. per ton, when hematite pig-iron was used, or £10 per ton when Swedish charcoal pig-iron was employed: my remuneration being a royalty of £2 per ton on all metal converted, the same as charged to all private manufacturers.

very silly observation of one of the members: "As regards the difficulty of the process, as well as the results of it, he thought the best thing for a member of a practical society to do was to *follow his example*, and go and see it for himself; nothing could be more simple or more perfectly under control." (Excerpt: *Minutes* of Proceedings of the Institution of Civil Engineers.)

CHAPTER XIV

THE BESSEMER PROCESS AND THE WAR OFFICE

I WAS kept for some time in daily expectation of a reply from the War Office accepting my tender, but no letter arrived, and at last I ventured on seeking an interview with the Minister of War, Mr. Sidney Herbert. He appeared to know very little on the subject. I took, however, the opportunity of explaining to him, in as clear and concise a manner as possible, the great national interests hanging on his decision. I told him that steel, the strongest of all known conditions of the metal iron, had hitherto been so costly as to considerably restrict its use; that by my process we produced it at a cost not exceeding £6 or £7 per ton, instead of £50 or £60, its ordinary market value; that instead of being made in small crucibles of 40 lb. or 50 lb. only in weight, we could make five tons of it in the short space of twenty minutes in a single operation; and, what was still more important, instead of being the hard and brittle material, such as is required to make cutting implements, the new steel possessed a toughness and tenacity far exceeding the very finest qualities of wrought iron known in commerce. I also endeavoured to impress on him the fact that Colonel Eardley Wilmot had seen the process in operation, had amply tested it, and had in his office at Woolwich pieces of gun-tubes that had been put to such unheard-of proofs as to afford to the meanest capacity overwhelming evidence of its fitness for the construction of ordnance. I also told him that in the chemical laboratory numerous analyses had been made by their own chemist; that in their rolling mill, bars had been rolled, and in their testing-house an immense number of most satisfactory tests had been made as to the tenacity and toughness. I said that the people at the head of each of these departments at Woolwich could adduce abundance of corroborative evidence of every statement I had made.

Mr. Sidney Herbert listened to all this, and remarked that it was a technical question which he was not prepared to deal with at that moment; but said that he would give the whole matter his most earnest attention, and that I might call on him that day week to hear his reply.

I waited impatiently for this second interview, in full confidence that Colonel Wilmot, and other heads of the chemical and testing departments, would have been called on to corroborate, or disprove, the statements I had made, and would have given him such proofs in favour of mild Bessemer steel as would at once have secured me the contract to erect at Woolwich the converting apparatus which Colonel Wilmot was so anxious to see in practical operation there. But Mr. Herbert did not examine or consult Colonel Wilmot, who could have told him all about it. He made no enquiries at the testing or other departments at Woolwich, nor did he take the trouble to look at the flattened gun-tubes, and other proofs, which would have irresistibly convinced any man of ordinary capacity and intelligence that this material was, at least, well worthy of being put to a practical proof in the interests of the State, by the immediate construction of a gun. He informed me that he had consulted Sir William Armstrong, who, he said, had at once declared that "steel was wholly inapplicable to the construction of ordnance;" and who, if Mr. Sidney Herbert's statements were true, had succeeded in convincing him that it would be a waste of time and public money to put up the Bessemer apparatus at Woolwich. It was quite evident that Mr. Sidney Herbert had made up his mind to fling to the winds all the labours and trials of Colonel Wilmot, and at the same time to utterly ignore me and the expense and trouble to which I had been put. The strongest protest on my part at this injustice, and my urgent request to have my process tried, failed to move Mr. Sidney Herbert one iota from his firm resolve to keep me and my process out of Woolwich, and to allow Sir William Armstrong, with his immensely more expensive welded iron gun, to have the field to himself. There was nothing for it but to submit, and I retired from this interview in deep disgust with Mr. Herbert and his arbitrary proceedings.

The event just recorded, although it had the effect of closing my connection with Woolwich Arsenal, did not in any way determine the fitness or otherwise of mild Bessemer steel for the construction of ordnance. I feel bound in honour, and in justice to my own name, to vindicate, not by mere words, but by an array of well-authenticated facts such as no intelligent person can lessen or deny, the perfect adaptability of this discarded material for that purpose.

It will be remembered by my readers that Bessemer steel, which is now used, and its value acknowledged, over the whole civilised world, was the direct outcome of my investigations in search of a more suitable metal than was at that time employed in the construction of ordnance. It is my present purpose to show that I had succeeded in attaining the result which I sought, and thus not permit the mere assertion of one man to obliterate from the page of history the fact that I originated a process and produced a material which, at the time the experiments were made by Colonel Wilmot at Woolwich, and for twelve years after that period,* and consequently during the whole tenure of office of Sir William Armstrong at Woolwich, stood unrivalled as a material for the construction of ordnance. No other known process could, at that period, produce steel with such marvellous rapidity, and at such an enormous reduction in price; no other known method could produce in large masses steel of such a degree of mildness as to pass, by almost imperceptible gradations, downwards until it became soft iron; nor did there exist any other known process by which large masses of almost chemically pure iron could be produced without weld or joint.†

Many persons who are not intimately acquainted with the early history of Bessemer steel have fallen into great error, and honestly believe that the Bessemer process was in itself uncertain and incapable of perfect control, and that the excellent material commercially produced at the present time has been the result of a long succession of improvements in the process since it left my hands. Nothing could be more absolutely erroneous or more historically untrue, as I shall show further on by incontestable proofs. No doubt all popular beliefs and prejudices

* The open-hearth process was patented in 1865, and practically introduced in 1869.

† I am speaking of a period of about twelve years prior to the manufacture of any kind of open-hearth steel, and when the production of mild crucible steel was extremely difficult and pure malleable iron in large cast masses was impossible by any known process but mine.

have some real, or supposed, good reason for their origin, and this particular popular error was, I admit, the outcome of circumstances only too well calculated to give rise to, and perpetuate, such a belief. The Bessemer process was sprung upon the iron trade suddenly, and in a moment, as it were, it excited the wildest hopes and the direst apprehensions. But it was very soon afterwards discovered that with ordinary phosphoric pig-iron it failed to produce iron or steel of any commercial value. It is almost impossible at this distant period to realise the sudden revulsion of feeling which then took place, and the utter disbelief in the whole scheme which followed, and, passing beyond all reasonable bounds, has not, even at the time I am now writing, entirely disappeared. When, after the labour of two years, I had succeeded in making "Bessemer Pig" from British hematite; when from that pig I had produced steel of excellent quality for all structural purposes; when I had manufactured a high-class tool steel from Swedish pig; and when also the tipping vessel was invented with the ladle provided with a bottom valve, the conical mould, and the hydraulic crane; when, in fact, the general system of the present day was a proved commercial reality at my own works in Sheffield; then, and not till then, did I again bring my process before the trade, when it still met with blank incredulity and distrust. But this time I was backed with proofs that could not be denied, for there, in the town of Sheffield, in the very heart of the great steel industry of the country, stood the Bessemer Steel Works, in daily commercial operation, underselling the old-established manufacturer, who still resisted its encroachment and obstinately refused to believe in it. But the temptation to the ironmaster to become a steel manufacturer at then existing prices was very great, and the adaptability of the process to the manufacturer of rails was self evident. Rail mills and steel works were established by people who had no previous knowledge or experience of steel and its peculiarities, and, what was still worse, there was not a manager or foreman, or even an ordinary workman, to be found who had any knowledge whatever of the new process. As an instance of this difficulty, I may mention a case in point. A very handy carpenter, whom I had employed to assist in the works, had acquired a certain amount of routine knowledge of the process. This made him a valuable man, and one of my licensees who had adopted my process bid a high price for this small amount of practical knowledge, and engaged this carpenter's services, under a five years' agreement, at £5 per week. It is only fair to say that he was quite worth it.

Thus it happened that those ironmasters who had adopted my process had to struggle against difficulties quite unknown in any old-established trade. Need we wonder, then, that the quality of their steel sometimes differed from day to day?

The ironmaster had been in the habit of making bar iron from every kind of pig, and he could not realise the fact that good steel by my process could only be made from a special quality of iron. This he did not like to buy from other makers; in those early days he did not fully understand how to make it himself, and hence he would use inferior hematite iron, or mix some of his own phosphoric pig with it, to eke it out and lessen the cost. The bad results so produced were all set down to the uncertainty of the Bessemer process; nor did the extreme jealousy of the steel trade prevent such unfavourable reports from being published with all the usual embellishments naturally arising from ignorance or prejudice.

This adoption of my process by the ironmaster for making rails went far to discredit it. If you told a steel maker that it was being largely used, he would say: "Well, perhaps it is good enough for rails; anything is good enough for rails." Indeed, it is true that in the case of rails moderate variations of temper were not fatal. The rail might be a little too hard or too soft, but in either case it was immensely superior to iron, and so it passed muster. But it was when boiler plates, ships' plates or crank-axles, were required, that the inexperienced ironmaster, with his inexperienced workmen, began to realise the fact that steel was wanted of a certain standard of quality for special purposes; and that he must train his men, who were little else than mere apprentices learning a new trade, to produce these several qualities with certainty. It is not at all surprising, under such conditions, that Bessemer steel acquired the character of being uncertain and not trustworthy. Hundreds of workmen who had never before worked a plate of steel in their lives, and were totally ignorant of its proper treatment, were engaged in the

c c

manufacture of steel boilers and in building steel ships. Such workmen had no hesitation in putting a hot steel plate down on the floor, with one end in a puddle of water; or in placing a mass of cold iron on a red-hot plate to keep it flat while cooling; or, on the other hand, in over-heating it in a furnace, quite unconscious that no steel would bear the same high temperature as iron. And when they had thus succeeded in spoiling a plate originally of good quality, they did not hesitate to lay all the blame on the Bessemer process, which they honestly believed was the sole cause of the mischief that their own want of experience as steel-smiths had occasioned.

When the investigation of the character and properties of Bessemer Steel, undertaken by Colonel Eardley Wilmot at Woolwich Arsenal, was completed, all the early difficulties of the process had been entirely removed. We had become intimately acquainted by use, and by analysis, with several brands of Swedish pig-iron, from which either soft ductile iron, or steel of any degree of carburation, could be—and in fact was—daily produced at our Sheffield works, on a commercial scale without the employment of spiegeleisen. We had command also of a practically unlimited supply of a very high-class non-phosphoric hematite Bessemer pig-iron, suitable for conversion into steel. We had also magnesian pig-iron from Germany, and Franklinite pig-iron from the United States, the latter containing about 11 per cent. of manganese, which was greatly preferred for deoxydising steel derived from British coke-made iron. We had our converting vessels at that time mounted on axes; and, in fact, the Bessemer process was so complete, and so under command, as to enable us to produce at will, pure Swedish steel of all tempers down to soft iron, and also mild hematite steel, as good in all respects as we are able to make at the present day. Above all, we had the advantage of the knowledge and experience of Mr. W. D. Allen, Managing Partner of the Bessemer Steel Works at Sheffield. In proof of my assertion that the Bessemer process was at that time as perfect in results as at any later date, I will give a few examples of our products, commencing at a period several months prior to the advent of Sir William Armstrong at Woolwich, covering the whole five years of his official power, and extending for some years after his departure from the Arsenal. Fig. 49, Plate XIX., is a photographic reproduction of some test specimens, to which I have already alluded, representing three out of four pieces of gun-tube tested at Woolwich, two of them made of mild steel, and the others being nearly chemically pure iron. It will be remembered that these cylinders were made at my works at Sheffield, and were crushed flat, in the presence of Colonel Wilmot and myself at Woolwich, while cold, under the heavy blows of a large steam hammer. In order to give a correct idea of the nature and appearance of these crushed gun-tubes and hoops, I refer my readers to the photographic reproduction, Fig. 49. The specimens illustrated were made of Bessemer hematite pig-iron, converted into steel by the Bessemer process, and of a quality precisely the same as we were, at that early period, daily using in the manufacture of railway-carriage axles, piston-rods of steam engines, and other general machine forgings.

In the illustration, A represents a portion of a gun-tube for a rifled gun, machined and finished ; B is one of these pieces, flattened, as shown, and C is a larger hoop, crushed flat with the heavy blows of the steam hammer. The two sides where the bend takes place are immensely stretched on the exterior surface, and also greatly compressed on their inner side, but at no point does the metal exhibit the smallest trace of fracture. The dimensions of these specimens will be readily seen by reference to the two-foot rule photographed with them. These examples of the toughness and endurance of Bessemer mild steel, after being subjected to violent and sudden strains, were exhibited in my large glass case at the International Exhibition of 1862, and must have been seen by hundreds and thousands of persons. When one reflects on the extent and prominence of my exhibit, covering an enclosed area of 1,225 square feet, and surrounded by a counter of more than 100 ft. in length, covered with steel exhibits, and having a 24-pounder gun forging on a pedestal at the central entrance, and an 18 pounder finished gun in the large central case, it is difficult to believe that this gun-hoop and these crushed gun-tubes were not seen during the time of the Exhibition by every engineer in London, and by every employé at Woolwich Arsenal, as well as by our Minister of War, who with a light heart excluded Bessemer process from Woolwich.

I desire to draw the reader's earnest attention to these crushed gun-tubes, for it is impossible, in my opinion, for any intelligent person to look at these marvellous proofs of the toughness and power of extension and distortion of the metal, and not be convinced that such a material was pre-eminently suited for the construction of ordnance. The two similar crushed cylinders which I gave to Colonel Wilmot were greatly prized by him, and were kept as trophies, with several other experimental proofs, on the writing-table in his private office at the Arsenal, where I saw them on several occasions prior to his vacating the office.

I may mention that when Colonel Wilmot inspected my Sheffield Steel Works, he happened to see on the scrap-heap a large mass of Bessemer malleable iron, which he wished to have for the purpose of experiment, and which, at his request, was sent to him to Woolwich. On May 24th, 1859, speaking of Bessemer iron and steel at a meeting of the Institution of Civil Engineers, he referred to this piece of iron during the discussion, and stated that a cylindrical piece of Bessemer pure iron, when only extended by forging to twice its original length, had a tenacity per square inch of 28 tons 13 cwt. 1 qr. and 2 lb., a tenacity which it possessed in all directions alike, as against the best Yorkshire iron, which was usually credited with a tenacity of 25 tons in the direction of its length, and very considerably less across the grain, even after being rolled and piled and again rolled into long bars. These bars, when welded together to form a large forging of any kind, will never afterwards possess the strength of the original bar, by two or three tons per square inch. The analysis given by Colonel Wilmot was issued from the Chemical Laboratory of the War Department, and can be fully relied on as showing that no impurity but sulphur existed in the specimen analysed in sufficient quantity to estimate, while no spiegeleisen or manganese was used in its production. (This metal was converted from Swedish pig.)

Colonel Eardley Wilmot's remarks are herewith reproduced from the *Proceedings* of the Institution of Civil Engineers.

Colonel Eardley Wilmot, R.A., said he had, from the commencement of these inquiries, taken a great interest in them, and had mechanically tested the products originally produced. A chemical examination was also made at the Royal Arsenal, Woolwich, and the result had indicated, and it had been stated at the same time, that the Bessemer process was perfectly

PLATE I.

Fig. 1. Copy in Relief of Raphael Cartoon

PLATE II.

Fig. 2. Copy of Oval Medallion

PLATE III.

Fig. 4.

The above is an example of an ordinary impressed stamp as now daily issued by the Stamp Office. The figures 21, 1, and 79, each in their respective circles, show the system of dating stamps communicated to Sir Charles Presley in 1833.

Fig. 5.

The above impression represents the Bessemer perforated stamp approved by the Stamp Office officials. The black portions show the perforations which were made by the die in a manner similar to the perforated word CANCELLED, below.

This word "cancelled" shows a mode of perforating paper introduced by Mr. J. Sloper a few years ago, and now much used in banks and public offices. No fraudulent ingenuity can efface this mark.

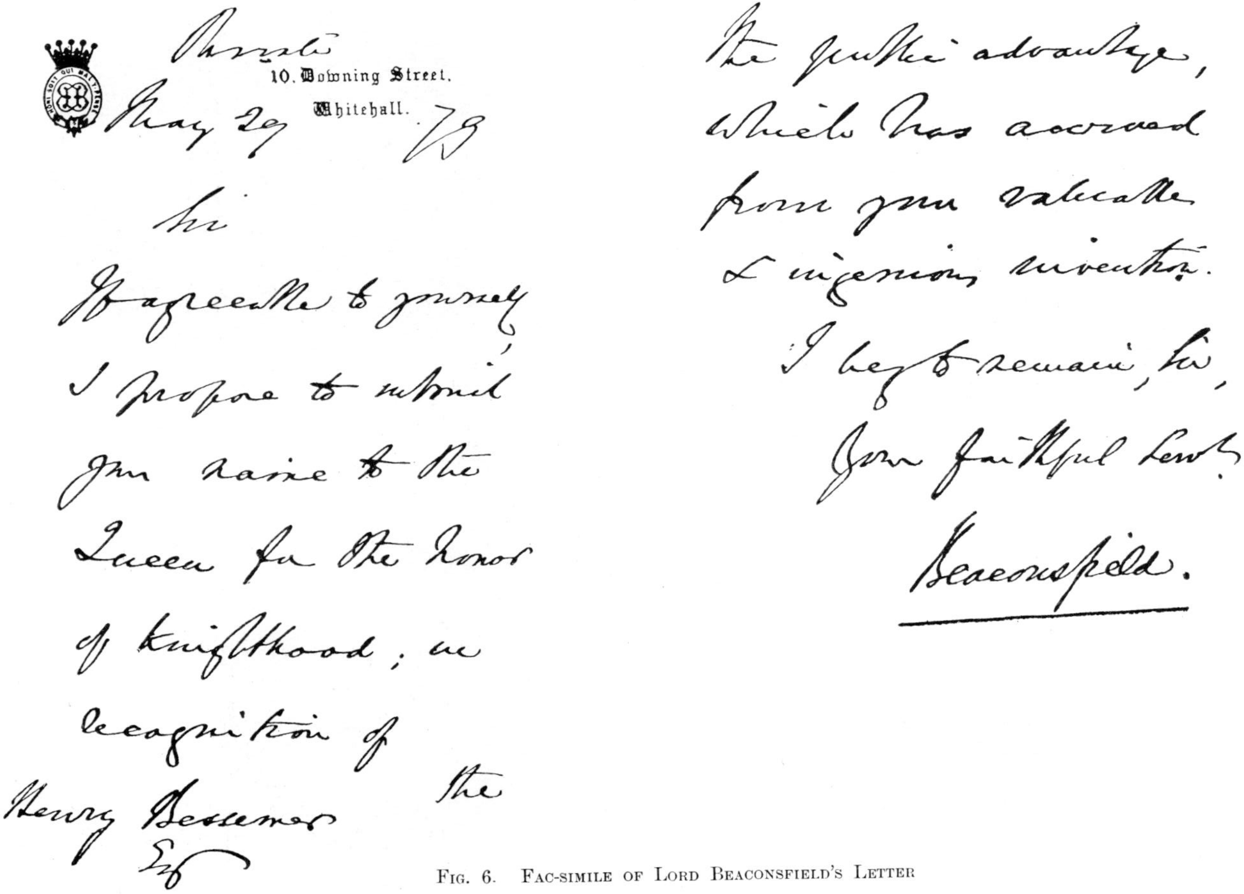

PLATE IV.

Fig. 6. Fac-simile of Lord Beaconsfield's Letter

PLATE V.

FIG. 9. REPRODUCTION OF THE FAMOUS "DOUBLE HEAD" MEDAL, NAPOLEON AND JOSEPHINE

PLATE VI.

FIG. 10. REPRODUCTION OF A NAPOLEON MEDAL.

FIG. 11. REPRODUCTION OF MEDALLION OF MINERVA'S HEAD

PLATE VII.

Fig. 15. Reproduction of Stamped Utrecht Velvet

PLATE VIII.

Fig. 18. Side Elevation of the Bessemer Sugar-Press

PLATE IX.

Fig. 24. Bessemer's Furnace for the Manufacture of Optical Glass

PLATE X.

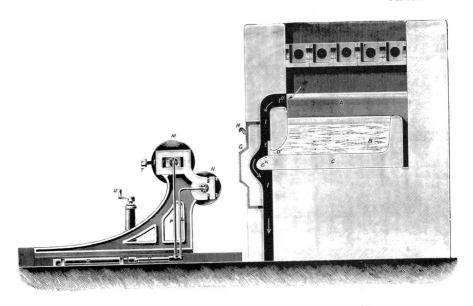

FIG. 25. BESSEMER'S FURNACE FOR THE PRODUCTION OF SHEET GLASS

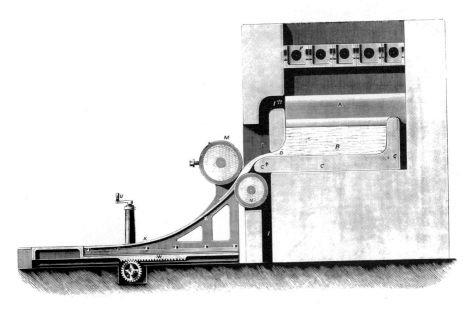

FIG. 26. BESSEMER'S METHOD OF ROLLING SHEET GLASS

PLATE XI.

FIG. 27.

FIG. 28.

DESIGN FOR BESSEMER PLATE GLASS WORKS

PLATE XII.

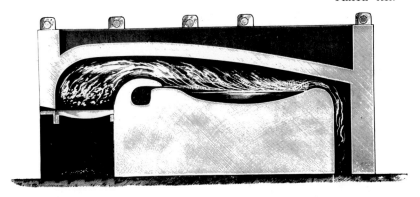

FIG. 35. VERTICAL SECTION OF FURNACE FOR MAKING MALLEABLE IRON

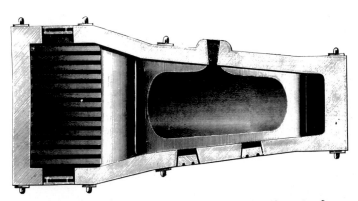

FIG. 36. HORIZONTAL SECTION OF FURNACE FOR MAKING MALLEABLE IRON

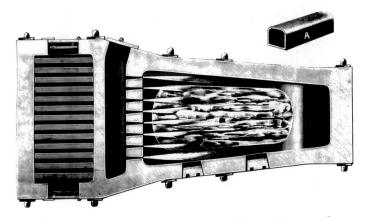

FIG. 37. HORIZONTAL SECTION OF FURNACE FOR MAKING MALLEABLE IRON

PLATE XIII.

FIG. 38. SECTION OF CRUCIBLE
WITH BLOW-PIPE

FIG. 39. SECTION OF VERTICAL CONVERTER

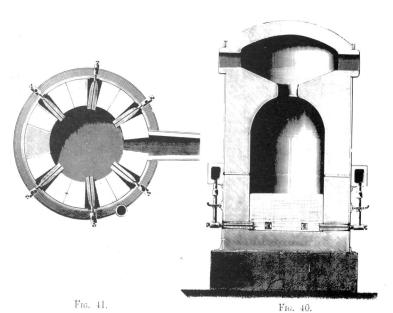

FIG. 41.

FIG. 40.

FIGS. 40 AND 41. SECTIONS OF VERTICAL CONVERTER WITH UPPER CHAMBER

PLATE XIV.

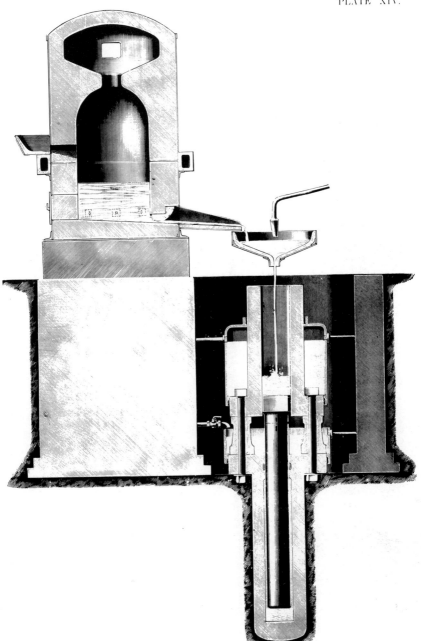

FIG. 42. SECTION OF CONVERTER, LADLE, AND HYDRAULIC INGOT MOULD

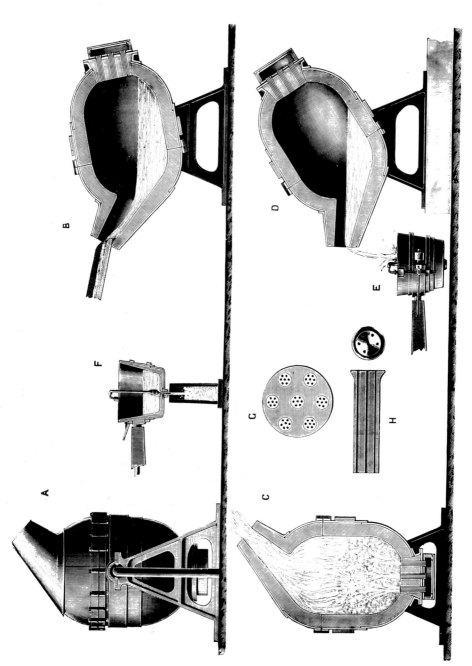

PLATE XV

Fig. 43. The First Form of Bessemer Moveable Converter and Ladle

PLATE XVI.

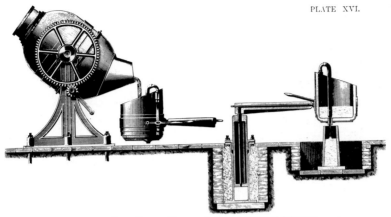

FIG. 44.   EARLY FORM OF BESSEMER CONVERTING PLANT AT SHEFFIELD

PLATE XVII

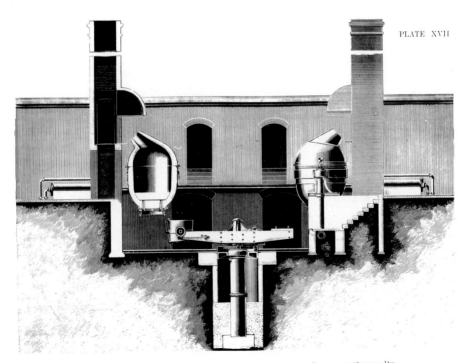

FIG. 45.   BESSEMER PLANT AT SHEFFIELD : CONVERTERS, LADLE AND CRANE, AND CASTING PIT

PLATE XVIII

Fig. 46. Plan of Bessemer Plant at Sheffield

PLATE XIX

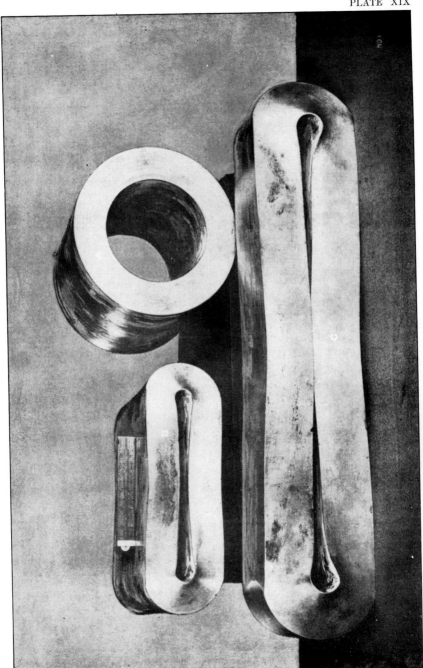

FIG. 49. SPECIMENS OF BESSEMER STEEL GUN TUBES, ETC

PLATE XX

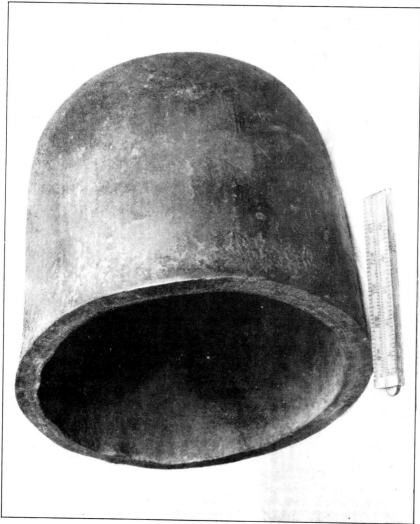

FIG. 56. BESSEMER STEEL BOILER PLATE PRESSED INTO A CUP, 1861

PLATE XXI

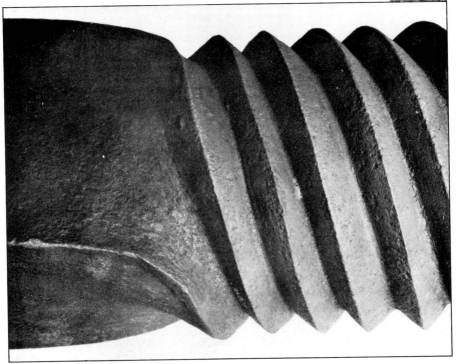

FIG. 57. SQUARE BAR OF BESSEMER METAL TWISTED COLD, AND SHOWN AT SHEFFIELD, 1861

PLATE XXII

FIG. 58. SQUARE BAR OF BESSEMER STEEL TWISTED COLD, AND SHOWN AT SHEFFIELD, 1861

PLATE XXIII

FIG. 59

FIG. 60

SQUARE BARS OF BESSEMER METAL TWISTED COLD, AND SHOWN AT SHEFFIELD, 1861

PLATE XXIV

FIG. 62. DISINTEGRATED WROUGHT-IRON BARS

PLATE XXV

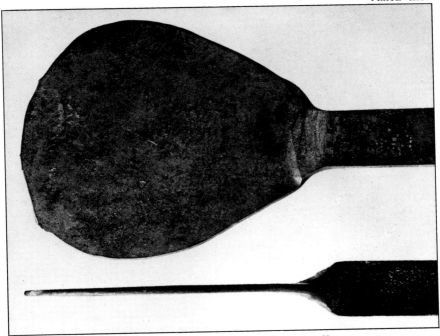

Fig. 63. Bessemer Mild-Steel Bar, Flattened under Hammer

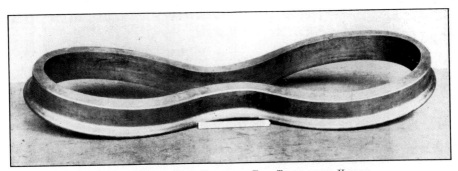

Fig. 65. Bessemer Steel Locomotive Tyre Tested under Hammer

PLATE XXVI

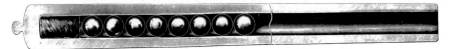

Fig. 66. Section of Bessemer Steel Gun Supplied to the Belgian Government, 1860

PLATE XXVII

FIG. 68. GROUP OF TEST-PIECES FROM BESSEMER STEEL GUN-FORGINGS

PLATE XXVIII

FIG. 69. BESSEMER STEEL GUN TEST-PIECE, SHOWN AT THE MEETING OF THE INSTITUTION OF MECHANICAL ENGINEERS AT SHEFFIELD, 1861

PLATE XXIX

Fig. 70. Bessemer Steel Gun Test-Piece, Shown at the Meeting of the Institution of Mechanical Engineers at Sheffield, 1861

PLATE XXX

Fig. 71. The Bessemer Display at the International Exhibition, London, 1862

PLATE XXXI

Fig 72. Alleged Faulty Bessemer Plate, 1875

PLATE XXXII

Fig. 74. Examples of Bessemer Steel Plate "Spun" into Vases, Etc.

PLATE XXXIII

Fig. 75.   Test Specimens of Bessemer Steel made at Sheffield, 1859-69

PLATE XXXIV

STEEL.

solid iron end of the press, made to resist great pressure; it is strongly bolted to the cylinder *a*, so as to resist the force of the ram ; *g, g*, iron rods, for bringing back the ram *b*, into its place after the pressure is over, by means of counter weights suspended to a chain, which passes over the pulleys *h, h ; i, i,* a spout and a sheet-iron pan for receiving the oily fluid.

STEEL. One of the greatest improvements which this valuable modification of iron has ever received is due to Mr. Josiah M. Heath, who, after many elaborate and costly researches, upon both the small and the great scale, discovered that by the introduction of a small portion, 1 per cent., and even less, of carburet of manganese into the melting-pot along with the usual broken bars of blistered steel, a cast steel was obtained, after fusion, of a quality very superior to what the bar steel would have yielded without the manganese, and moreover possessed of the new and peculiar property of being weldable either to itself or to wrought iron. He also found that a common bar-steel, made from an inferior mark or quality of Swedish or Russian iron,

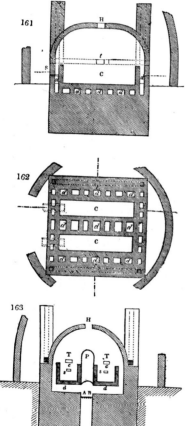

would, when so treated, produce an excellent cast steel. One immediate consequence of this discovery has been the reduction of the price of good steel in the Sheffield market by from 30 to 40 per cent., and likewise the manufacture of table-knives of cast steel with iron tangs welded to them ; whereas, till Mr. Heath's invention, table-knives were necessarily made of shear steel, with unseemly wavy lines in them, because *cast* steel could not be welded to the tangs. Mr. Heath obtained a patent for this and other kindred meritorious inventions on the 5th of April 1839; but, strange and melancholy to say, he has never derived any thing from his acknowledged improvement but vexation and loss, in consequence of a numerous body of Sheffield steel manufacturers having banded together to pirate his patent, and to baffle him in our complex law courts. I hope, however, that eventually justice will have its own, and the ridiculously unfounded pretences of the pirates to the prior use of carburet of manganese will be set finally at rest. It is supposed that fifty persons at least are embarked in this pilfering conspiracy.

The furnace of cementation in which bar-iron is converted into bar or blistered steel is represented in *figs.* 161, 162, 163. It is rectangular and covered in by a groined or *cloister* arch: it contains two cementing chests, or sarcophaguses, c, c, made either of fire-stone or fire-bricks: each is 2½ feet wide, 3 feet deep, and 12 long ; the one being placed on the one side, and the other on the other of the grate, A B, which occupies the whole length of the furnace, and is from 13 to 14 feet long. The grate is 14 inches broad, and rests from 10 to 12 inches below the inferior plane or bottom level of the chests; the height of the top of the arch above the chests is 5½ feet ; the bottom of the

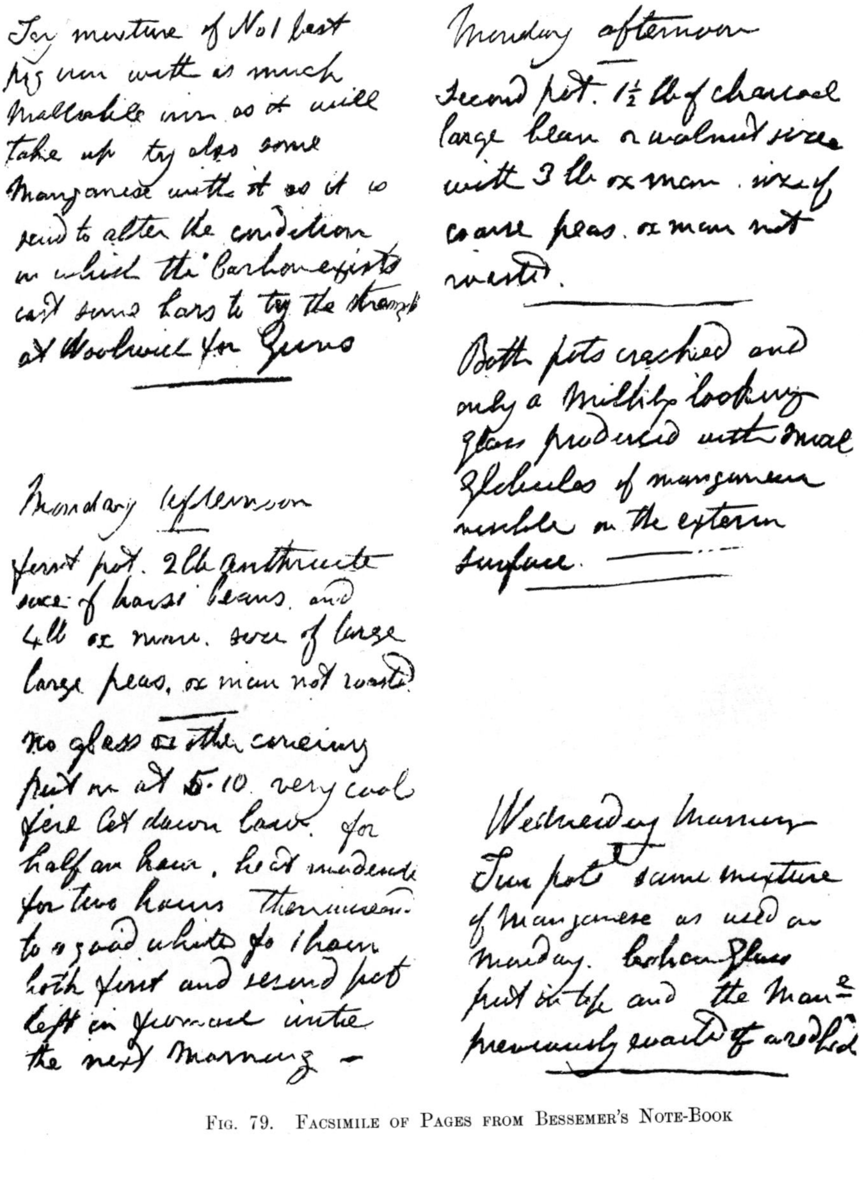

PLATE XXXV

Try mixture of No 1 best
pig iron with as much
malleable iron as it will
take up try also some
Manganese with it as it is
said to alter the condition
in which the Carbon exists
cast some bars to try the strength
at Woolwich for Guns

Monday afternoon
First pot. 2 lb anthracite
size of horse beans. and
4 lb ox man. size of large
large peas. ox man not roasted

No glass or other covering
put on at 5.10 very cool
fire let down low. for
half an hour. heat moderate
for two hours thermometer
to a good white po them
both first and second pot
left in furnace until
the next morning —

Monday afternoon
Second pot. 1½ lb of charcoal
large bean or walnut size
with 3 lb ox man. size of
coarse peas. ox man not
roasted.

Both pots cracked and
only a millily looking
glass produced with small
globules of manganese
visible on the exterior
surface. ———

Wednesday morning
Ten pots same mixture
of Manganese as used on
Monday. broken glass
put on top and the Man²
previously roasted & added

FIG. 79. FACSIMILE OF PAGES FROM BESSEMER'S NOTE-BOOK

PLATE XXXVI

FIG. 80. STATUARY AND CLOCK IN SIR HENRY BESSEMER'S HALL AT DENMARK HILL

PLATE XXXVII

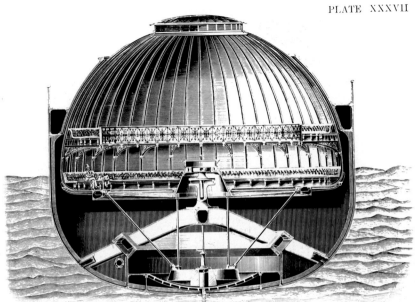

FIG. 81. SECTION THROUGH EARLY FORM OF BESSEMER SALOON, IN STILL WATER

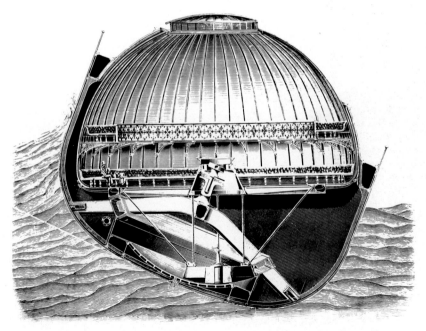

FIG. 82. SECTION THROUGH EARLY FORM OF BESSEMER SALOON, WITH VESSEL ROLLING

PLATE XXXVIII

Fig. 83. Model of Bessemer Saloon, with Hull in Horizontal Position

PLATE XXXIX

Fig. 84. Model of Bessemer Saloon, with Hull Inclined

PLATE XL

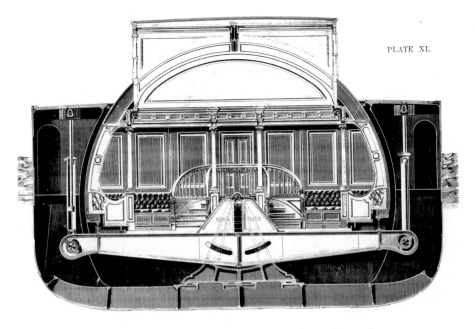

FIG. 85. TRANSVERSE SECTION OF SALOON ON CHANNEL STEAMER "BESSEMER"

PLATE XLI

FIG. 86. THE CHANNEL STEAMER "BESSEMER"

PLATE XLII

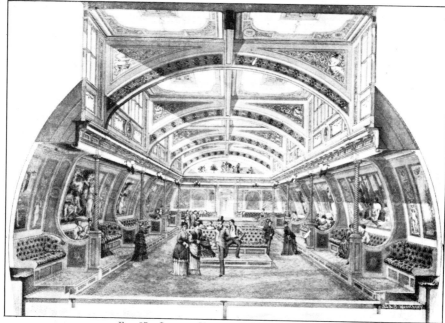

Fig. 87.   Interior View of the Bessemer Saloon

PLATE XLIII

Fig. 88.   View of Sir Henry Bessemer's Residence at Denmark Hill

PLATE XLIV

FIG. 89. THE CONSERVATORY AT DENMARK HILL

PLATE XLV

FIG. 90. THE GROTTO AT DENMARK HILL

Dear Sir Henry Bessemer

I am amusing myself by writing a Narrative of my life, which has been to me, a very interesting one, on account of the numbers of remarkable Persons and circumstances that I have met with during the long and happy course of it — This must be eminently so with you! and long may it continue to be so —

In my Narrative I have touched upon some of my most favorite Hobbies and also have made reference to some of my Mechanical contrivances and "Notions" some of which have borne good fruit, others have fallen aside from this tree — among the latter class is my Patent of 1854 for Improving the Process of "Puddling" by the Employment of Steam as the oxidizing and agitating Agent by introducing it "Beneath the Surface of the Molten Iron" by means of a Steam pipe "Rabblers" - This Patent was so far successful in greatly reducing the time and Labor of Puddling, and also, in yielding a very superior quality of Iron, but as "the waste" was said to be greater than the Enhanced value of the Iron (in a commercial sense) my Patent made little progress, and was soon after Eclipsed by your grand Invention!

The Evening before the meeting of the British Association at Cheltenham in 1856 You, accompanied by Mr Clay of the Mersey Iron works, did me the favor to show me a strikingly beautiful specimen of Iron made by your process, and (If my memory serves me right?) you told me that I was among the first who ought to see this specimen, Because, (as I understood you to say) That my Patent for Steam Puddling had in some degree led your thoughts into the line of reasoning That had resulted in your invention

(86258)

FIG. 93.   FAC-SIMILE OF A LETTER FROM MR. JAMES NASMYTH

PLATE XLVI

In the discussion that followed your description of your
Process before the Mechanical Section next day
I had the boldness to Express my anticipation of
the vast importance and value of your invention
and Holding up your Specimen before the members
of the Section as " a true British Nugget "! all of
which anticipations have been so lately and substantially
confirmed by the magnificent results of your invention
and accessories. —

Impressed with the remembrance of your kind and candid
remarks on my system of Steam Pudding and its possible
Homoeopathic Influence in leading your thoughts in the
direction of your grand invention. If I might furth your
sanction I refer to your kind remarks, above alluded to,
I should feel much honoured by being allowed by you to
do so in my reference to my patent for steam Pudding
in my narrative.

I trust all goes on to your Entire satisfaction with
your noble astronomical Enterprize

Believe me I am
Your faithfully yours
James Nasmyth

H 🔩 M

Hammerfield
Penshurst Kent
Oct 31. 1881

PLATE XLVII

FIG. 96. THE OBSERVATORY, DENMARK HILL

PLATE XLVIII

FIG. 97. GALLERY FLOOR OF OBSERVATORY

PLATE XLIX

FIG. 98. INTERIOR OF OBSERVATORY, GROUND FLOOR

PLATE L

Mr. Anthony Bessemer   Mrs Anthony Bessemer

effectual for removing the silicon from iron, but that it did not operate upon the phosphorus or the sulphur. Acting on this knowledge, which was corroborated in many quarters, Mr. Bessemer had wisely dealt with such iron as yielded the desired result. As regarded the difficulties of the process, as well as the results of it, he thought that the best thing for a member of a practical society to do, was to follow his example, and to go and see it for himself. Nothing could be more simple, or more perfectly under control; and having, by a few trials, ascertained the particular kind of treatment required, with the sample of iron to be dealt with, it was operated upon with certainty. It was said that there was nothing new in the process; but it might be fairly asked, was it, or was it not, a new result, that a bar of iron 4 in. in diameter could be bent cold into a perfect contact, without any sign of flaw? As regarded the particular product in which he was most interested, namely, a cast metal for cannon, projectiles, iron plates for shot-proof ships, and all military purposes, a circumstance had not being mentioned which he would name as being peculiarly instructive; while the metal, after having being operated on to the extent required to make it malleable iron, was in the ladle ready for pouring into the moulds, an accident occurred to the tapping-hole of the ladle, and the metal was allowed to get cold in it, instead of being poured out. Here was the ordinary condition of a common casting in a gun mould, with, however, this important difference, that in this case it was very shallow, as compared with the gun mould, and there was, therefore, no condensation of the material from fluid pressure. A cylinder of 2 in. diameter was taken out of this mass, and gave a tenacity of 42,908 lb. on the square inch, and a specific gravity of 7.626. A similar cylinder was drawn out under an ordinary smith's hammer to twice its length, and then gave a tenacity of 64,426 lb., and a specific gravity of 7.841. This portion of metal was examined by the Chemist to the War Department, and was found to contain—

Silicon	.	.	.	.	0.00
Graphite	.	.	.	.	0.00
Combined Carbon	.	.	.	.	minute quantity
Sulphur	.	.	.	.	0.02
Phosphorus	.	.	.	.	trace
Manganese	.	.	.	.	trace

This appeared to him to approach more nearly to true iron than any he had seen. The ordinary iron of the market was, in that sense, not iron, but a compound of iron and certain other ingredients. The ordinary re-melting would remove, or combine, the graphite only; the Bessemer process would remove the silicon, and when applied to an iron having but little phosphorus and sulphur, would do all that was required. If an additional process was discovered for removing these, all the iron ores of England, instead of only a very large portion of them, could be converted into pure iron.

As regarded the steel, he had been using it for turning the outside of iron guns, cutting off large shavings several inches in length, and he had found none superior to it, although much more costly. It was only necessary to witness the operation of the manufacture by the Bessemer process to be satisfied that the expense of converting the pig-iron into any of the products involved scarcely any cost beyond the labour, and that for a very short period of time. And as far as the price went, Mr. Bessemer had offered to supply such sizes as it was worth his while to make at the prices stated.

The above quotation serves to corroborate what I have previously said as to the deep interest taken by the Superintendent of the Royal Gun Factories on this subject, and I much regret that the numerous analyses made from time to time at the Arsenal have not been preserved. But I find that I gave in my Paper, which I read in May, 1859, at the Institution of Civil Engineers, when Colonel Wilmot was present, a number of official tests of the tensile strength of soft Bessemer malleable iron in its cast unhammered state, and also when hammered. There were also several tests of highly-carburised Bessemer steel, in hammered and unhammered condition.

The extreme limits of tensile strength of the converted metal are shown in the following Tables, which give the results of many trials made at different times at the Royal Arsenal at Woolwich, under the superintendence of Colonel Wilmot :—

TESTS MADE AT WOOLWICH OF BESSEMER STEEL.

Tensile Strength per Square Inch.

Bessemer Steel.	Various Trials.	Mean Tensile Strength.
In the cast unhammered state.	lb. 42,780 48,892 57,295 61,667 64,015 72,503 77,808 79,223	63,023 lb. = 28.13 tons per square inch.
After hammering or rolling.	136,490 145,512 146,676 156,862 158,899 162,970 162,974	152,912 lb. = 68.26 tons per square inch.

TESTS MADE AT WOOLWICH OF BESSEMER IRON.

Tensile Strength per Square Inch.

Bessemer Iron.	Various Trials.	Mean Tensile Strength.
In the cast unhammered state.	lb. 38,197 40,234 41,584 42,908 43,290	41,243 lb. = 18.41 tons per square inch.
After hammering or rolling.	64,059 65,253 75,598 76,195 82,110	72,643 lb. = 32.43 tons per square inch.
Flat Ingot rolled into Boiler Plate without piling.	63,591 63,668 72,896 73,103	68,319 lb. = 30.50 tons per square inch.

CHAPTER XV

THE late Ebenezer Parkes, of Birmingham, a well-known metallurgist and tube manufacturer, conceived the bold idea that copper tubes for locomotive boilers of, say, 2 in. in diameter and 12 ft. in length could be formed without a seam or joint from flat circular plates of copper of 27 in. in diameter and about $\frac{3}{16}$ in. in thickness. He forced these plates through an opening 11 in. in diameter, in a die under an hydraulic press; they thus became short cylinders. These cylinders were afterwards drawn out longer and less in diameter on steel mandrils, which were made for him at our Sheffield Works. He, however, found that the strain on ordinary sheet copper was so severe that many plates cracked and failed, and it was not until he obtained chemically-pure copper—the result of electrolysis—that his manufacture was a commercial success. On one occasion I met Mr. Parkes at my Works at Sheffield, and, in speaking of the extreme toughness of our mild steel, he said he had no doubt that he could force plates of it through his dies, as he was doing with copper. I must confess that I did not think this possible; but on his persisting in his assertion, I arranged to return with him to Birmingham the same evening, taking five discs of our mild steel, varying from $\frac{1}{4}$ in. up to $\frac{3}{4}$ in. in thickness. I was anxious also to try a very stout plate, and there happened, at the time, to be some locomotive boiler tube-plates (ordered by the Lancashire and Yorkshire Railway Company) in course of construction at our works. One of these was found to be sufficiently large to allow us to cut off a disc from one end, 27 in. in diameter, without spoiling the plate. Taking these discs, Mr. Parkes and I proceeded to Birmingham, and on the next morning we commenced operations. We succeeded in making these steel plates into deep cylinders of 11 in. in diameter. They were quite cold when operated on: had they been red-hot, those parts in contact with the cold dies would have become cooled, and stretching unequally with the hot parts, would inevitably have failed. Figs. 50 to 55 illustrate the mode of operation.

In Fig. 50, A represents the ram of an hydraulic press, and B a circular punch, the lower angles of which are slightly rounded; C is a circular ring, or die, having a trumpet-shaped mouth, shown in

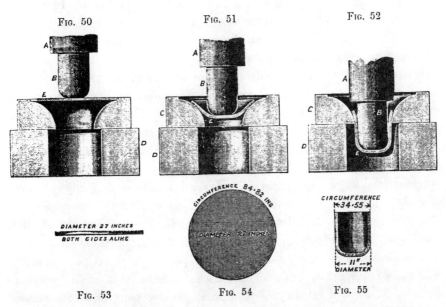

FIG. 50   FIG. 51   FIG. 52

DIAMETER 27 INCHES
BOTH SIDES ALIKE

CIRCUMFERENCE 84·82 INS
DIAMETER 27 INCHES

CIRCUMFERENCE
←34·55→

←--11"--→
DIAMETER

FIG. 53   FIG. 54   FIG. 55

FIGS. 50 TO 55. BESSEMER STEEL BOILER PLATE BEING PRESSED INTO A CUP (1861)

section, and resting on the hollow bed D, of the hydraulic press. A circular recess of 27 in. in diameter was made on the upper side of the die to receive the plate of steel to be operated on; E, Fig. 50, shows the cold plate of steel placed in the die ready for bulging. The descent of the ram forced the plate into a dished form, shown at E in Fig. 51. The further descent of the ram, as shown in Fig. 52, drove the plate nearly through the die: it, however, still had its mouth slightly splayed. Another movement of the ram pushed the plate entirely through the die, and made it into a plain parallel cylinder, with a slightly-rounded

D D

bottom, as represented at Fig. 55. In spite of this marvellous trans-
formation, in form and dimensions, the metal remained at all parts wholly
uninjured, as was incontestably proved by the fact that the cylinder
became a beautiful sonorous bell, in which the critical musical ear could
not detect any fault in tone, due to crack or injury of any kind.

Now let us for one moment consider what changes the solid cold
steel underwent, as it flowed like a piece of plastic clay, and suffered so
great a change in the position of all its constituent particles. In Fig. 53
we have the original disc seen on edge ; it was $\frac{3}{4}$ in. in thickness, 27 in. in
diameter, and $84\frac{3}{4}$ in. in circumference ; both its sides were originally
of the same area. When made into a cylinder or cup it measured on
the outside $34\frac{1}{2}$ in. in circumference, and on the inside 29 in. only ; the
metal which originally formed its outer circumference had been reduced
to $34\frac{1}{2}$ in. Such a change of form and flow of cold steel from one part
of the mass to another, required enormous force, and yet so great was
the toughness and resilience of this mild steel that the changes of form
and dimension were possible without producing a symptom of rupture. I
fearlessly challenge any person of ordinary intelligence to study, however
slightly, these diagrams, and then to cast his eye on the accompanying
illustration, Fig. 56, Plate **XX.**, which is a reproduction of this
steel cup, without coming to the conclusion that in these early days
of the Bessemer process we could, and did, produce a metal pre-eminently
adapted to the construction of ordnance : a metal that could be manufactured from Swedish charcoal pig-iron in homogeneous, unwelded masses
of from 5 to 20 tons in weight, at less than one-half the price paid for
Lowmoor iron bars, from which the Armstrong gun coils were made.
I cannot tell the precise date of the actual production of the cup illustrated,
but I know it was many months before the great Exhibition of 1862.
I can trace it back to that period by evidence that cannot be disputed.
The Engineer newspaper of the first week in May, 1862, describing my
exhibit of Bessemer steel, says :—

There are also some extraordinary examples of the toughness of Bessemer steel made from
British coke-made pig-iron, among which may be enumerated two deep vessels of one foot in
diameter, with flattened bottoms and vertical sides ; at the top edge, one of them is $\frac{5}{8}$ in. and
the other $\frac{7}{8}$ in. in thickness. These are formed up in a press from flat circular discs of steel.

They can now be drawn into long tubes, either of their present diameter, or they may be reduced to locomotive boiler tubes of 2 in. in diameter; there is also shown an attempt to raise a piece of the best Staffordshire iron plate by the same tools; this only went about as deep in proportion as an ordinary soup plate before it fractured all around the punch, and almost fell into two pieces. It may be remembered that Mr. Parkes, who invented this beautiful system of making unwelded tubes, has been obliged to use the very highest quality of copper for that purpose; the ordinary copper of commerce generally cracks, but the Bessemer steel, as seen by these examples, stands this fearful ordeal with perfect safety.

On the closing of the Exhibition of 1862, I presented this cup to Dr. Percy, who placed it in the gallery of the Geological Museum in Jermyn Street, whence it was, many years ago, transferred to the South Kensington Museum. The Curator kindly allowed me to have a photograph taken of it, and from this photograph the engraving on Plate XX. has been made.

Since writing the above, I have called to memory an earlier date on which one of these deep cups was exhibited. I refer to the occasion of Sir William Armstrong's visit to Sheffield, as President of the Institution of Mechanical Engineers, which held its summer meeting there on July 31st, 1861; in proof of this I refer to the copy of my Paper as printed and issued by that Institution. In the *Proceedings* of the Institution the Secretary interpolated, between the reprint of my Paper and the report of the discussion thereon, the announcement which is here reproduced.

Mr. Bessemer exhibited an 18-pounder gun made of the Bessemer steel cast in a single ingot of the required size and subsequently hammered, with a variety of specimens of the metal, broken to show the quality of the fracture; also some piston rods, a boiler plate flanged for a locomotive firebox, and a plate bulged in a die without cracking or tearing; a plate of thin metal punched with a number of small holes very close together, and a tube of the metal which had been crushed flat without the surface of the metal cracking. He showed also one of the fireclay tuyères used for blowing the melted metal in the converting vessel, and specimens of the ganister used for lining the vessel and ladle, both new and after use.

The "variety of specimens of the metal broken to show the quality of the fracture" should have been described as "specimens crushed to show the toughness of the steel." "A plate bulged in a die" is the deep cup made from a flat piece of boiler plate 27 in. diameter, and already mentioned as being illustrated in Fig. 56, Plate XX. The tube of metal crushed flat without cracking (see c, Fig. 49, Plate XIX.) was similar to the crushed gun-tubes so many years exhibited in the South Kensington Museum, and now in the possession of the Iron and Steel Institute. Figs. 57 to 60, on Plates XXI., XXII., and XXIII., show other specimens exhibited at the meeting.

It is unnecessary to multiply examples, since those already given cannot fail to convince any unprejudiced person that in these early days of the Bessemer process all those manufacturers who understood it, and took the amount of care which is necessary in all properly-conducted manufacturing operations, were able to produce steel of high quality with as great a degree of regularity as is common with any other modes of

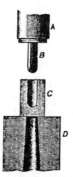

FIG. 61.   PRESSING BESSEMER STEEL BLOCK FOR RIFLE BARREL

production.   I, however, cannot refrain from giving yet another instance of the wonderful tenacity and endurance of this metal when subjected to the most violent strains.

About the year 1862, a Mr. Thompson, of Bilston, took out a patent for a novel and ingenious mode of manufacturing Enfield rifle-barrels, and after many trials he chose Bessemer mild steel as the material most suitable for this purpose.   Our works at Sheffield supplied him with large quantities of mild steel, in the form of round bars 3 in. in diameter.   These were afterwards sawn into lengths of about 6 in., and when made red-hot were placed on end under the steam-hammer, which carried a cylindrical steel punch of 1 in. in diameter, having a conical end resembling an armour-piercing shot, as shown in Fig. 61.   The hammer A had projecting from it the punch B, beneath which was placed the steel piece c, shown partly pierced by one or two blows. This piece was placed over an opening in the anvil block D, and after two or three more blows it was pierced from end to end, forming a short tube from which no metal has been removed. This violent treatment did not split or injure the steel in any way, but was well calculated to show any defect if the metal operated upon was not absolutely sound. After the operation of punching, the short tubular piece was rolled between a pair of rollers having a series of tapering grooves formed on them, and also an enlarged recess to form the breech part out of the solid, so that a barrel in one piece without welding was produced. This was afterwards finished in the usual way. The severe test to which these mild steel barrels were subjected at the Proof House, Birmingham, is shown in the annexed tabular statement, which is taken from a Paper read by me at the Royal United Service Institution on May 2nd, 1864, and published in the *Transactions,* from which the Table herewith given is copied.

TRIAL OF TWO STEEL GUN-BARRELS (ENFIELD PATTERN), AT THE PROOF-HOUSE, BIRMINGHAM.

Barrels made from Bessemer Steel by Thompson's Patent process.

Barrels, 1853 Infantry Pattern, .577 bore. Bullets used, 715 grains. Diameter, .551

Length, 1.043. Ratio of length to diameter, 1.893.

Result of Experiments :

1st round, charge 205 grains, 7½ drachms powder, 1 bullet.

2nd round, charge 224 grains, 8¼ drachms powder, 1 bullet.

3rd round,   ,,   ,,   2 bullets.

4th round,   ,,   ,,   3 bullets.

5th round,   ,,   ,,   4 bullets.

6th round,   ,,   ,,   5 bullets.

The barrels were now examined and found intact.

7th round, charge 224 grains, 8¼ drachms powder, 6 bullets.

8th round,   ,,   ,,   7 bullets.

9th round,   ,,   ,,   8 bullets.

10th round,   ,,   ,,   9 bullets.

11th round,   ,,   ,,   10 bullets.

12th round,   ,,   ,,   11 bullets.

13th round,   ,,   ,,   12 bullets.

14th round,   ,,   ,,   13 bullets.

15th round,   ,,   ,,   14 bullets.

16th round,   ,,   ,,   15 bullets.

Barrels found intact.

17th round, charge 224 grains, 8¼ drachms powder, 16 bullets.

The firing was now continued with one barrel only, the nipple having been blown out of the other, which, still retaining its charge of 16 bullets, remained intact.

18th round, charge increased to 269 grains, 9¾ drachms powder, 17 bullets.

19th round,   ,,   ,,   ,,   ,,   18 bullets.

*20th round,   ,,   413 ,,   15   ,,   ,,   and 25 bullets.

Length of each bullet, 1.043.

The barrel was then examined and found intact. Further test was deemed unnecessary.

Proved by Mr. Samuel Hart, Assistant Proof-Master, *in the presence of Ezra Millward, Esq., Proof-Master at Birmingham, December 23rd, 1863.

With these examples of the extraordinary toughness and tenacity of both pure Bessemer iron and Bessemer steel, no one, with any knowledge of the violent strains to which the test pieces were subjected, can doubt the fact that between the copper-like toughness of the pure Bessemer iron, and the great tenacity of the more highly carburised steel which we were at that time supplying to engineers, for making every description of cutlery and cutting tools, there did exist, and could easily have been found by trial, the precise quality of steel most suitable for the construction of ordnance. It must be borne in mind that it was not until some ten years later, that is, in the year 1869, that any Siemens-Martin, or open-hearth steel, was made, and consequently that the only varieties of cast-steel then available for guns were crucible cast-steel and Bessemer cast-steel. The fact must also be recognised that both the difficulty and the cost of producing large masses of crucible steel increased greatly whenever the metal was required to be of the very mild quality known as low carbon steel, which is most difficult to fuse in crucibles, as well as to retain in fusion during the time occupied in filling a large mould from hundreds of separate small vessels. Hence the strong temptation the steel manufacturers had to supply a more carburised, and consequently a more easily fusible and *less tough*, steel than was specified; while the price of this crucible steel was greatly augmented as the ingot became larger, increasing to over £100 per ton. It is equally notorious that not one of these disadvantages applied to the Bessemer metal; it was, in fact, cheaper to produce a single mass of 10 or 20 tons in weight than to make the same weight in a number of small batches of 3 tons to 5 tons. Nor was there any greater difficulty in making the mildest possible quality of steel, because we always began by making pure soft iron. From the zero point of decarburisation the hardest qualities of steel could be made, differing by almost imperceptible gradations, and depending on the number of pounds of rich carburet of iron added to the pure iron for that purpose.

The material had been proved in all respects suitable for the manufacture of ordnance, and, as I have already said, Colonel Eardley Wilmot and I had arranged, under contract, to erect a Bessemer plant in the old gun foundry at Woolwich, which was amply large enough for that purpose. This project, had it been carried out, would have rendered wholly unnecessary the erection of a second arsenal at Elswick, built under the guarantee of the British Government at a cost of £85,000. It must also be borne in mind that by my process we had the advantage of being able to make, if desired, malleable iron guns in a single piece *without a weld or joint*, by founding, or by the combined processes of founding and forging, with or without hoops; so that if malleable iron, and not steel, had in reality been the best material for the construction of ordnance, such guns could have been produced at Woolwich Arsenal, either as complete gun-castings, or as ingots to be forged, at a cost not exceeding £6 or £7 per ton if made of British iron, and not exceeding £10 per ton if made of Swedish charcoal pig-iron; whereas the Lowmoor iron bars used to make the coiled guns cost over £20 per ton, and were the mere raw material to start with. Nor did the Bessemer pure malleable iron, if used for guns, admit of any of the charges that had been made to depreciate the value of steel for that purpose, namely, that it was very uncertain in quality, and could not be obtained of the precise degree of carburisation and toughness required. Such a charge could not possibly be made in reference to pure iron, which was wholly decarburised, a condition which it was impossible to mistake during its manufacture, for the huge white flame issuing from the converter suddenly drops when all the carbon is burnt out, a result which occurs with unerring certainty. At all events, if Bessemer steel could not be depended upon at Woolwich, Swedish charcoal pig-iron, wholly decarburised, could have been made in masses of 10 to 20 ton, at a cost not exceeding £10 per ton, and of, at least, 5 tons per square inch greater tensile strength than Lowmoor bars, as was proved by Colonel Wilmot's experiments at Woolwich Arsenal; while the cost of the huge unwelded mass would have been less than half the cost per ton of the bar-iron used to make a welded coil with its many imperfect junctions.

I should like to say a few words here about the broad distinctive characters of the two materials, wrought or bar-iron, and cast homogeneous iron or steel. I need scarcely remind the reader that bar-iron making begins with the process of puddling, which produces a ball or mass of iron that, in every case, is mechanically mixed with fluid scoria, and sometimes with sand and dry oxide or iron scale. From this crude material, puddle bars are made, and these are cut into lengths of 2 ft. or 3 ft., and formed into a bundle or pile, which is brought up to a welding heat in a suitable furnace, and then rolled into a merchant bar. This process of rolling and piling is repeated more than twice, and a bar is in this way produced, which to the eye appears, and is supposed, to have all its separate parts welded or united so as to form an undivided and indivisible mass. But this is not so. I have never seen a bar of wrought iron produced by puddling that, in two or three minutes, by a very simple treatment, I could not separate more or less perfectly into its component bars, which are in reality never thoroughly united, although they adhere more or less soundly. In fact, so imperfect is this adhesion called "welding," that whenever bar-iron is worked under the hammer, it is necessary to forge it at such a degree of heat as will continue the welding process; for by working it much below this temperature, the imperfectly coherent mass begins at once to separate at all the junctions between the several bars of which it is composed, and tumbles to pieces.

I will describe an experiment clearly illustrating this fact. Two pieces of ordinary commercial bar-iron of 1 in. square were heated to a blood-red heat, and put under a small steam-hammer, where they received several blows on alternate sides; the result was a complete disintegration of the mass, as shown in Fig. 62, Plate XXIV. The lower example was similarly treated on alternate angles, instead of on the flat sides. It may be supposed that the far-famed Lowmoor and other

Yorkshire irons are exempt from this defect, but this is not so, the simple fact being that "best-best" iron has been piled more times than common iron, and the result of working it at a temperature that will not continue the welding process, only divides it into more numerous filaments than a bar of common iron. I may mention the fact that, on one occasion, during a short stay at my works at Sheffield, I had the honour of a visit from an active partner in one of the great Yorkshire firms which stand so deservedly high among bar-iron makers. I mentioned this fact of imperfect welding, and the consequent disintegration of bar-iron by simply working it at a temperature below welding heat. My visitor laughed outright at the possibility of such a thing happening to any bar-iron that his firm had ever turned out. I said : "If you will wait while one of my people goes to an iron warehouse in the town and purchases a bar of your iron, I will convince you that I am right." Well, he patiently waited until the bar was procured, and admitted at once that the brand stamped on it was his own. A short length was then cut from it and heated in his presence. It was put under one of the rapidly-moving tilt hammers at that moment being used in forging our bar steel at the same low heat. The result was that the Yorkshire iron bar divided, under this simple treatment, for about a foot of its length into a mass of fibres forming a veritable birch-broom, to the utter astonishment of the manufacturer.

At the time when the two bars of 1-in. square iron, shown in Fig. 62, Plate XXIV., were hammered, a similar bar of Bessemer mild steel was treated at the same temperature under the same hammer. The illustration, Fig. 63, Plate XXV., shows how it simply became extended into a flat undivided surface, without crack or rift in the material. These examples of forging below a welding heat serve to show the imperfection inevitable in all puddled or welded iron; while the steel example also shows the continuity of parts resulting from the Bessemer steel or homogeneous iron being formed into an ingot while the metal is in a fluid state, hence producing an undivided and indivisible mass, however much it may be hammered, hot or cold.

It will be readily understood how deeply interested I was in the application of my invention to the construction of ordnance, and how

E E

much I felt encouraged by the high appreciation of what I had achieved by so competent a person as Colonel Eardley Wilmot. Although I saw that there was an almost endless variety of applications in industry to which this cheap and superior metal could be advantageously applied, I nevertheless felt a strong desire to see it used in the manufacture of guns. Its summary rejection at Woolwich, however, without even a trial, furnished me with yet another proof of the utter foolishness of relying on Government, and made me throw up all idea of following that branch of manufacture as a speciality. With a still lingering desire to put my material to the test of gun-making, I had looked pretty deeply into the subject, in order to see what had already been done by others, and how far the road was still open to me as a gun-manufacturer. On searching at the Patent Office I found the specification of Captain Blakeley, dated February 27th, 1855; in this specification, Captain Blakeley described his invention as consisting of certain improvements in the construction of ordnance, in which an inner tube or cylinder of steel, gun metal, or cast iron, was enclosed in a case or covering of wrought iron or steel, which casing was made in parts, either shrunk on to a cylindrical tube, or forced cold on to a tube, the exterior surface of which was slightly conical, so as, in either case, to tightly grip the inner tube, adding materially to its strength and power of resisting internal pressure. This casing, whether made of cast-iron or steel, might itself be further supported or strengthened by one or more outer layers of rings or hoops, also put on under tension. Here we had clearly and distinctly laid down the vital principles embodied in all modern built-up guns, in this and in other countries—that of external compression of the inner tube by an outer one; and, unless it can be shown that this patent of Captain Blakeley was anticipated by a prior invention, he must stand before the world as the originator and father of modern built-up artillery. From this patent I saw at once that it would be impossible for me to manufacture *built-up guns* having an *internal steel tube, without direct infringement.* Captain Blakeley, at this early period (February, 1855), had the sagacity to see that a steel tube or lining was an *indispensible condition* of a perfectly built-up gun : not only because of its homogeneous character and freedom from welded joints, and its greater cohesive strength, but also because of its greater hardness and power to resist the severe abrasion of its inner surface, caused by the studs on the projectile moving along the rifled grooves under immense lateral pressure. Although he knew that steel was the best possible material for the lining of the gun, he, nevertheless, thought it prudent to claim also the use of gun-metal and cast-iron, lest he should have his invention evaded by the substitution of either of these last-named homogeneous metals. He, however, evidently thought it unnecessary to guard himself against the possible evasion of his patent built-up gun by the substitution of a welded wrought-iron tube in place of a homogeneous steel one. This doubtless arose from his knowledge of the great inferiority of wrought-iron, as compared with steel, for such a purpose, and also from his practical experience, as an artillerist, of the searching and highly corrosive nature of the intensely-heated powder gases, which, sooner or later, find out and deeply corrode the numerous imperfectly-welded joints inevitable in a wrought-iron gun tube.

The natural effects of corrosion on wrought-iron bars must have been commonly observed. Take, as an example, an old pump-handle, and see how the once smooth and even surface is eaten into deep grooves and furrows by corrosion, commencing at, and following, all the lines where the several parts, of which the bar is composed, are imperfectly welded together. Or examine an old chain cable, the links of which were made of smooth round iron rods, and see the indented shape it has acquired, the once smooth surface of each link being grooved by corrosion of the metal where the parts were imperfectly welded in the original formation, even of the high-class iron used for cables. This is the effect of water only on ordinary wrought iron. If any one doubts the destructive effects of fluids more corrosive than water, let him put a bright, well-finished piece of bar-iron into water containing only one-tenth of its weight of sulphuric acid, and he will find that in less than one hour he will have a perfect picture of the arrangement of parts of which the bar is composed, showing all the imperfectly-welded fibres, like a beautifully engraved map. What, then, must be the result from the union of the oxygen in the saltpetre with the sulphur in gunpowder, producing sulphuric acid gas, acting under enormous heat and pressure within the gun, and searching out and attacking all its welded joints?

In my search at the Patent Office, I also found the provisional petition of Mr. William George Armstrong (afterwards Lord Armstrong), dated February 11th, 1857, being two years less sixteen days after the patent of Captain Blakeley, which is dated, 27th February, 1855. Annexed is a copy of Mr. Armstrong's provisional specification, issued under the authority of the Commissioners of Patents :—

(This Invention received Provisional Protection only.)

PROVISIONAL SPECIFICATION left by William George Armstrong at the Office of the Commissioners of Patents, with his Petition, on the 11th February, 1857.

I, WILLIAM GEORGE ARMSTRONG, of Newcastle-upon-Tyne, in the County of Northumberland, Civil Engineer, do hereby declare the nature of the said Invention for " Improvements in Ordnance," to be as follows :—

The improvements relate, firstly, to forming guns with the internal tube or cylinder of wrought iron or gun metal in one piece, surrounded by one or more cylindrical casings of wrought iron or gun metal shrunk upon the internal cylinder.

It will be seen that this proposal of Mr. W. G. Armstrong differs from the invention set forth in Captain Blakeley's prior patent, by substituting a wrought-iron internal tube for a steel one. As I could not lawfully make a built-up gun with collars or rings shrunk or forced on to a steel tube, and as I had no intention of evading Captain Blakeley's patent by using an inferior material for the inner tube of the gun, I abandoned all idea of the manufacture of built-up guns, and contented myself with supplying Captain Blakeley with steel tubes, or with forged steel guns complete in one piece, with the trunnions formed thereon out of the solid ingot. This manufacture I commenced as early as February, 1861; between that date and February 5th, 1863, I had manufactured at my works in Sheffield no less than seventy forged steel guns for foreign service, not one of which was ever returned to me, or was reported to be in any way defective.

All these orders for guns came to me spontaneously, and were never sought for by travellers, advertisements, circulars, or otherwise. But not one gun, or gun-block, was ever ordered of me by the British Government to test the qualities of this new steel, which at that period was the subject of the deepest interest and most careful examination by intelligent engineers in every State in Europe.

In the early part of the year 1859, the Bessemer Steel Works at Sheffield had regularly embarked in the manufacture of high-class steel for tools, and also for cutlery. For this purpose I had investigated the whole question of the supply from abroad of pure charcoal pig-iron, and had practically tried the famous Algerian iron from Bône and other mines, and also Indian, Nova-Scotian, Styrian, and Swedish pig-irons. Among the latter, I found on analysis, to my astonishment, that certain brands of charcoal pig, which, when delivered in Sheffield, cost only £6 to £7 per ton, were, when decarburised by my process, superior in purity to some of the highest brands of Swedish bar-iron, costing in Sheffield from £16 to £24 per ton. One Swedish brand of pig-iron in particular, costing £6 10s. delivered in Sheffield, was capable of making malleable iron by my process more pure than the far-famed Danemora L bar-iron, worth £30 per ton in Sheffield, and with which particular brand the small malleable iron gun which I exhibited in May, 1859, at the Institution of Civil Engineers, was made. The analysis of this by Mr. Edward Riley has already been given (page 182 *ante*).

It will be conceded that if we obtained malleable iron of this extreme purity by my process, steel of very high quality could also readily be produced from that particular class of pig-iron.

Thus fortified by practical working and by actual analysis, and also by the purchase of a large consignment of this pure charcoal pig, we laid ourselves out at the Sheffield works for the production of high-class tool steel, which we put on the market at 15s. or 20s. per hundredweight below the ordinary trade prices for this article. My process, so admirably adapted for the production of large ingots, was not so well fitted to make a great number of the $2\frac{3}{4}$ in. square ingots generally used in the Sheffield steel trade for tilting into small bars, which particular size of ingot had all its long-established trade rules and prices connected with it. So we determined to convert our pig into steel in large quantities, and to pour the converted metal into an iron cistern filled with water, in order to granulate the whole charge, and avoid all costs of moulds, casting, etc. By this means, and by the blending of different charges in definite proportions, we insured the production of steel of any desired temper, or degree of carburation, with an accuracy wholly unattainable by the old crucible system. For it must be borne in mind that in the ordinary crucible process the steel melter has to deal with bar-iron that has been subjected for several days, in a very large closed box or chamber, to the action of charcoal powder at a high temperature, during which treatment the iron bars absorb about one per cent. of carbon, more or less, dependent on time and on temperature. The amount of absorption depends also on the relative positions of the bars in the converting-box; hence, when the bars are thus converted into blister steel, it is almost impossible that the ends and the middle of any particular bar should be equally carburised, or that bars occupying different positions should absorb an equal quantity of carbon. After the withdrawal of the bars from the converting-chest, they are broken into short pieces for the melting crucible. Now the only mode of telling how far each piece of the broken bars has been carburised is to examine the crystalline fracture by the eye, and thus class and assort the various fragments for each quality of steel. It is wonderful how accurately a clever practised steel melter will judge of the state of carburation of the metal; but his judgment, after all, can only be approximate. Such visual determination is not like measuring or weighing the constituents of a mixture. Crucible steel is made in separate pots of from 40 lb. to 50 lb. each, and the steel maker cannot afford to make forty-five separate quantitative analyses of every ton of steel he turns out. Even if he could do so, after he had made the metal into ingots, he would not be more secure, since he could not alter the ingot when once cast. As a matter of fact, the precise degree of carburation of each 50 lb. of steel produced in the old crucible process depended on the judgment of a man looking at the crystallised fracture of each piece he put into his crucible; and all must agree that it is highly creditable to those engaged in this mere guesswork that they got as near as they did to the quality required.

In the manufacture of tool steel, on the system which I laid down at my Sheffield works, we entirely eliminated this source of inequality, by dispensing with ocular examination of a crystalline fracture, which is subject to numerous modifications in character, from causes other than its precise degree of carburation. We converted five tons of pig iron at one charge, and having granulated it by pouring the molten steel into a cistern of water, we had this quantity of shotted metal in a condition that was still practically fluid as far as the power of mixing was concerned. If each granule weighed, on an average, seven grains, we had in our 50 lb. crucible 50,000 separate pieces, the precise degree of carburation of which had been ascertained by careful quantitative analysis of the whole five tons, which analysis we could afford to make—and did make—very carefully.

We produced, as nearly as practicable, three qualities of converted metal, say A, with half per cent. of carbon, B, with one per cent., and C, with one and a-half per cent. ; we also made pure iron upon which we could absolutely rely. These four qualities, accurately analysed, were kept in separate bins ; the analyst who gave the order to the steel melter to make two or three tons of steel of any precise and predetermined degree of carburation would, say for example, weigh $41\frac{1}{4}$ lb. out of bin A, and put $8\frac{3}{4}$ lb. from bin B into it, thus making the 50 lb. charge, always using the nearest of the three qualities to the one required, and making it a little milder or a little more highly carburised as desired. Most minute differences could thus, at all times, be made with unerring certainty by the simple fusion in a crucible of two metals, the carburation of which had, in each case, been tested by careful analysis. The mixing of these accurately-ascertained qualities in definite weights while in the granulated state resulted in the production of a quality the exact mean of the known constituents of the two qualities mixed. It gave a more certain and a more accurate result than could possibly be obtained on the old system of crucible steel-making, where judgment by the eye took the place of accurate analysis and the weighing machine, as used in my system. Hence it was an undeniable fact that we could, and did, produce commercially crucible cast steel of great purity, and of any precise and predetermined degree of carburation, with greater accuracy than was obtained by the method employed to produce crucible steel in Sheffield.

CHAPTER XVI

BESSEMER STEEL GUNS

THE course of events now brings me to an incident connected with Woolwich Arsenal, which I would fain pass over in silence, but, if history is to be written at all, the historian must speak the truth. In 1859 the firm of Henry Bessemer and Company, of Sheffield, had qualified themselves to receive proposals to tender to Woolwich Arsenal, for the supply of steel for cutting tools, and on June 3rd of that year, we tendered unsuccessfully, under a form of contract sent by the War Office, at the same price as we were obtaining from several first-class engineers—namely, £42 per ton, the ordinary trade price in Sheffield for such tool steel varying from £50 to £60 per ton. We tendered again for another lot of tool steel on July 8th, at £40 to £42 per ton; again our offer was not accepted. We tendered also on September 5th, at prices still lower, viz., from £32 to £40 per ton; and again, on September 7th, for some bars at £40, and for some (the greater part) at £32 per ton. But this low tender also failed to secure us the order, and, as we could make the highest quality of tool steel by my process from Swedish pig-iron at an extremely low cost, we were determined on the next occasion to get the order, or know the reason why. On December 7th, 1859, forms of tender were sent us for two different sizes of steel bars, and we quoted as low as £20 per ton for each of them ; our tender was then accepted for the first time, and we commenced at once to make the steel. Bars of each quality were carefully tested by us in our own works, so as to prevent the possibility of a single bar being sent out of any but the very highest quality, my managing partner personally taking charge of these special tests. This rigid inspection at our works was considered by our firm to be absolutely necessary in this case, because we felt assured that our former tender of £32 to £42 was far below that of any Sheffield house, although it was not accepted; hence our belief that the steel about to be sent would undergo the most severe and rigid tests.

In due course the steel was delivered to the carriage department at Woolwich Arsenal, as directed, but, after several days, we were informed that it was useless, and that we must take it back. Now, the conditions of the tender were such that the Government officials were the sole judges of the fitness of the material, and had absolute power of rejection if not satisfied with it. In case of the steel not proving satisfactory, the Government had also power to purchase a like quantity of any other manufacturer, and charge the difference in price to the person whose steel was rejected. Thus the Government could send back to us all the steel which had been tendered for at £20 per ton, and purchase a like quantity at £50 or £60, making our firm pay the difference of £30 or £40 a ton. Under these circumstances I was determined to investigate this matter for myself. I accordingly went down to the Arsenal, and was shown into the office of the head of the carriage department. I asked him in what way the steel was defective. Before replying, he got up from his chair, opened a drawer, and took out ten or dozen "chipping chisels," which were made, as usual, out of an octagon bar of steel known in the trade as $\frac{7}{8}$ in. "octagon chisel steel." All but two of the chisels were broken; they were very slender and delicate, and had been a good deal punished by the prover's hammer. Notwithstanding this, I was much astonished at such a result, and on attentively examining the fractured parts I became convinced that they were not made of the quality known as "chisel steel," which is invariably used for this purpose. I then looked over the written contract that had been sent to us, and found that among the specified shapes and sizes of steel bars therein described, there was not one single bar of octagon steel. I handed the list to the gentleman who received me, and asked him to point out octagon steel, which, of course, he could not do. In order that there should be no possible mistake on this point, I have had the entry made by my clerk at the time, in his rough order book at Sheffield, photographed, as shown in Fig. 64, thus furnishing unquestionable evidence of the absence of any octagon bars in the contract.

F F

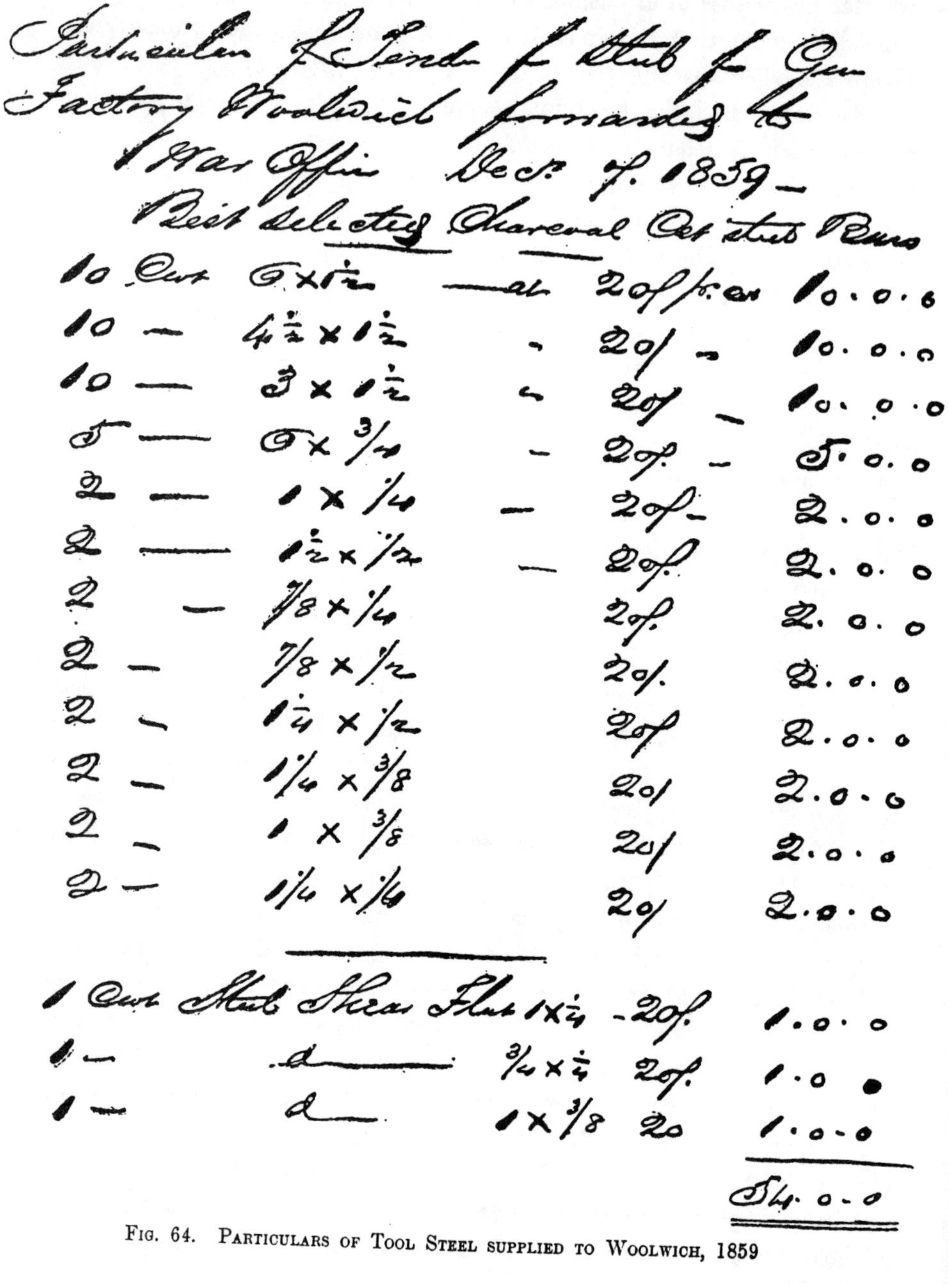

Particulars of Tender for Steel for Gun
Factory Woolwich forwarded to
War Office   Dec. 7. 1859 —
    Best selected Charcoal Cast steel Bars

10	Cwt	6 × 1½	— at	20/ p. cwt	10 . 0 . 0
10	—	4½ × 1½	"	20/ "	10 . 0 . 0
10	—	3 × 1½	"	20/ "	10 . 0 . 0
5	—	6 × 3/4	—	20/ —	5 . 0 . 0
2	—	1 × 1/4	—	20/ —	2 . 0 . 0
2	—	1½ × 1/2	—	20/	2 . 0 . 0
2	—	7/8 × 1/4		20/	2 . 0 . 0
2	—	7/8 × 1/2		20/	2 . 0 . 0
2	—	1¼ × 1/2		20/	2 . 0 . 0
2	—	1¼ × 3/8		20/	2 . 0 . 0
2	—	1 × 3/8		20/	2 . 0 . 0
2	—	1¼ × 1/4		20/	2 . 0 . 0

1	Cwt	Steel Shear Flat 1 × 1/4	— 20/	1 . 0 . 0
1	—	d°	3/4 × 1/4 20/	1 . 0 . 0
1	—	d°	1 × 3/8 20	1 . 0 . 0

54 . 0 . 0

FIG. 64. PARTICULARS OF TOOL STEEL SUPPLIED TO WOOLWICH, 1859

On my pointing out the absence of octagon steel in the contract, the gentleman touched the bell, and told the messenger to send the store-keeper to him. On the arrival of this person, his chief said : " I told you to make a dozen octagon chipping chisels, in order to test the Bessemer steel, and now I find that we had not ordered any ; what did you do ? " " Oh," said the man, " I gave out one of the larger bars, and had it drawn down to octagon, and brought you the chisels." Now, the nearest bar in size in the whole list that could be made into $\frac{7}{8}$-in. octagon bars was in cross-section 3 in. by $1\frac{1}{2}$ in., or more than six times the area of the $\frac{7}{8}$ in. octagon chisels made from it, and it was, as the fractures showed, of much too hard and highly carburised a quality to be made into chipping chisels ; not to mention the damage it must have received from the excessive heating in a common black-smith's forge. Instead of being tilted down to the proper size, as in a steel works, it was worked with a smith's hammer by an ordinary blacksmith, and not a steelsmith—a fact in itself enough to endanger this highly carburised steel, which must not be overheated or " burnt." Hence it must be clear that this so-called test of the quality of Bessemer steel, supplied under this contract, was, even in the case of chisel steel, no test at all of its quality. Under these circumstances, any fair and impartial person would have apologised for such a gross mistake and wholesale condemnation, and would have said that the other bars should be carefully tested as to their suitability for the several purposes for which they were required. But, on the contrary, the chief, who never even pretended that any other tests had been made, insisted on condemning the whole of the bars embraced under this contract. I said : " I will take back the steel which you have power under the *words* of the contract to reject so unfairly, and will wash my hands of Woolwich for all time ; but let me tell you that, having condemned this steel, it is your duty to your employers to purchase an equal quantity of some other manufacturer, and make our firm pay the £30 to £40 difference in price. But this is just what you dare not do, because I should resist such a claim, and that would bring the question into a Court of Law, where your conduct would become known to the world." The whole

of this steel was returned to our Sheffield works. We were at that
time regularly supplying this kind of tool steel to the most eminent
engineers in this country, among whom may be mentioned Sir Joseph
Whitworth, Messrs. Sharp, Stewart and Co., Sir William Fairbairn,
Messrs. Beyer, Peacock and Co., etc., who paid us £42 per ton for the
same quality for which we had quoted £20 per ton in the Woolwich
contract, in order to force the Arsenal authorities to accept it. Every
bar of this steel, so shamefully rejected at Woolwich, was marked
in the centre by a special punch, and sent as required to the eminent
firms above referred to, and not one of the bars was ever returned to
us or complained of.

In contrast with this summary rejection of Bessemer steel at
Woolwich, I may mention that we had, during the time when Colonel
Eardley Wilmot was Superintendent of the Royal Gun Factory, supplied
him with tool steel, which had given him every satisfaction. Indeed,
he was so pleased with it that, during the discussion which followed
the reading of my paper on May 24th, 1859, before the Institution of
Civil Engineers, he incidentally made the remarks which I reproduce
below from the printed Minutes of the Proceedings of the Institution.
He said :—

> As regarded the steel, he had been using it for turning the outside of iron guns,
> cutting off large shavings several inches in length, and he had found none superior to
> it, although much more costly. It was only necessary to witness the operation of the
> manufacture by the Bessemer process, to be satisfied that the expense of converting the
> pig iron into any of the products involved scarcely any cost beyond the labour, and that for
> a very short period of time. And, as far as the price went, Mr. Bessemer had offered
> to supply such sizes as it was worth his while to make, at the prices stated.

So exceptionally heavy were the cuts and sizes of the shavings he
referred to, that he placed on the table a box full of them, to show
their unusual character.

In the latter part of the year 1859 important changes in the control
and management of the Arsenal took place, and on November 4th
Sir William Armstrong was appointed " Superintendent of the Royal
Gun Factory for Rifled Ordnance." It was on December 7th of the
same year that Henry Bessemer and Company, as one of the authorised

contractors to the Government, supplied a quantity of tool steel at the low price of £20 a ton, which was summarily rejected under the circumstances before described. It was quite clear to me that neither I, nor my steel, was wanted at Woolwich, and I made up my mind to leave the place severely alone in future.

In the year 1858 we were getting fairly into commercial working at Sheffield, and on September 8th of that year we supplied a first sample order of steel boiler-plates to Sir William Fairbairn, of Manchester.

It was deemed desirable to communicate these facts to the world, through the Institution of Civil Engineers, whose members could not fail to be deeply interested in the production of a new kind of homogeneous cast steel, having greater toughness and cohesive strength than the best wrought iron, and at a cost considerably less than that of cast steel made by any other known process. I, therefore, wrote a paper " On the Manufacture of Malleable Iron and Steel," which was illustrated by many interesting examples of the metal that had been subjected to various tests of the most severe description. This paper I submitted to the Council of the Institution about the end of December, 1858. It was accepted, and read at a crowded meeting on May 24th, 1859.

Now, I had no intention whatever to ask Sir William Armstrong, as a favour to myself, to adopt and use this wonderfully tough and rapidly produced metal, for the manufacture of gun-tubes, in lieu of the weaker, and much more costly, coiled iron employed by him for that purpose. But, I felt that, notwithstanding the summary rejection of Bessemer steel and Bessemer iron by Lord Herbert, it was a public duty which I owed to my country to give him a further opportunity, both of hearing and seeing what was daily being done with welded masses of Bessemer iron and with Bessemer mild steel. I knew that Sir William Armstrong had been, for several years, a member of the Institution of Civil Engineers ; he was, when my paper was accepted, also a Member of Council, and, therefore, was one of the persons by whom all communications submitted to the Institution were examined, criticised, and finally voted worthy—or otherwise—of being read before a public meeting of their members, and of being published in their Proceedings. In the ordinary course of events, my paper would, I knew, be examined by Sir William Armstrong, and that this would be so appeared to me the more certain, because the careful and punctual secretary, Mr. Forrest, was in the habit of sending the actual paper that was to be examined to the private residences of all Members of Council who might be absent from the Council meetings. It was also his custom to invite important persons, who were supposed to be specially interested in the subject, to attend and take part in the discussion which follows the reading. Here again it seemed certain, if everything else failed, that Sir William Armstrong would be invited to come and join in the discussion of a subject in which he, as a paid servant of the State, must, or should, take the deepest interest. It was in this way that Colonel Eardley Wilmot was invited, and was present during the reading of my paper. But the one man in all Great Britain who was—or who ought to have been— most deeply interested in the subject, was not present at this important meeting; and thus I lost the unique opportunity I so much desired of bringing before him, while in the presence of the most eminent engineers of Great Britain, the proofs of the fitness of my metal for the construction of ordnance. But, such was the impression made on the other members of the Council of the Institution by the facts I brought before them, and by the marvellous proofs afforded by the specimens exhibited, of the value of this new kind of mild steel for constructive purposes, that they voted me the Telford gold medal; later, they made me a member of the Institution, and they also, " as the originator of the greatest improvement in the Iron Manufacture of Great Britain during the preceding five years," presented me with the Howard Quinquennial Prize, a massive gold cup, intrinsically worth 120 guineas. Finally, when advancing years rendered my duties as a Member of Council too arduous, they further conferred on me the great and distinguished position of Honorary Membership.

I will not trouble my readers with any lengthy abstracts from this paper, but it may be of interest to show some important portions of it. The following is one of the extracts referred to, which has been reproduced from the report of my paper, and the discussion thereon, printed by the Institution of Civil Engineers, and sent to all its members.

In the early part of this Paper it was shown that the process of puddling unavoidably introduces into the metal more or less cinder, and other mechanically-mixed impurities ; also, that the different degrees of refinement and decarbonization of the numerous lumps of metal which compose a puddle ball, render the production of a homogeneous mass, by that means, a desideratum not yet achieved. It has likewise been pointed out how, in the working of the other malleable metals, all these difficulties are avoided by casting the metal in a fluid state into moulds. Now this is precisely what the Bessemer process proposes to accomplish—that is, to bring malleable iron, or steel, into the same category with the other malleable metals, and by its purification, in a fluid state, to avoid the diffusion of cinder throughout the mass ; so that when cast into an ingot, or into a single homogeneous mass of any desired form, or size, a metal of equal hardness in every part may be produced, without the necessity of welding or joining of separate pieces. That this can be accomplished, is shown by the specimens exhibited. The iron bars of 3 inches square, which have been bent and doubled-up cold, the twisted bars, and the collapsed cylinders which do not split, but yield like copper to the blows of the hammer, prove this. If assurance be required, that there are no hard ribs, or sand cracks, the examples of the malleable iron gun, or the iron and steel cylinders may be taken. With reference to the tensile strength of iron bars, or boiler plate, so made from English coke pig metal, the careful testing of plates made of puddled iron, according to Mr. W. Fairbairn, has given an average of 45,300 lbs. per square inch for Staffordshire plates, 45,000 lbs. for Derbyshire, and 57,120 lbs. for Yorkshire plates. Now, four samples of the Bessemer iron plate, tested at the Royal Arsenal, Woolwich, according to the report of Colonel Eardley Wilmot, gave an average of 68,314 lbs., or 63,591 lbs. as the least, and 73,100 lbs. as the highest proof for boiler plates $\frac{3}{8}$ths of an inch in thickness. Here, then, is a result showing a greater amount of tensile strength above Low Moor, or Bowling iron boiler plates, than those plates possess above the ordinary quality of Staffordshire plates.

Here there is proof that Bessemer *iron* plates, tested at Woolwich Arsenal by Sir William Armstrong's immediate predecessor in office, gave an average tensile strength of 68,314 lb. per square inch = 30½ tons, quite five tons over the best Yorkshire plates. Also, the fact is demonstrated that this superior iron could be made from Swedish charcoal pig iron at about one-half the cost of Yorkshire iron bars, and that it could be made with great rapidity into masses of any form of several tons in weight without welding.

Again I quote from the paper :—

In order to show the extreme toughness of such iron, and to what a strain it may be subjected without bursting, several cast and hammered cylinders were placed cold under the steam hammer, and were crushed down, without the least appearance of tearing the metal. Now these cylinders were drawn from a round cast-iron ingot, only 2 in. larger in diameter than the finished cylinder, and in the precise manner in which a gun should be treated. They may, therefore, be considered as short sections of an ordinary 9-pounder field gun. Iron so made requires very little forging; indeed, the mere closing of the pores of the metal seems all that is necessary. The tensile strength of the samples, as tested at the Royal Arsenal, was 64,566 lb. per square inch, while the tensile stress of pieces cut from· the Mersey gun gave a mean of 50,624 lb. longitudinally, and 43,339 lb. across the grain; thus showing a mean of 17,550 lb. per square inch in favour of the Bessemer iron.

If it be desired to produce ordnance by merely founding the metal, then the ordinary casting process may be employed: with the simple difference that the iron, instead of running direct from the melting furnace into the mould, must first be run into the converting vessel, where in from ten to twenty minutes it will become steel, or malleable iron, as may be desired; and the casting may then take place in the ordinary way. The small piece of ordnance exhibited will serve to illustrate this important manufacture, and is interesting in consequence of its being the first gun that ever was made of malleable iron without a weld or joint. The importance of this fact will be much enhanced when it is known that conical masses of this pure tough metal, of from five to ten tons in weight, can be produced at Woolwich at a cost not exceeding £6 12s. 0d. per ton, inclusive of the cost of pig iron, carriage, re-melting, waste in the process, labour, and engine power.

It will be interesting to those who are watching the advancement of the new process to know that it is already rapidly extending itself over Europe. The enterprising firm of Daniel Elfstand and Company, of Edsken, who were the pioneers in Sweden, have now made several hundred tons of excellent steel by the Bessemer process. Another large works has since started in their immediate neighbourhood, and two other companies are making arrangements to use the process. The authorities in Sweden have most fully investigated the whole process, and have pronounced it perfect. The large steel circular saw-plate exhibited was made by Mr. Goranson, of Gefle, in Sweden, the ingot being cast direct from the fluid metal, within fifteen minutes of its leaving the blast furnace. In France, the process has been for some time carried on by the old-established firm of James Jackson and Son, at their steel works, near Bordeaux. This firm was about to go extensively into the manufacture of puddled steel, and indeed had already got a puddling furnace erected and in active operation, when their attention was directed to the Bessemer process. The apparatus for this was put up at their works last year, and they are now greatly extending their field of operations by putting up more powerful apparatus at their blast furnaces in the Landes. There are also in course of erection, four other blast furnaces in the South of France, for the express purpose of carrying out the new process. The long and well-earned reputation of the firm of James Jackson and Son is in itself a guarantee of the excellent quality of the steel produced by this process. The French samples of bar steel exhibited were manufactured by this firm. Belgium is not much behind her neighbours in the race, as the process is being put in operation at Liége. While in Sardinia preparations are being made to carry it into effect, Russia has sent to London an engineer and a professor of chemistry to report on the process, and Professor Müller, of Vienna, and M. Dumas and others, from Paris, have visited Sweden to inspect and report on the new system in that country.

These facts will serve to show how, on the Continent of Europe, the fame of this new metal was spreading, and its manufacture extending. It will be seen from the foregoing that Colonel Wilmot fully corroborated what I have previously stated, and gave the results of some experiments of his own with a mass of iron he happened to see lying with other waste scrap at my works at Sheffield. This mass of iron (see page 196 *ante*) he desired to be sent to Woolwich, and from it were cut the two cylindrical pieces which he described to the meeting; he proved that Bessemer pure iron, only slightly hammered, showed in the proving-house a tenacity of 64,426 lb., or 28.76 tons per square inch.

Another year or more slipped away, almost unnoticed in the ardour and excitement created by the rapid development and progress of my invention. Our own works were crammed with orders for locomotive double-throw cranks, which had hitherto been exclusively made at Lowmoor, or at some other of the justly-celebrated Yorkshire ironworks, but which were now being constructed of Bessemer steel. We were also busy with plain engine and carriage axles, marine engine and screw-propeller shafts, steel guns and gun blocks, locomotive engine and carriage tyres, etc. Our works were daily engaged in superseding welded Lowmoor tyres, and we were turning out, as fast as the mills could roll them, mild steel weldless tyre-hoops from 4 ft. 6 in. to 5 ft. in diameter, to be shrunk on to locomotive engine driving-wheels, and also 3 ft. tyres for carriage wheels, of which many thousands were ceaselessly running on our railways. All these hoops were tightly shrunk on to the wheels with a firm grip, just in the same manner as hoops are shrunk on to built-up guns. These thousands of hoops were daily responsible for the lives of tens of thousands of passengers seated immediately above them. Every train of twenty-five carriages would have a hundred of these steel tyres supporting their heavy load of wood and iron, and their still more valuable living freight, rushing over the steel rails at a high speed, and tending, by their rolling motion and heavy pressure at a single point of their circumference on the steel rail, to become elongated and loosened from the wheel, a tendency which this strong elastic steel most successfully resisted. It must be borne in mind that the loosening of this firm grip on only one of these hundred hoops, or the fracture of any one of them, might have wrecked a whole train, and killed more people than the bursting of a gun—an instrument that may be required to do duty for a few hours,

at intervals of many years, or, perhaps, never be used at all. That
these thousands of Bessemer steel tyres did not fail in constant
service, and did not lose their grip upon the wheels, furnished no
proof to those obtuse intellects who could only recognise the virtues
of welded iron. Bessemer steel hoops, so extensively used with the full
sanction of the eminent engineers of our British railways, found,
however, no favour at Woolwich or at Elswick. They were, nevertheless,
employed by Captain Blakeley, the original inventor of built-up guns,
and also by the Blakeley Ordnance Company of London, for the manufacture of built-up guns which were being made for Russia, and other foreign
governments, while Woolwich and Elswick were rapidly manufacturing
welded iron guns with welded iron hoops, for home use.

As a practical proof of how far weldless steel tyres would resist
fracture under the most severe trials, a locomotive engine-tyre, turned
and finished, was placed up on edge under a steam hammer, and received
blow after blow until its two opposite sides touched each other, when
its elasticity again allowed it to spring back a few inches. This large
tyre was thus formed into a long flat loop (see Fig. 65, Plate XXVI., in
which its dimensions are indicated by the foot-rule lying in front of it).
With all this ill-usage it showed no sign of cracking or fracture. This
tyre has for the last thirty-five years been exhibited in South Kensington
Museum, and is undeniable evidence of the toughness and endurance of
Bessemer steel under the most violent and abnormal strains. It also affords
a good example of the tough mild steel manufactured at our Sheffield
works at that early date.

In the summer of 1861, the Institution of Mechanical Engineers
held a provincial meeting at Sheffield, and, as a member of this Institution,
it was only natural that I should read a paper on the occasion of their
visit to the town where my steel works were located. I was still most
anxious that my own countrymen should use Bessemer steel for the
manufacture of ordnance : for this, as my readers are aware, was the
express purpose to which I had devoted myself for so long a period,
and striven so earnestly to accomplish.

The fact that I had succeeded in making a special mild steel, in
every way adapted for the purpose, was proved by a report of the

Belgian Government, which had spontaneously applied to me to make them a trial gun, thirteen months before the date on which I read my paper before the Sheffield meeting: a meeting which was presided over by Sir William Armstrong. This gun was made at our works, and sent to the Fort, at Antwerp, on the 16th June, 1860, its receipt being acknowledged in the following letter.

Brussels,
August 19th, 1860.

Sir,

I have the honour to inform you that the conical steel forging, rough from the forge, which was manufactured in your establishment, and of which you advised the shipment in your letter (stated in the margin), was received by the Commander of Artillery, in the Fort of Antwerp. Being submitted to the examination of a commission composed of officers of the cannon foundry of Liége, it was found to weigh 840 kilos. (equal 16 cwt. 2 qrs. and 22 lbs.), and to be of good quality of steel.

Be pleased, Sir, to accept the assurance of my distinguished consideration.

(Signed) THE MINISTER OF WAR.

This gun-block was bored and finished under military inspection, at Antwerp, and went through the regulation proofs in a perfectly satisfactory manner. It was afterwards determined to bore it to a much larger size, viz., 4.75 in. in diameter, suitable for 12-pounder spherical shots, and to fire larger charges of powder and to increase the number of shots, each of such additions being repeated three times, until the gun should at last give way, the charges of powder rising from 2 lb. up to $6\frac{3}{4}$ lb., and the shots from one to eight. On firing the second round of eight shots the gun gave way, apparently by the over-riding of the spherical shot.

I have annexed an accurate scale engraving of the gun as altered to a 4.75 in. bore, suitable for 12-pounder spherical shot (see Fig. 66, Plate XXVI.). In re-boring, the gun was reduced to $9\frac{1}{4}$ cwt., only about ten times the weight of the eight shot, the thickness of metal at the breech being $2\frac{3}{8}$ in., and $1\frac{3}{8}$ in. at the muzzle. In fact, it was little more than a mere gun lining, but it nevertheless afforded the most incontestable proof of the extraordinary endurance of this metal under conditions of extreme severity. The fact that the Belgian Government should seek out a foreign manufacturer, and put this new

material to the test, only makes it more extraordinary that our own
Government should have passed it by.

Nor was this Belgian gun an isolated case, for, up to the date of
which I am writing (Midsummer, 1861), several agents of foreign Governments had spontaneously applied to the Bessemer Steel Works, at Sheffield,
for steel guns.   But our firm could not manufacture built-up guns with
a steel barrel or inner tube, because this would have manifestly been a
direct infringement of Captain Blakeley's patent of February, 1855; and
knowing that iron, in any welded form, would be vastly inferior to steel
for the inner tube of a gun, we declined to manufacture such an inferior
article, and confined ourselves to making simple solid-forged steel guns

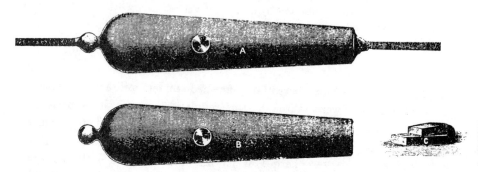

FIG. 67.   FORGED BESSEMER STEEL GUN WITH TEST PIECES

and gun-tubes.   Up to this time we had supplied twenty-eight guns,
consisting of 12-, 18-, and 24-pounders, forged, and ready for the boring
mill, at £45 per ton, a price about three times their actual cost, but
still very considerably below that of crucible steel forgings.

I may here mention that every gun, after being forged by our firm,
had its quality tested in the following simple and practical manner.
The gun when being forged had a part of both ends drawn down
under the hammer, into a flat bar of some 12 in. or 15 in. in length
and 3 in. wide by 2 in. in thickness—this was our standard test.   A
gun so forged is shown in the annexed engraving, Fig. 67.   In this
illustration the view A shows the gun with these test pieces still
projecting from each end; they were cut off and bent, when cold, into

the form shown in c, while B shows the gun-block ready to be turned and bored.   A group of these test pieces is reproduced to a scale of half the actual size in Fig. 68, Plate XXVII., and this engraving— prepared from a photograph — clearly shows how wonderfully these pieces bore the enormous strain due to the cold bending of so large a mass, the metal in each case bulging out laterally on the inside of the bend, and contracting in width on the outside of it, thus supplying the material forming the greater length of the outer surface.   Notwithstanding this interchange of parts, not a sign of tear or breaking is visible in any one of their sharply-defined angles.

Mr. A. L. Holley's remarks on our steel guns, published in his book on Ordnance in 1863, are subjoined, and form an independent testimony to their value.

141. *Bessemer Steel Guns.*—The Bessemer process of making steel direct from the ore, or from pig-iron, promises to ameliorate the whole subject of Ordnance and engineering construction in general, both as to quality and cost.   This product has not yet been used for guns to any great extent, although Mr. Krupp, the leading steel-maker, has introduced it.   Captain Blakeley and Mr. Whitworth have also experimented with it, and expressed their faith in its ultimate adoption.   Messrs. John Brown & Co., Sheffield, have made over 100 gun-forgings, some of them weighing above 3 tons, from solid ingots of this steel.   During the present year, their production of Bessemer steel will exceed 400 tons per week.   With the two new converting vessels then in operation, solid ingots of 20 tons weight can be fabricated.   A large establishment about to be started in London, with a 50-ton hammer, and a capacity to pour 30-ton ingots, will afford the best possible facilities for the development of this process.

As a point of special interest in connection with the paper I was going to read at the Sheffield meeting of the Institution of Mechanical Engineers, I determined to take strict account of the time occupied in making, at my steel works, an 18-pounder gun, and to put the finished weapon on the table in front of the Presidential chair.   By this means the Superintendent of the Royal Gun Factory at Woolwich could not help being placed in possession of all the facts and arguments I was going to put forward in my paper, and which I intended should be illustrated with plenty of actual specimens.   I have reproduced here pages 144 and 145 from the published *Proceedings* for 1861 of the Institution of Mechanical Engineers, in order to show what words Sir William Armstrong actually heard, and what facts were brought to his knowledge at that meeting, and also what mechanical proofs of the marvellous toughness of Bessemer mild steel were placed on the table immediately in front of him.

The special aim of the author during the first year of his labours, which throughout the last six years has never been lost sight of, was the production of a malleable metal peculiarly suitable for the manufacture of ordnance. By means of the process that has been described solid blocks of malleable cast steel may be made of any required size from 1 to 20 or 30 tons weight, with a degree of rapidity and cheapness previously unknown. The metal can also with the utmost facility be made of any amount of carburation and tensile strength that may be found most desirable: commencing at the top of the scale with a quality of steel that is too hard to bore and too brittle to use for ordnance, it can with ease and certainty be made to pass from that degree of hardness by almost imperceptible gradations downwards towards malleable iron, becoming at every stage of decarburation more easy to work and more and more tough and pliable, until it becomes at last pure decarbonised iron, possessing a copper-like degree of toughness not found in any iron produced by puddling. Between these extremes of temper the metal most suitable for ordnance must be found; and all qualities are equally cheap and easy of production.

From the practice now acquired in forging cast steel ordnance at the author's works in Sheffield it has been found that the most satisfactory results are obtained with metal of the same soft description as that employed for making piston rods. With this degree of toughness the bursting of the gun becomes almost impossible, its power of resisting a tensile strain being at least 15 tons per square inch greater than that of the best English bar iron. Every gun before leaving the works has a piece cut off the end, which is roughly forged into a bar of 2 inches by 3 inches section, and bent cold under the hammer in order to show the state of the metal after forging. Several test bars cut from the ends of guns recently forged are exhibited.

The power of this metal to resist a sudden and powerful strain is well illustrated by the piece of gun muzzle now shown, which is one of several tubular pieces that were subjected to a sudden crushing force at the Royal Arsenal, Woolwich, under the direction of Colonel Wilmot; the pieces were laid on the anvil block in a perfectly cold state, and were crushed flat by the falling of the steam hammer, but none of them exhibited any signs of fracture when so tested. Probably the best proof of the power of the metal to resist a sudden violent strain was afforded by some experiments made at Liége by order of the Belgian government, who had one of these guns bored for a 12 lbs. spherical shot of $4\frac{3}{4}$ inches diameter, and made so thin as to weigh only $9\frac{1}{4}$ cwts. This gun was fired with increasing charges of powder and an additional shot after each three discharges, until it reached a maximum of $6\frac{3}{4}$ lbs. of powder and eight shots of 12 lbs. each or 96 lbs. of shot, the shots being thus equal to about one tenth of the weight of the gun. It stood this heavy charge twice and then gave way at about 40 inches from the muzzle, probably owing to the jamming of the shots. The employment of guns so excessively light and charges so extremely heavy would, of course, never be attempted in practice.

Some idea of the facility of this mode of making cast steel ordnance is afforded by the time occupied in the fabrication of the 18 pounder gun now exhibited, which was made in the author's presence for his experiments on gunnery. The melted pig iron was tapped from the reverberatory furnace at 11.20 A.M., and converted into cast steel in 30 minutes; the ingot was cast in an iron mould 16 inches square by 4 feet long, and was forged while still hot from the casting operation. By this mode of treating the ingots their central parts are sufficiently soft to receive the full effect of the hammer. At 7 P.M. the forging was completed and the gun ready for the boring mill.

The erection of the necessary apparatus for the production of steel by this process, on a scale capable of converting from crude iron enough steel to make forty of such gun blocks per day, will not exceed a cost of £5000, including the blast engine; hence the author cannot but feel that his labours in this direction have been crowned with entire success: the great rapidity of production, the cheapness of the material, and its strength and durability, all adapt it for the construction of every species of ordnance.

Sir William Armstong had thus another opportunity of seeing and trying, if he chose to do so, a quality of steel which he himself told the meeting that he had never tried; a kind of steel that for constructive purposes had attracted the serious attention of the most eminent engineers in every country of Europe; a kind of steel invented and perfected expressly for the manufacture of ordnance; a kind of steel that was much sought after abroad for military purposes, and from which I had, up to that period, made twenty-eight guns for foreign governments; a kind of steel that could be made in masses of 5 to 10 tons in less than half an hour, at a cost of £10 per ton, if made from pure Swedish charcoal pig-iron. These important facts were not new facts—they were known to thousands of people. But this was the one opportunity that was left, after many others had failed, when by force of circumstances, I had Sir William Armstrong before me face to face, and also in the presence of a public audience; and I there made him look at these things, and hear my statements, which were backed with substantial proofs on the table before him, such as could not be denied or set down as exaggerations. But my efforts were again entirely fruitless.

In the early days of the Bessemer steel manufacture, many persons who had no love for steel, and saw in it a most formidable rival to iron, had with much perverted ingenuity raised a bogey to scare and alarm the uninitiated. They asserted that although many splendid specimens of steel were produced, the metal was very uncertain in its quality, and reliance could not be placed on it, as it had the fault of failing unexpectedly. Like all other trade prejudices, or mere creations of the imagination, this only required looking at steadily in open day, and in the light of well-ascertained commercial facts, to show how hollow and without foundation it really was. In fact, this crusade against steel was entirely unsuccessful in influencing engineers who took the trouble to inquire into the real facts. It did not prevent the use of thousands of steel railway tyres, which, by their great superiority, rapidly displaced the Lowmoor welded tyres previously almost exclusively relied on. It did not prevent hundreds of steam boilers being made of Bessemer steel for private establishments, nor did it stand in the way of our locomotive engine-boilers being made of this material, in place of the high-class Yorkshire iron previously used for that purpose. Those clever people who set up this bogey of "uncertainty" in the quality of steel, simply for self‑protection, dared not assert that occasional bars of bad iron were unknown in commerce. The same persons who so strenuously advocated the building up of heavy masses of wrought‑iron could not pretend that the welding of many parts to form a whole was exempt from uncertainty and failure. It was even then a well-known fact that the welding of large masses of wrought iron involved more risk and uncertainty in its results than any other of the processes used in the manufacture of iron.

The question of the uncertainty in quality of the Bessemer mild cast steel simply resolved itself into a question of cost, because the quality was easily ascertainable in the earliest stages of its manufacture, and thus the loss of working up bad material into a costly finished article could be most easily avoided. To show this fact, I will take as an example the production of a Bessemer steel gun-tube, suitable for a 40-pounder gun of 4.75 in. calibre. Such a forging would simply be a plain solid steel cylinder, 8 in. in diameter and 10 ft. long, weighing 15 cwt. and 20 lb., and, with a flat test piece formed on each end, it would weigh 15½ cwt. A 10-ton converter would cast eleven ingots of 1 ft. square, weighing 18½ cwt. each, and if 3 cwt. were cut off the top end of each of these ingots to ensure absolute soundness of the part used, we should then have the requisite weight in each ingot to make the gun-tube, and 3 cwt. of scrap metal worth something, but which may be discarded in this case. Now, if this forging, when tested by bending the flat bars formed at each end for analysis, should turn out not to be of the precise standard quality for use as a gun-tube, let us see what would be the loss. The highest quality of Swedish charcoal pig-iron would be used, costing from £6 10s. to £7 per ton (say £7), and with a small quantity of ferro-manganese, the 10 tons of steel ingots would not cost £10 per ton, and could be utilised for engine or tender axles, steam engine shafts, piston rods, plates or other articles. As the ingots were made of this pure Swedish iron, they could be sold for more than than their prime cost, at a time when steel axles and engine shafts, made from British iron smelted with coke, were sold at £16 to £20 per ton. But suppose, for the sake of argument, and to give no excuse for rejecting these figures, that 20 per cent. reduction was necessary to ensure the ready sale of the ingots, there would then be a loss of £20 on the 10 tons. Now, all experience showed that not one out of every ten charges converted was made of the wrong quality, and it is almost inconceivable that a converting-house could be so grossly mismanaged as to make one charge out of every five of the wrong quality. But if it had been so mismanaged, it would simply have diminished the output of the converting house 20 per cent. ; and at a period when railway bars made from British coke-iron were selling at £12 per ton, such Swedish steel ingots would surely have realised £8 per ton, entailing a loss of £2 on one-fifth of the steel made, thus bringing the cost per ton of ingots up from £10 to £10 10s. per ton.

It must be borne in mind that this particular manufacture of Bessemer steel had one most important element of certainty as to its composition or quality not possessed by any other iron or steel known in commerce at that period, viz., the contents of the converter when poured into the casting ladle, and well stirred by the revolving agitator, would cast ten separate ingots of a ton weight each that were absolutely identical in quality, so that after testing one of them, the other nine could be used with certainty. This absolute identity in quality was unattainable by any other system : a fact which none of those persons who watched with dismay the daily encroachment of steel on the domain of iron were able to deny.

H H

The 18-pounder gun exhibited on the occasion of Sir William Armstrong's visit to Sheffield sufficed to show that in the short period of eight
hours a gun-bock of forged steel could be obtained from pig-iron. The
gun-ends bent cold, which were placed on the table to illustrate my paper,
bore testimony to the quality and toughness of the steel of which this
gun, and many others, had been made. Some of these I have already
dealt with, and I have selected for illustration, in Figs. 69 and 70,
Plates XXVIII and XXIX, two more striking specimens from among
the number I displayed.

Month after month rolled on, and no application came from Woolwich
for any of the Bessemer steel, which Sir William Armstrong admitted
he had never tried for guns. Nevertheless, we continued making guns
to go abroad.

The managers of the International Exhibition of 1862, fully appre-
ciating the importance of this new steel process, allotted me a very large
space, measuring no less than 35 ft. by 35 ft., equal to 1225 square
feet area, with a free passage 8 ft. wide all round it. A photographic
reproduction of my exhibit is given in Fig. 71, Plate XXX; it was
taken from an imperfect print made in the dark days near the close
of the Exhibition. It will be seen, however, that on a pedestal in front
of the central case is a rough forging of a 24-pounder gun with trunnions
formed out of the solid; inside the case is a finished 18-pounder gun, a
large and massive gun-hoop, etc., etc. There were also shown an
embossed steel shield, a star formed of bayonets, a group of revolvers,
cavalry swords and sheaths, military rifles, projectiles, a model breech-
loader, etc. On the external counter was placed a 4-inch diameter bright
steel shaft, 35 ft. long, in one piece, steel hydraulic press cylinders,
railway axles and carriage and engine tyres, a circular saw, 7 ft. in
diameter, every size of steel wire for ropes, steel bars and rods of all
sizes, and, in fact, an immense number of other interesting objects that
would fill a long catalogue.

The enumeration of these objects may seem commonplace enough
at the present day, but at that time they were undoubtedly marvellous
industrial results, and an immense excitement was caused by this display
of the new steel, which attracted engineers, ironmasters, and steel manu-

facturers from every part of Europe and America. Indeed, I exhibited beautiful specimens of steel made, under my patents, both in France and in Sweden.

I cannot refrain from comparing the small effect which my exhibit made upon the stolid inertness and indifference of the War Office, with the results it produced on the active mind and business instincts of one of the most important and most intelligent Lancashire engineers, an employer of some 5000 workmen. I refer to the late John Platt, M.P. for Oldham, where his large works were situated. This successful engineer visited the Exhibition on the opening day, and at once grasped the importance of my steel process from an engineering point of view; he pointed out its value to some of the heads of departments in his own works, who made the same high estimate. Mr. Platt, on the fourth or fifth day after the opening of the Exhibition, had a long interview with me, and said that he himself, and nine of his immediate friends and connections, wished to join in the purchase of one-fourth share of my patent. It was very natural that I should entertain an offer to recoup me for my large expenditure, and at the same time to afford a handsome profit, thus avoiding some of the risks to which all patent property is subject. But I had so strong a faith in the great future of my invention—a faith based on proved facts—that I felt bound to decline his offer, as I desired myself and my friend and partner, Robert Longsdon, to retain the absolute control of the patents, and thus be able at any time to raise or lower my royalties as I thought best. Mr. Platt, however, approached me again on the subject a few days later, saying that he and his friends were prepared to waive all right to control the patents so long as I retained one half, trusting that in the interests of that half I should do what was best for myself, and consequently what was best for them. This proposal quite met with my approval, in principle; that is, I was willing to enter into a bond with these gentlemen to hand over to them five shillings out of every pound paid to me by way of royalty by my licensees, the patents, price of royalties, etc., being governed by myself and my partner, Longsdon, just as though no such bond were in existence.

It therefore became only a question what the purchase price should be. To fix this, these ten gentlemen met us by appointment at the

Victoria Hotel, Westminster, about ten days after the opening of the
Exhibition. Mr. Longsdon left this delicate negotiation entirely to me,
and at the meeting I pointed out the peculiar difficulties we had met
to discuss. The thing to be purchased could neither be measured nor
weighed ; there was no analogous case to use as a guide or precedent ;
the patents might bring in a very large sum of money, or a
quibble of the law, or some other invention, might render them of little
value. Thus I had to propose a sum which might fairly be estimated
as a very profitable purchase for them, if all went well. At the same
time I was to realise a considerable present profit, while my future
action was wholly untrammelled, as my partner and I still retained
three-fourths of the whole property intact. Having thus briefly reviewed
the position, I said : "Gentlemen, we have thought this matter thoroughly
over, and I have come to a fixed resolution to accept a certain sum
in cash for this one-fourth part of the proceeds of my invention ; or,
otherwise, I will keep the whole and run my course uninsured. I must
therefore beg you to give me a distinct " Yes " or " No " to my offer. I
cannot haggle, for no one can demonstrate it is worth so much more
or so much less. I have fixed on an easily divisible sum among ten
gentlemen, which has all the advantages of round numbers. I have
fixed on £50,000 as the purchase price." Mr. Platt, who occupied the
chair, said : " We have heard your definite proposal, and if you will be
so good as to go into the adjoining room for a few minutes, we will
discuss your proposition, and give you a reply."

I then left the meeting ; after a lapse of not more than ten
minutes I was called back, when the chairman said they had talked
the matter over, and had unanimously agreed to accept my offer.

In the course of a few days, a formal and satisfactory document
was prepared by the joint industry of the solicitors on both sides, and
Mr. Longsdon and I were invited to dine with Mr. Platt and his friends
at the Queen's Hotel, Manchester. This was about three weeks after
the opening of the Exhibition. We had a very pleasant and friendly
dinner ; we were all mutually pleased with our bargains, and in a
bumper the company drank to the success of the new steel process, and
long life to the inventor, a toast to which I had the pleasure of

responding. Then came the formal reading of the bond, and its signature, after which there was still another interesting ceremony, which was performed in a genuine Lancashire fashion, each gentleman producing from the depth of his pocket a neat little roll of Bank of England notes of the value of £5,000, which was handed to us in the proportion of our respective shares, viz., £40,000 to myself and £10,000 to my partner Longsdon. The meeting then broke up in a most cordial manner, and the friendly feeling thus inaugurated was never for one moment clouded by a single expression of dissent or dissatisfaction in the whole ten years of our business intercourse, during which time I had the great pleasure of handing over to my friends their 5s. in the £, amounting on the whole to something over £260,000. As a further testimonial of our mutual friendship and regard, Mr. Platt presented to Lady Bessemer, in his name and those of our Manchester friends, a portrait of myself painted by Lehmann, and exhibited in the Royal Academy.

I have mentioned these facts because it is almost impossible to conceive higher testimony to the value of my processes than this purchase of a share of the invention with all its risks; a testimony which was justified by the results obtained, while our War Office officials did not venture to purchase even a few ingots of our steel sufficient to make half a dozen 40-pounder gun-tubes.

At last there came a time when the British Government abandoned welded-up iron gun-tubes, and they and Sir William Armstrong parted company (on February 5th, 1863), the Government paying the Elswick Ordnance Company £65,534 4s. as compensation for breaking the contract with that Company, as well as paying the other sums which are given at page 5 of the Report of the Select Committee on Ordnance, ordered by the House of Commons to be printed, July 23rd, 1863. The following copy taken from that Report accurately gives these amounts.

The whole supply of Armstrong guns and projectiles has been obtained from the Royal Arsenal at Woolwich and the Elswick Ordnance Factory.

1st. The sum of 965,117l. 9s. 7d. has been paid to the Elswick Ordnance Company for articles supplied.

2nd. After giving credit for the value of plant and stores received from the Company, a sum of 65,534*l.* 4*s.* has been paid to the Elswick Ordnance Company as compensation for terminating the contract.

3rd. The outstanding liabilities of the War Office to the Elswick Ordnance Company, for articles ordered, amounted on the 7th May last to the sum of 37,143*l.* 2*s.* 10*d.*
The whole of these payments and liabilities amounts to the sum of 1,067,794*l.* 16*s.* 5*d.*

4th. The sum of 1,471,753*l.* 1*s.* 3*d.* has been expended in the three manufacturing departments at Woolwich on the Armstrong guns, ammunition, and carriages, making altogether a grand total of 2,539,547*l.* 17*s.* 8*d.*

On May 4th, 1862, Sir William Armstrong was examined by the Select Committee on Ordnance, on which occasion the Right Hon. William Monsell occupied the chair; in reply to his question, No. 3163, Sir William Armstrong gave a somewhat lengthy description of his system of making guns of coiled iron tubes, etc. He also gave his reasons for not using steel instead of iron, which he admitted was too soft for that purpose.

The reason which Sir William Armstrong gave to the Ordnance Committee for not using the superior metal quite astounded me when I saw the printed report of his evidence before that Committee. I read it over and over again, each time with increasing astonishment; a feeling which will, I doubt not, be shared by every person who has read the preceding pages.

The three quotations herewith reproduced are part of Sir William Armstrong's evidence, as printed in the Report of the Select Committee of the House of Commons, 1863.

From the very first I saw, and I still feel, that steel is the proper metal for the barrel of a gun, if it can be obtained, and my only reason for not persevering in the use of steel was the difficulty of getting it of suitable quality. There can be no question that wrought iron is too soft, and that brass is still more objectionable than wrought iron, and if we can only obtain, with certainty and uniformity, steel of the proper quality, there can be no question as to the expediency of using it.

5004. Then, in speaking in the answer to which I have referred you, of "the gun with the barrel of steel," you did not intend to rely on that as the difference between the two guns?—I merely stated it as the fact. We could not get steel suitable for the barrels; the steel was not to be had; I would have used it without hesitation if I could have got it. I am quite sure that no patent Captain Blakeley held would have been adequate to prevent my using steel.

5007. Then am I right in inferring, that your system of construction "as it was then and is now," involved an internal lining of steel, with twisted cylinders of wrought-iron tightly contracted ? When the steel is to be obtained. I do not think I can possibly be more explicit than I have been already; I have stated that if the steel can be obtained, it is unquestionably the best material, and it is the proper mode of construction; but if steel cannot be obtained, the alternative is to use coils for the barrels.

It was only natural that I should be astonished at such a declaration, for I could not forget the numerous proofs of the fitness of Bessemer mild steel, which I had given to Sir William Armstrong's immediate predecessor, Colonel Wilmot, at Woolwich; nor could I forget the display I had made of crushed gun-tubes, the malleable iron gun produced, *in one piece* without weld or joint, and other examples of steel, on the occasion of the reading of my Paper on the manufacture of iron and steel, at the Institution of Civil Engineers; to say nothing of the indisputable proofs of the suitability of Bessemer mild steel for the manufacture of ordnance, brought before the Institution of Mechanical Engineers at their meeting at Sheffield, on July 31st, 1861.

With regard to the reasons assigned by Sir William Armstrong, in his evidence before the Ordnance Select Committee, for persisting in the use of welded-iron gun-tubes, I must remain absolutely silent; such admissions and declarations as he there made do not admit of discussion, and hence I dismiss for ever this unsatisfactory episode in the long struggle I had maintained to induce the British Government to avail themselves of the immense advantages which my invention offered.

In closing this portion of my history, I have the satisfaction of feeling that I have done my duty to my country, untainted by personal and selfish motives; and in this hard struggle I have had the satisfaction of seeing the survival of the fittest successfully demonstrated by the universal acceptance of mild cast steel for the construction of ordnance.

CHAPTER XVII

CAST STEEL FOR SHIPBUILDING

AMONG the almost endless variety of useful purposes to which Bessemer mild cast steel has been applied, there is none more important than its employment in the construction of steam ships for the conveyance of passengers and merchandise, and also of ships of war and fast cruisers. The great strength of this material, as compared with the best brands of iron; its even and homogenous character; its great power of elongation before rupture; and its unequalled amount of elasticity under severe strains; all combine to form a material not only admirably adapted for the plates, beams, and angles of the ship itself, but equally suitable for the construction of her masts and spars, her boilers and her machinery; and for the still more important manufacture of the heavy armour-plates necessary to protect ships of war from the assaults of the enemy.

From a very early period I had become deeply impressed with the importance of the application of my new steel to shipbuilding, and my first impulse was naturally to try and force my own conviction on the British Admiralty, and induce them to employ it in the construction of ships of war. But the remembrance of my treatment at Woolwich came upon me as a warning, for there I had given, at much cost and labour to myself, the most irrefutable proofs of the perfect applicability of my mild steel to the manufacture of ordnance, and all these proofs had been overlooked and thrown aside by the Minister of War in favour of an inferior substitute for steel. This experience determined me not to be foiled a second time by attempting to convince the "How-not-to-do-it" Government official. I therefore preferred to await the more certain and reliable action of mercantile instinct. Private shipbuilders, I had no doubt, would soon find out the merits of steel, and feel a personal interest in its adoption. Boiler-makers, I also felt assured, would recognise its value, and use it instead of iron, many years before the Admiralty officials would wake up and become conscious of the advantages it possessed over the weaker material. Nor did I have long to wait for the verdict of practical men on the value of Bessemer mild cast-steel plates, as applied to the construction of steam boilers; an application which in itself is a sufficient guarantee of their high quality, and their superiority over plates made of the highest brands of British iron. Every person connected with the iron trade is well aware that the articles known to the trade as boiler-plates are superior in quality to those known as ship-plates; in fact, iron ships were never built with the high-class iron used for boilers.

I have already stated that, on the occasion of the Institution of Mechanical Engineers holding one of their annual meetings at Sheffield, in July, 1861, under the presidency of Sir William Armstrong, I read a paper on "The Manufacture of Cast Steel and its Application to Constructive Purposes." I now refer again to that paper, simply to quote a few lines from the speeches made in its discussion, by two eminent practical Lancashire engineers, in order to show what had been done *up to that early date* in the application of the new steel to the construction of steam boilers. This discussion, be it observed, took place no less than fourteen years prior to the date on which Sir Nathaniel Barnaby, then the Chief Naval Architect at the Admiralty, read his paper before the Institution of Naval Architects, in which he criticised adversely the use of Bessemer steel plates for shipbuilding and boiler-making. Hence it will be interesting to see how far this material had already been employed for boiler-making.

At this meeting of the Institution of Mechanical Engineers above referred to, Mr. Daniel Adamson,* the well-known engineer and manufacturer of steam boilers, whose works were at Hyde, near Manchester, exhibited some beautiful specimens of deep and difficult flanging in some fire-boxes for locomotive boilers. Mr. Adamson said he had already used 200 tons of boiler-plates made from the new steel, and

* Died January 13th, 1890.

I I

was about to procure a further supply of 70 tons. He found the
metal of excellent quality, and of regular character throughout, and
it was an admirable material for working. The flanged fire-box plates
shown were duplicates of a number that he had used in the manufacture of boilers for very high pressure, with the most satisfactory
results. The metal flanged beautifully, and was like copper in this
respect,* but with the advantage that it was not so liable as copper
to be damaged by overheating. He could fully confirm the statements
given as to its strength, having tested it severely. As a precaution
every plate had been ordered with a 1-in. margin all round, which
was sheared off, and bent double, as a test of the quality of the plate.
The metal was found to stand this test well, and bent double, like the
specimens exhibited, without cracking at any part of the surface.

The other engineer referred to, who took part in the discussion of
my paper, was Mr. William Richardson, the active practical partner
in the firm of Messrs. John Platt and Company, Engineers, Oldham,
in which firm Mr. Richardson had, for over twenty years, the direction
and supervision of some five thousand workmen. In the course of the
discussion on my paper, Mr. Richardson said, " He had made trial
of the Bessemer steel plates for some time in boilers at Messrs. Platt's
works at Oldham, where, some years ago, a higher pressure of steam
was adopted than was then usual. At that time they frequently found
distress at the joints of the boilers, and had adopted double riveting ;
the furnace plates were frequently blistered, though of a good make
of iron. Subsequently three boilers were made of plates of ' homogeneous
metal,'† which had been at work three years, but since the Bessemer
steel had been produced at a cheaper rate and equally reliable in strength
and quality, they had used it extensively, and had now six boilers constructed of the new plates. They had no more trouble from blistered
plates and strained joints, while a great saving was effected, owing to
the reduced thickness of the metal requiring less fuel to produce the

---

* Copper is thus frequently referred to by metallurgists as an example of extreme
toughness.

† A beautifully tough, but very expensive kind of iron, made of charcoal bar-iron melted
in crucibles, and first introduced by Messrs Howell and Company, of Sheffield.

same heating power. * * * They had had only two years' experience of the new plates, but during that time the results had proved thoroughly satisfactory."

This latter remark of Mr. Richardson shows the high opinion formed, from personal observation, of the new steel, at least two years prior to the date at which it was spoken. Thus, as far back as July, 1859, Mr. Richardson had erected, at the works of Messrs. John Platt, of Oldham, no fewer than six Bessemer steel boilers, of 6 ft. 6 in. in diameter by 30 ft. in length, each having one flue-tube of 3 ft. 10 in. in diameter, with plates $\frac{5}{16}$ in. thick, and working at a pressure of 85 lb. per square inch.

These facts will serve to show the high reputation acquired by these mild cast-steel plates, even at this early period: a reputation that steadily increased throughout the country, and which, in the early part of 1863, had so fully convinced the firm of Messrs. Jones, Quiggins, and Company, shipbuilders, of Liverpool, of the suitability of steel as a shipbuilding material, that they determined to put it to a practical test by building a small steam-ship. For this vessel the firm of Henry Bessemer and Company, of Sheffield, produced the steel, which was afterwards rolled by Messrs. Atkins and Company, of Sheffield, this being the first of many extensive orders given us by this enterprising firm for the Bessemer mild cast-steel ship-plates.

I am indebted to the Chief Surveyor of Lloyd's for the following list of Bessemer steel ships, classed by them during the years 1863, 1864 and 1865.

Name of Vessel.	Tonnage.	Built in
Screw steam-ship, "Pelican"	329	1863
Screw steam-ship, "Banshee"	325	1863
Screw steam-ship, "Annie"	330	1864
Paddle-wheel steam-ship, "Cuxhaven"	377	1863
Sailing-ship, "Clytemnestra"	1,251	1864
Paddle-wheel steam-ship, "Rio de la Plata"	1,000	1864
Paddle-wheel steam-ship, "Secret"	467	1864
Screw steam-ship, "Susan Bernie"	637	1864
Paddle-wheel steam-ship, "Banshee"	637	1864
Screw steam-ship, "Tartar"	289	1864
Paddle-wheel steam-ship, "Villa de Buenos Ayres"	536	1864

Name of Vessel.				Tonnage.		Built in
Sailing-ship, "The Alca"	...	...	...	1,283	...	1864
Paddle-wheel steam-ship, "Isabel"		...	...	1,095	...	1863
Paddle-wheel steam-ship, "Curlew"		...	...	1,095	...	1865
Paddle-wheel steam-ship, "Plover"		...	...	410	...	1865
Screw steam-ship, "Soudan"	...	...	...	184	...	1865
Paddle-wheel steam-ship, "Midland"		...	...	1,622	...	1865
Paddle-wheel steam-ship, "Great Northern"			...	1622	...	1865

At the time when the "Clytemnestra," a steam sailing-ship of 1,251 tons, was in course of construction, it was found by the builders that want of capital would prevent it being finished, and result in the shutting-up of the shipyard. I was so anxious that the application of my new steel to shipbuilding should not receive a sudden check, that I was induced to lend the firm £10,000, to put their financial affairs in order. This, however, did not effect the desired object, and, unfortunately for me, the prior claims of secured creditors converted my loan into an absolute loss. It had, however, one good effect; it enabled the firm to continue for a while; and by the end of 1865 no less than eighteen steel ships, aggregating 13,489 tons, had been built of Bessemer steel, classed at Lloyd's, and duly placed on the Register. Every person connected with shipping is fully aware that the careful examination of Lloyd's experienced surveyors is an absolute guarantee of the strength and structural good qualities of all ships passed by them. But these steel ships had more than the ordinary credit of going through this ordeal, for, on a thorough investigation of the whole subject, Lloyd's surveyors became so satisfied of the much greater strength and reliability of Bessemer steel, compared with ordinary commercial iron ship plates, that they considered it unnecessary for shipbuilders to use the same thickness of steel that was required for iron; therefore, they permitted a reduction of 20 per cent. to be made in the weight of steel used in the construction of every steel ship: a concession of vast importance for high speed or great carrying capacity. Thus, if a ship of certain size and form would require say, 1,000 tons of iron for the construction of its frames and shell, Lloyd's would give the same class to a steel ship of precisely the same form and dimensions, containing only 800 tons of steel, and therefore capable of carrying 200 tons more merchandise than could an iron ship of the same form and size. It is difficult to conceive a higher testimonial to the strength and fitness of Bessemer steel for shipbuilding than is afforded by this reduction of 20 per cent. by Lloyd's. Prior to the construction of steel ships at Liverpool, in 1863, I had introduced the last of the important improvements in my steel process, by inducing Mr. Henderson, of Glasgow, to manufacture ferro-manganese for me, so as to produce steel of exceptional mildness for plates and rivets. Hence, at that date, 1863, Bessemer steel was regularly made of as high a quality as it ever has been, or can be, made. Thus I established my claim to have successfully introduced the use of mild cast steel for the construction of ships of every class and description no less than thirteen years prior to the construction of the first Siemens-Martin steel-built ship, the sailing vessel "Stormcock," 466 tons, built in 1878, and registered at Lloyd's.

It will be seen from the foregoing that I had formed a pretty accurate estimate of the inertness and inactivity of the British Admiralty, when I decided on not wasting my time in endeavouring to awaken them to a sense of the vast national importance of employing mild cast steel for shipbuilding.

Private shipbuilders and shipowners had, as I felt assured they would, availed themselves largely of the many advantages possessed by this material, and had set an example of alertness and activity to the officials of the Admiralty, an example which they wholly disregarded. Thus, year after year rolled by, and still there were no signs of the Admiralty waking up to the consciousness of the great metallurgical revolution that was rapidly spreading over Great Britain and the whole continent of Europe, and that had already extended in full force to the energetic people of the United States. In fact, everywhere steel was replacing iron for innumerable structural purposes, varying from viaducts and bridges of large span, down to such small items of domestic hardware as milk-cans and saucepans.

After ten years of indifference on the part of the Admiralty, it was discovered that, notwithstanding the fact that the Bessemer process was a British invention, the more active and more enterprising officials of the French Admiralty had fully recognised the value of steel for the construction of ships of war, and that the French Government were far advanced with the large iron-clad, "Redoubtable," then being built of steel at L'Orient, and that they were also pushing forward two other large steel vessels of war, the "Tempête" and the "Tonnerre," which were then being built of steel in French ports. When this important fact came upon our quietly-sleeping Admiralty officials, then, and not until then, did they rub their eyes, and wake up sufficiently to recognise their position. They knew that this important fact could not long be concealed from the public press, and would thus come to the ears of John Bull, who is apt to demand a scapegoat when he finds that his country has allowed itself to be beaten in the race with other nations. Possibly it was felt by the Admiralty that some reason or other ought to be advanced for their not having commenced to build a single steel war ship, while our nearest neighbour had nearly completed three magnificent steel ironclads. Whether this surmise be accurate or not, it is certain that, with the consent of the Admiralty, Sir Nathaniel Barnaby, then the Chief Naval Architect of the Royal Navy, read, in 1875, a paper on "Iron and Steel for Shipbuilding," before the Institution of Naval Architects, in which paper the alleged "uncertainties and treacheries of Bessemer steel in the form of ship and boiler plates" were explained to the public. This comprehensive summing up of the uncertain quality and undesirable characteristics of the material was still further emphasised by Sir Nathaniel Barnaby holding up to the meeting an isolated example of the failure of a thin piece of plate metal, said to be a part of a Bessemer steel ship-plate, which had cracked when it was bent to a very small angle. As represented in Fig. 72, Plate XXXI., this shocking example proved too much ; it was, in fact, so bad a plate that, if originally made of such an unheard-of quality, it could never have been either rolled or sheared in the makers' works without proclaiming its utterly valueless character to every workman engaged in its manufacture. It must not be forgotten that it is physically impossible for the Bessemer process to produce a single isolated plate of such a bad quality, for the simple reason that Bessemer steel is never made in less than 5-ton batches, every part of each "blow" being equally good or bad. Now, after deducting 20 per cent. for waste in shearing, these five tons of homogeneous fluid steel will produce twenty-three ship-plates, 8 ft. long by 3 ft. wide and ⅜ in. in thickness. All of these twenty-three plates must, therefore, be equally good or bad, so that one bad plate alone could not be made, though any number of good plates may be spoiled by an ignorant, careless, or designing workman. The exhibition at a public meeting of such an unheard-of specimen of steel plate, and the proclamation of the "uncertainties and treacheries" of Bessemer steel, together with other damaging statements, by a person holding high authority, compels me to discuss the above-named paper at some length, and in justice to myself, to show that Bessemer steel is now and was then, in reality, a metal immensely superior to ordinary puddled iron, and that the example exhibited at the meeting in no way represented its true character and properties.

In order to clearly understand this question of bad plates, it is important to bear in mind that the iron plates used by shipbuilders were infusible in any of the heating furnaces that were to be found in ship-yards at that date. Hence an iron plate worker could leave an iron plate in the furnace, and make it very hot with impunity. But cast steel, as its name implies, has undergone fusion, and if ever it again be subjected to an unnecessarily high temperature, approaching its point of fusion, its molecules rearrange themselves, and the valuable qualities conferred on the cast ingot by hammering and rolling are lost in proportion to the amount of overheating it may have been subjected to ; so that, at a temperature quite possible to be given to it by a careless or ignorant workman, it becomes almost like the normal unwrought ingot from which it was formed. But this property of cast steel is so well understood by the practised *steel*-smith that he will pass hundreds of plates, or other articles, through any of the processes of heating in the furnace, tempering, hardening, or annealing, without the smallest injury to any one of them. It is the unpractised *iron*-worker, who does not understand the properties and mode of working steel, who makes mistakes of this kind.

It must also be observed that neither at the date about which I am writing, nor at any subsequent date, has it been possible to make cast steel which could not, either by ignorance, carelessness, or design, be rendered unfit for use by overheating it. Such liability to damage is not peculiar to steel made by the Bessemer process, since this quality is common to cast steel, however manufactured. When the molten cast iron in the Bessemer converter has been decarburised by blowing air through it, and has been poured into an ingot mould, the Bessemer process is complete; and such an ingot, like every one made in crucibles, or by the Siemens or open-hearth process, may be treated properly and make an excellent plate, or it may be treated improperly and be rendered worthless. The Bessemer process, like all others, may also make bad steel, if raw material of inferior quality be used in its manufacture. Sir Nathaniel Barnaby neglected to use the most perfect, and, at the same time, the only possible, means at his disposal of proving beyond dispute if the particular piece of plate, which he held up to the meeting, owed its bad quality to the Bessemer process, or to improper treatment after it had left the converter in a pure state. If he had had this sample of steel carefully analysed before he condemned it publicly, he and his audience would have known whether it contained such an amount of phosphorus, sulphur, or any other deleterious matter, as would account for the extraordinary cracking at so slight an angle, or whether the steel was free from these deleterious matters; or if it was of excellent quality when it left the converter, and had been spoiled afterwards by its treatment in the shipyard. Unfortunately, nothing was told us in this incomplete paper as to how, or by whom, this little sample was prepared for exhibition. Was the workman who made it a *steel-smith*, or was he an *iron-worker*, ignorant of the nature and proper treatment of cast steel?

If an actual plate, which had failed in the course of shipbuilding, had been shown at the meeting, it would have been much more satisfactory than a sample-piece, by whomsoever made, and such an actual plate could have been most easily produced, if such plates were common enough to justify what was said of the material in Sir Nathaniel Barnaby's paper. In the early part of this paper, the author damned Bessemer steel with faint praise; he said, "No doubt, excellent steel is produced in small quantities by the converter." Quite so; the small quantity of Bessemer steel made in England alone was, during the year in which this paper was read, over 700,000 tons, or more than *one hundred times* the total production of cast steel in Great Britain prior to the introduction of the process. These 700,000 tons were worth £6,000,000 or £7,000,000 sterling; so that the great commercial importance that Bessemer steel had attained at the date when Sir Nathanial publicly denounced it as a treacherous material, could not be hidden by calling it a "*small quantity.*" Or did Sir Nathaniel Barnaby desire his hearers to understand that only very little of this 700,000 tons was good steel? One per cent. of this *small quantity* would have supplied the Admiralty with 7000 tons, or enough to build two of the largest ships of war ever—up to that time—constructed; so the smallness of the quantity was no excuse for not using it.

Again, Sir Nathaniel Barnaby said: "Our distrust of it is so great that the material may be said to be altogether unused by private shipbuilders, except for boats, and very small vessels, and masts and yards." This statement was absolutely unwarranted.

We were also told that "Marine engineers appear to be equally afraid of it." Every Englishman who reads this will be surprised at this confession of want of courage, on the part of our marine engineers. However this may be, it was very gratifying to know that we had among us eminent practical engineers in Great George Street, who had the courage of their opinions, and under whose sanction and advice hundreds of thousands of tons of Bessemer steel were at that time being used for structural purposes. At the meeting, when this paper was read, there was present Mr. Francis William Webb, the well-known Chief Mechanical Engineer of the London and North-Western Railway, who was kind enough to bring for exhibition several test-pieces illustrative of the tests to which every plate of the locomotive boilers made under his supervision at Crewe was subjected before it was used. These test-pieces consisted of strips of boiler-plate, doubled up quite into close contact while cold; and other pieces of plate, each having a hole $\frac{3}{4}$ in. in diameter punched into it, which hole was then expanded or "drifted" out to $2\frac{1}{2}$ in. in diameter, by driving a conical punch or "drift," with a hammer, into the small hole first made.

Mr. Webb told those present at the meeting, that in their testing-house at Crewe they had 11,000 sets of these test-pieces, all duly

K K

stamped and numbered, each one referring to a corresponding number
stamped on 11,000 Bessemer steel plates that had been worked up into
locomotive boilers at Crewe, all of which had stood the ordeal of these
bending and "drifting" tests. Further, he said that Bessemer steel
had entirely superseded iron plates for boiler-making at Crewe, although
his company had previously bought the best iron that could be found
in this country. He also said that the London and North-Western
Railway Company had, at the time this paper was read, no less than
three hundred locomotive boilers in daily use, and that they were
building at Crewe rather more than six steel boilers every week. All
the steel plates were punched and worked, and then flanged into various
shapes with steel hammers; they were not tickled with copper hammers,
as Sir Nathaniel Barnaby had told his audience was a necessary
precaution in French shipbuilding.

I may add that the London and North-Western Railway Company
had, at that date, established extensive Bessemer steel works at Crewe, and
made their own steel; thus demonstrating what could be accomplished
for a great commercial company, advised by a thoroughly practical
engineer, not given to fear and doubting.

Now, I would ask any reasonable man what there was to prevent
the Admiralty from using such a simple and infallible mode of testing
every steel plate brought into the shipyard, the responsible officials
thus assuring themselves, beyond the possibility of doubt, that every
plate in their ships was of the high standard quality contracted for,
and so ending all the ridiculous suspicions of the treacherous nature
of a material that was being daily used so successfully? The simple
mode of testing used by the London and North-Western Railway
Company in 1875 is illustrated by Fig. 73, where (1) shows the irregular-
shaped plate as it leaves the rolls; (2) shows it when sheared on three
of its sides, a dotted line indicating where the fourth side is to be
sheared; and (3) shows the plate sheared on all four sides. Now, if
the Admiralty had ordered every plate delivered to them from the steel-
maker to have one side left unsheared, as shown in (2), their own people
could have sheared this one side, and cut three pieces, numbered respectively
5, 6, and 7, as marked on the sheared-off piece shown on an enlarged scale

at (4). Having done so, the prover would have taken (5) and hammered it into close contact while quite cold, as shown in (8); he might then have taken the piece marked (6), made it red-hot, and while at the proper

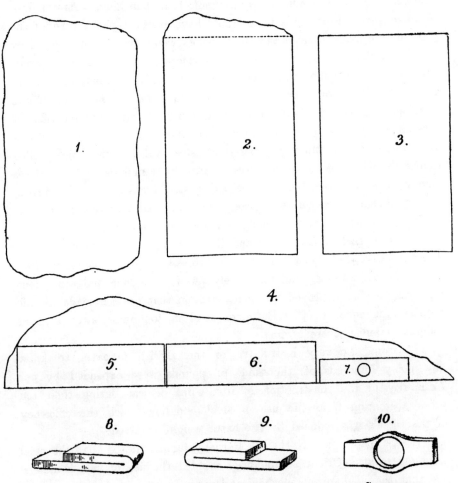

FIG. 73. SYSTEM OF TESTING BESSEMER STEEL PLATES ADOPTED AT CREWE BY MR. F. W. WEBB

temperature for working, hammered it into close contact, as shown in (9); these two tests would have proved or disproved the workable quality of the plate, both hot and cold. The piece marked (7) would then have had a $\frac{3}{4}$-in. hole punched in it, and a conical steel plug, or "drift," would have been driven into this hole until it was expanded to a given standard size, as shown at (10); this would have proved whether the plate would, or would not, bear punching. Any failure to stand these three usual tests would have justified the return of the plate to the manufacturer, and thus no loss would have been incurred by the Admiralty. With the certainty of perfect safety which these proofs afforded, the London and North-Western Railway Company, acting under the advice of their engineer, and under the responsibility of the directors, did not hesitate to stake the lives of many thousands of persons every day, for whole years together, daily transporting them over hundreds of miles of Bessemer steel rails, over which rolled thousands of Bessemer steel tyres, drawn by hundreds of locomotives having Bessemer steel boilers, steel axles, steel cranks, steel piston-rods, steel guide-bars, steel connecting-rods, etc., etc. All this went on hourly, weekly, and for years, and had been going on for ten years under the eyes of the British Admiralty and their officials. Mr. Webb and his directors were fully justified in this extensive use of Bessemer steel, for they had carefully and tentatively put it to a long and continuous practical test, and proved to demonstration that no iron made in this country was equal to this Bessemer steel in toughness, strength, and endurance under severe strains.

It would be very instructive to the British taxpayer to know how many hundreds of thousands of pounds were expended by our Admiralty in the construction of iron ships of war during their ten years' abstention from the use of steel, and how much the efficiency of the vessels was reduced by the extra weight involved.

In his paper, Sir Nathaniel Barnaby further stated that the steel shipbuilders at L'Orient scrupulously avoided the use of iron hammers, and that they had various mechanical devices for "coaxing and humouring this material." Why did not the author give the meeting some account of what had been done nearer home? Why did he steer clear of Liverpool, where the material of eighteen steel ships had been shaped and fashioned with steel hammers wielded by the powerful arms of the practised steelsmith, without any "coaxing and humouring?" The meeting was also informed that the ordinary steel angles in use at L'Orient cost £27 per ton, and the double-tee bars about £41 per ton ; and to this there was to be added the cost of such careful labour as he had described. But private shipbuilders and ship-owners were not deterred by the price of Bessemer steel from using it even ten years before the date at which this paper was written, when Bessemer steel was at least 30 per cent. dearer than in 1875. Would it not have been far better to have quoted the then prices of Bessemer steel in England, instead of giving the absurdly high prices said to obtain in France?

I was present at the reading of Sir Nathaniel Barnaby's paper, when he held up to the meeting the piece of steel plate, which he called " the treacherous Bessemer steel," illustrated in Fig. 72, Plate XXXI. I invite my readers to compare this illustration with the various examples I have had photographed of Bessemer steel tests of gun-forgings (see Figs. 69 and 70, Plates XXVIII. and XXIX.) and with the 11,000 test pieces then accumulated at Crewe. But even more striking than these were the specimens I had prepared thirteen years before. Few would believe, without ocular demonstration, the extraordinary fact that a thin steel plate, 11 in. in diameter and $\frac{1}{16}$ in. thick, can be brought without rupture into the forms shown in Fig. 74, Plate XXXII., while Fig. 75, Plate XXXIII. shows various pieces of Bessemer steel, of our regular daily manufacture at Sheffield, tested cold. The former are examples of what is called " spinning ; " the cold steel plate is made to revolve in a lathe, and is pressed heavily upon by a blunt instrument as it revolves, just as a piece of soft clay revolving on a potter's wheel is pressed upon by his thumb and fingers, and is fashioned into a vase. As the thin cold steel plate revolves it yields to the pressure exerted upon it by the blunt instrument forced dexterously against it, and by degrees its particles are expanded in some directions and contracted in others, the solid cold steel flowing, like its prototype the potter's clay, and forming almost any variety of circular form which the work-man desires to give it. This wondrous change of position of the several parts of the original flat plate takes place without the smallest symptom of a crack or failure at any part of its surface. These examples demonstrate the marvellous toughness of the Bessemer cast steel when manipulated by a skilful workman.

The small vase on the left, $4\frac{1}{4}$ in. in height and $3\frac{1}{2}$ in. in diameter (Fig. 74, Plate XXXII.), is by no means a solitary example. It was one of a group of vases of various forms exhibited by me at the International Exhibition of 1862, that is, thirteen years before Sir Nathaniel Barnaby held up to the public meeting an isolated example of a maltreated plate as a representation of the "treacherous Bessemer steel," which he seemed to think was sufficient to excuse the British Admiralty for their ten years' indifference and apathy. During those ten long years, twenty-four Bessemer steel works had been erected in England alone, having 112 converting vessels with their powerful blast engines, steel-rolling mills, and other expensive plant and buildings, producing annually 700,000 tons of Bessemer steel.

At the time at which I write (1896), when we look into the present state of British shipbuilding, we find that merchant sailing-ships and passenger steam-ships are, in all cases, built of mild cast steel, which is admitted to be the most suitable of known materials for their construction. The way in which mild cast steel (Bessemer and open-hearth) has absolutely superseded iron is proved by the annexed extracts from Lloyds Register of British Shipbuilding for the year 1895.

During 1895, exclusive of war ships, 579 vessels of 950,967 tons gross (viz., 526 steamers of 904,991 tons and 53 sailing vessels of 45,976 tons) have been launched in the United Kingdom. The war ships launched at both Government and private yards amount to 59 of 148,111 tons displacement. The total output of the United Kingdom for the year has, therefore, been 638 vessels of 1,099,078 tons.

As regards the material employed for the construction of the vessels included in the United Kingdom returns for 1895, it is found that, of the steam tonnage, nearly 98.8 per cent. has been built of steel and 1.2 per cent. of iron. The iron steam tonnage is practically made up of trawlers, and comprises no vessel of more than 425 tons. Of the sailing tonnage, 97.0 per cent. has been built of steel, and 3.0 per cent. of wood. No iron sailing vessel appears to have been launched during the year.

Can any evidence more clearly show how the opinions of shipbuilders and shipowners, including the great passenger steam-ship owners and the Admiralty itself, have practically condemned iron as a shipbuilding material, with the consequent adoption of mild cast steel in its stead? In considering this evidence it must not be forgotten that mild Bessemer steel has not undergone the smallest alteration in manufacture, or any improvement in quality, since the completion of the eighteen Bessemer steel ships which were built at Liverpool. All that we did then we do now, and consequently the steel was as well adapted for the building of ships at that period as it is at the present day. From 1875 up to 1896—that is, a period of twenty years—the London and North-Western Company have built no less than 4000 Bessemer steel locomotive boilers, and during these twenty years of constant wear and tear, not one of these has ever been treacherous enough to burst. It may further be recorded that the London and North-Western Railway Company made all the Bessemer steel plates used for building their splendid fast Dublin and Holyhead passenger boats, which have so long been in constant use.

Although I have unavoidably used words of censure in speaking of that abstraction, the British Admiralty, no one can doubt that its officials are gentlemen of honour and integrity. They are liable, like the rest of humanity, to errors of judgment, while the traditions of the office, and the conditions under which they work, must tend to develop the conservative side of their character, and render them averse to experiment. But the course they pursue, whether it be technically the wisest or not, represents, I am sure, their honest opinion, and under no circumstances whatever would they stoop to the meanness of attempting to escape the consequences of any errors of judgment by making a scapegoat of the man through whose energy and perseverance the construction of mild cast-steel ships was rendered commercially possible, and whose invention has so greatly benefited the nation generally, and the British Admiralty in particular. Although that great department of the State failed for so long to recognise the merits of my steel, I have received the most ample recognition of the value of my inventions, alike from reigning sovereigns, from the learned societies, and scientific institutions of every State in Europe, all of which I acknowledge with every expression of profound gratitude.

CHAPTER XVIII

MANGANESE IN STEEL MAKING

IN giving a brief account of the more salient points of my life's history, I have deemed it desirable in some cases not to keep strictly to the chronological order of events, which would so entangle different subjects with each other as to render each incident difficult to be understood. I have therefore preferred sometimes to follow up the details of a series of connected events, and thus trace each subject to its natural conclusion, afterwards retracing my steps to recall other incidents which have thus been unavoidably displaced and left to some extent in the background. In accordance with this plan, I now go back to August, 1856, the month in which I read my—to me—memorable, paper at the British Association. I have mentioned on another page* that one of the immediate results of that paper was the application for a large number of patents by various people, either *bonâ-fide* though unpractical inventors, or others who deliberately planned to take advantage of the premature publication of my invention, by obtaining patents which should hedge me round and force me to divide with them the fruits of my labours. I think I have already made it clear that none of these efforts, *bonâ-fide* or otherwise, ultimately interfered with the triumphant development of my own patents. I am treading on very delicate ground, and although the events I have to refer to occurred many years ago, and are entirely done with so far as I am concerned, I feel that even now I may not be able to write without prejudice, much as I should desire to do so. I shall therefore confine myself entirely to a narrative of facts, and keep my own individuality and personal feelings as far as possible in the background.

As all I have to say in this Chapter bears intimately upon the

* See page 166 *ante.*

employment of manganese in the manufacture of cast steel, it will be in the natural order of things if I commence with a short review of the use of manganese in this industry.

In all the old published accounts of steel making, we find that steel works were located in places where manganesian iron was found. The ancient steel manufacturers of Styria produced the famous German "Natural Steel," which was so much used in this country before Sheffield had achieved its present high reputation. The manganesian iron ore, known here as spathose, or white carbonate, was in Germany known as stahlstein, a term indicative of its well-known special aptitude for the production of steel from the pig-iron known in Styria as spiegel eisen, then and now so much used in steel making. Towards the end of the eighteenth century and the beginning of the nineteenth, efforts were made in this country to combine the metal manganese with our British iron, and thus obtain pig-iron so alloyed with manganese as to give it those qualities which enabled the Germans to produce with their manganesian iron ores the finest steel in the market in those early days.

The first in the long list of inventors and patentees is one William Reynolds, who, in December, 1799, obtained a patent in this country "for a new method of preparing iron for the conversion thereof into steel," by employing oxide of manganese, or *manganese* (that is, metallic manganese), which was to be mixed either with the *material for making the pig* or cast-iron, *or with the cast iron*, to be converted into malleable iron in the finery, bloomery, puddling furnace or otherwise.

In either case, ordinary British pig-iron would be converted into manganesian pig-iron, or spiegeleisen, by the employment of Reynolds's patent process of preparing cast-iron "for its conversion into steel;" a process that has, at the time I am writing, now been public property for a period more than eighty years. Thus I had acquired, in common with all other persons in this country, the right to put oxide of manganese into the blast furnace with the iron-making materials, and so produce manganiferous pig-iron of any desired quality for conversion into malleable iron or steel. By the falling into public use of this long-expired patent I had, in common with all other persons, also acquired

L L

the right to add *manganese* (that is, the metal manganese) to cast-iron in order to render it more suitable for conversion into steel. I had the full right to use such alloyed cast-iron for making steel by my process; and by my patent, bearing date October 17th, 1855, I had the right, after the blowing process, to recarburise, or alter the state of carburation of, the converted metal by the addition thereto of molten pig-iron : a right of which no subsequent patent could deprive me.

This patent of Mr. Reynolds' started a host of imitators, who all laid claim to improve iron for steel making, or to improve steel when made, by alloying it with manganese. In case any of my readers should desire to see how these very " numerous inventors " tried to claim this valuable material for their own special use and advantage, I give below a list of most of them for easy reference to their respective specifications.

MANGANESE PATENTS.

Reynolds, Wm., A.D. 1799. " For a New Method of Preparing Iron for Conversion thereof into Steel." Oxide of manganese is to be mixed, either with the materials for making the pig, or cast iron, or with the cast iron, to be converted into malleable iron, in the Finery, Bloomery, Puddling Furnace or otherwise.

John Wilkinson, A.D. 1808. " Making Pig, or Cast Metal, from the Ore for the Manufacture into Bar Iron equal to Russian or Swedish," by manganese, or ores containing manganese in addition to iron-stone.

John Thompson, A.D. 1819. " Extracting Iron from Ore." The inventor smelts a mixture of iron ore and oxide of manganese.

Charles Schafhautl, A.D. 1835. " Manufacturing Malleable Iron," by using oxide of manganese.

Josiah Marshall Heath, A.D. 1839. "Manufacture of Iron and Steel." Manufacture of cast steel in a furnace with deficient fuel ; uses oxide of manganese. " Carburet of manganese may be used in any process for the conversion of iron into cast steel."

William Vickers, A.D. 1839. "Manufacture of Cast Steel." Wrought-iron borings and scraps are melted with oxide of manganese and carbon in crucibles to produce cast steel.

Charles Low, A.D. 1844. "Manufacture of Iron and Steel." Uses oxide of manganese and charcoal in pots.

John D. M. Stirling, A.D. 1846. "Alloys and Metallic Compounds, and Welding the same to other Metals." Molten cast iron and malleable iron and metallic manganese are used.

Moses Poole, A.D. 1847. "Manufacture of Cast Metal, Iron and Steel." Chromate of iron, oxide of manganese, etc., are used.

Alexander Parkes, A.D. 1847. "Manufacture of Metals containing Iron and Steel." To improve iron, some metallic manganese may be melted with it, etc.

John D. M. Stirling, A.D. 1848. "Manufacture of Iron and Metallic Compounds." Molten iron is mixed with 5 to 30 per cent. of scrap and one per cent. of manganese in a reverberatory furnace.

Josiah Marshall Heath, A.D. 1848. "Manufacture of Cast Steel." Granulated de-oxydised pure iron, mixed with manganese and carbon.

Richard A. Brooman, A.D. 1853. "Producing Castings in Malleable Iron." Manganese is used with wrought scrap in crucibles with carbon.

J. Leon Talabot, A.D. 1853. "Manufacture of Cast Steel." Blister steel is melted with oxide of manganese.

John D. M. Stirling, A.D. 1854. "Manufacture of Steel." Cast iron is repeatedly melted with iron oxides containing manganese.

C. A. B. Chenot, A.D. 1854. "Manufacture of Steel, Iron, and different Alloys." Iron ore is roasted, pulverised, and converted into a "sponge." It is then mixed with manganese, and fused.

Auguste E. L. Bellford, A.D. 1854. "Manufacture of Steel and Wrought Iron directly from the Ore." Iron ore is mixed with manganese and other substances, and is roasted. It is then melted in crucibles.

Charles Sanderson, A.D. 1855. "Manufacture of Iron." Sulphate of iron and manganese are added to molten iron.

Abraham Pope, A.D. 1856. "Manufacture of Iron." Iron ore, boghead coke, and oxide of manganese are melted in a reverberatory furnace.

Richard Brooman, A.D. 1856. "Manufacture of Cast Steel." Manganese and other materials are added to wrought iron to make steel.

John D. M. Stirling, A.D. 1856. "Manufacture of Steel." Manganese is used in the manufacture of steel from cast iron and iron ore.

Joseph Gilbert Martien, A.D. 1856. "Manufacture of Iron." Manganese is blown into molten iron.

William Clay, A.D. 1856. "Manufacture of Wrought or Bar Iron." Uses manganese.

Abraham Pope, A.D. 1856. "Manufacture of Steel." Manganese is used in the cementation process.

From the foregoing long list of claimants to the use of manganese in various ways in steel making, it must be evident that a knowledge of of its beneficial effect was widely known and highly appreciated nearly a century ago; but the most prominent, and the most practically successful, of all these patentees was a Mr. Josiah Marshall Heath, a civil servant under the Indian Government, who, noticing in the native Wootz steel-making of India the marvellous effect of manganese, conceived the idea of producing steel of superior quality from inferior brands of British iron by its use in the cast-steel process then extensively carried on in Sheffield. Heath came over to this country, and obtained a patent, bearing date the 15th of April, 1839, for the employment of carburet of manganese (that is, manganese in the metallic state) in the manufacture of cast steel: an invention of very great utility, as by its use cast steel of excellent quality could be produced from British iron that had been smelted with mineral fuel. Such steel possessed the property of welding either to itself or to malleable iron. The Sheffield cutlers were thus enabled to weld iron tangs on to the cast-steel blades of table-knives, and also to weld many other similar articles: a process which was not successfully carried on previous to the use of metallic, or carburet of, manganese under Heath's patent.

Mr. Heath, in his specification, does not confine his claim to the use of carburet of manganese in crucible steel melting, but distinctly claims "the use of carburet of manganese in any process whereby iron is converted into cast steel." All that Heath claimed lapsed and became public property when his patent expired, and the right to use carburet of manganese "in any process whereby iron is converted into cast steel" became common property by this publication, even if the patent were invalid. Heath was fully justified in making this general claim, because the results obtained depended on an inevitable chemical law, viz.: whenever metallic manganese, with its powerful affinity for oxygen, is put into molten iron containing disseminated or occluded oxygen, a union of the oxygen and the manganese follows as an inevitable consequence of their strong affinity for each other, wholly irrespective of the process employed in the manufacture of the iron or steel so treated.

In consequence of this successful invention of Heath's, no British iron that has been smelted with mineral fuel is ever made into cast steel in Sheffield without the employment of carburet of manganese. In the early days of Heath's invention, he supplied the carburet in small packages to his licensees; he made this by the deoxydation of black oxide of manganese mixed with coal-tar, or other carbonaceous matter, in crucibles heated in an ordinary air furnace. This was a costly process, and as the demand increased he suggested to his licensees that it would be cheaper to put a given quantity of oxide of manganese and charcoal powder into their crucibles, along with the cold pieces of bar iron or steel to be melted. These materials would, when sufficiently heated, chemically react on each other, and produce the requisite quantity of carburet of manganese in readiness to unite with the steel as soon as the latter passed into the fluid state. But Heath's licensees said, " This is not precisely your patent, Mr. Heath," and they claimed the right to carry out this suggestion without paying him any royalty. This was the cause of some eight or nine years of litigation, by which poor Heath was ultimately ruined, although his patent was established by a final decision of the House of Lords —alas! only too late; for Heath died a broken-hearted, ruined man, wholly unrewarded for his valuable invention.

Thus we see that both in the use of a carburet, and also by the use a mixed powder, consisting of oxide of manganese and carbon, Heath's process has been successfully and commercially carried on from the date of his patent, in 1839, up to the present hour.

Now, as my converting process was specially intended to deal with iron that had been smelted with mineral fuel, it will be readily understood how disastrous it would have been to me, if, by the action of another patentee, I had been prevented from using manganese; for if manganese, in some form or other, were absolutely necessary for the production of steel of good quality from iron smelted with mineral fuel, it would follow that if the use of manganese, in all its known forms and combinations when applied to the Bessemer process, could be patented, thus becoming the exclusive property of some other persons, then I should have been rendered utterly powerless, and my invention could not have been worked without the permission of the holders of these patents, and I should consequently have been wholly at their mercy.

This part of my narrative turns upon a patent obtained by Mr. Joseph Gilbert Martien, on September 15th, 1855, about a month before I took out my first steel patents. Mr. Martien's invention referred to improvements in the manufacture of iron and steel. He was at that time engaged at the Ebbw Vale Works, either on the staff of that company or as an independent experimenter. There would have been no need for me to refer to Mr. Martien's patent of 1855, but for subsequent events with which it was associated. It was really a valueless patent, and one which found no practical application; nevertheless, I must describe it briefly here, and I cannot do better than reprint some passages from Mr. Martien's specification.

Specification.   A.D. 1855.—No. 2082.

Martien's Improvements in the Manufacture of Iron and Steel.

This Invention has for its object the purifying iron when in the liquid state from a blast furnace, or from a refinery furnace, by means of atmospheric air, or of steam, or vapour of water applied below, and so that it may rise up amongst and completely penetrate and search every part of the metal prior to the congelation, or before such liquid metal is allowed to set, or prior to its being run into a reverberatory furnace in order to its being subjected to puddling, by which means the manufacture of wrought iron by puddling such purified cast iron, and also the manufacture of steel therefrom in the ordinary manner, are improved.

In carrying out my Invention, in place of allowing the melted iron from a blast furnace simply to flow in the ordinary gutter or channel to the bed or moulds, or to refinery or puddling furnaces, in the ordinary manner, I employ channels or gutters, so arranged that numerous streams of air, or of steam, or vapour of water may be passed through and amongst the melted metal as it flows from a blast furnace.

Thus we are distinctly told that the crude metal, after treatment in the gutter, is made into malleable iron or steel, by puddling in the ordinary manner, and not by the action of the steam, air, or vapour of water blown through it. In evidence of this I give another quotation from Mr. Martien's printed specification.

In treating the liquid or melted metal as stated, either as it directly comes from a blast furnace or from a finery fire, it is left in the form of pigs, plates, or in a granulated state, as may be desired ; or it may be conducted after such treatment directly and without material loss of heat to a reverbatory or other furnace or furnaces, and there subjected to intense heat and manipulation, and speedily converted into balls of malleable metal of iron and steel.

Martien was under the impression that he could, in part, supersede the ordinary finery fire, and render the crude iron more suitable for puddling, there being no new method or process of making malleable iron or steel described, or even in the most remote manner suggested, in this patent. In fact, in the last quotation, he tells us "the metal is left in the form of pigs, plates (that is, I presume, finer's plate metal), or in a granulated state, and if it be desired to make it into malleable iron or steel, the old process of puddling must be resorted to.

Possibly—I think probably—we should never have heard any more of Mr. Martien's invention had it not been for my Cheltenham paper of August, 1856. This paper, as we have seen, was fertile in suggestions to many would-be inventors. Amongst them in the records of the Patent Office we find, on September 16th, 1856, the applications for two patents connected with the manufacture of steel; one of them was taken out in the name of Robert Mushet and the other by Joseph Gilbert Martien. Six days later—that is, on September 22nd—two other patents were applied for by Robert Mushet, all four of the patents named being for the use of manganese in the manufacture of steel; and therefore they were, intentionally or otherwise, obstructive patents from my point of view. It must be remembered that these patents were applied for in the fourth and fifth weeks immediately following the reading of my paper at Cheltenham, at which period the whole iron trade of this country was in a state of extreme agitation and excitement in reference to my invention, which, at that moment, it was believed would effect a complete revolution in the iron industry.

Now, at this period, hundreds of men in Sheffield knew perfectly well that cast steel made from iron that had been smelted with mineral fuel was so much improved in quality by being alloyed with manganese, that such iron was never made into cast steel in Sheffield without the addition thereto of oxide of manganese and carbonaceous matter in the form of powder, which was put into the crucible or vessel in which cast steel was made. I have, however, already dwelt at length on Heath's invention, and have shown that his patents, which had expired long years before, had given to the world the free use of manganese in steel-making, and that its general application was a matter of universal knowledge.

Mr. Mushet's specification commences, "Now know ye that I, the said Robert Mushet, do hereby declare the nature of my said invention, and in what manner the same is to be performed to be particularly described in and by the following statement. When cast iron, including grey and white pig iron and refined metal, has been decarburised or purified by forcing air through or amongst its particles, either in the manner described in the specification of Letters Patent, dated the 15th day of September, 1855, granted to Joseph Gilbert Martien, or in any other convenient manner, with a view to convert it into malleable iron, etc." Now, it is clear that Martien did not blow air through molten iron, in order to convert it into malleable iron, but simply in order to prepare such cast iron for the after-process of puddling, by which process, and not by the air blown through it, it was to be converted into malleable iron. Further, any addition of pitch and oxide of manganese could not possibly convert into steel iron treated in the manner described in this patent of Martien so specifically referred to. There was at that time no commercially-known process of converting pig iron direct into malleable iron or steel, while still retaining its fluidity, except that patented by me, to which alone Mr. Mushet's patent could possibly be applied.

Any attempt to carry into practice Mr. Mushet's process, in the manner described in his patent of September 16th, 1856, would have been attended with great danger, and failure must have inevitably followed. In the manipulation of cast steel a small quantity of oxide of manganese and charcoal in the form of powder is put into the bottom of covered crucibles, nearly filled with cold broken-up steel bars. In such crucibles only a very small amount of atmospheric air is present, consequently the charcoal at the bottom of the covered crucible is not consumed. But as soon as a very high temperature is attained the carbon present gradually deoxydises the manganese, producing a fluid carburet of that metal, which unites with the steel as soon as the latter is fused. Now, Mr. Mushet proposed a somewhat different method of procedure. In this first patent for improvements in the manufacture of steel he stated that he preferred to use pitch as the carbon element, and having melted it, to put into the fluid pitch an equal weight of oxide of manganese in the form of powder, and to stir them well together. This mixture was to be allowed to cool, after which the brittle mass was to be reduced to a state of powder, and a quantity equal to one-fifth, or to one-tenth, the weight of the converted metal was to be used before, during, or after the conversion. Now, I have found on testing the specific gravity of this fine powder that a cubic foot of it weighs, as near as may be, $62\frac{1}{2}$ lb. (the same as water); hence the minimum charge of one-tenth of the weight of the contents of an ordinary 5-ton converter, or 10 cwt., would have a bulk of 13.9, or nearly 14, bushels—we may call it 13 bushels —while the maximum charge would be 26 bushels. Let us see how such an addition would behave if put into a Bessemer converter: a vessel with an interior lining brilliantly red-hot, and containing about 90 to 100 cubic feet of atmospheric air, at a temperature of about 1000 deg. Fahr. Certainly the first shovelful of such a highly-combustible powder thrown into this red-hot chamber filled with heated air would result in a dangerous gas explosion, and the instant rejection of the unreduced manganese powder present in the mixture. How, then, were the 13 bushels, or the 26 bushels, of this explosive powder to be got into the red-hot vessel? For even if it were possible to put in only the smaller quantity of 13 bushels, of this powder, it would form for a few minutes a huge bath of molten pitch, and it would require a very bold man to pour into it 5 tons of molten iron. The whole proposition is so absolutely unpractical that it requires no further comment.

Six days later (September 22nd, 1856), Mr. Mushet applied for another patent, which did not differ from the use of carburet of manganese as patented by Josiah Marshall Heath in 1839, for years used by Sheffield steel manufacturers, and in which patent Mr. Heath claims, fourthly, " the use of carburet of manganese, in any process whereby iron is converted into cast steel," to which I have previously referred. Now, it is obvious that this use of carburet of manganese, even if it could not have been claimed by Heath in his patent of 1839, had—as I have already stated—become, by mere publication, common property for a period of no less than sixteen years prior to Mr. Mushet's patent of September, 1856. The only plea that could possibly be advanced to justify Mushet's claim to a long-ago expired patent, which had been extensively used, was that the steel into which this carburet of manganese was to be put had been made by a different process. Now, let us see to what a deadlock all improved manufactures would be reduced if once we admit such a claim. Let us take an example which is strictly analogous. Some fifty or more years ago a great discovery was made by Mr. Pattinson, of Newcastle, who invented a most ingenious mode of extracting metallic silver from ordinary commercial pigs of argentiferous lead. Previous to this, silver had been almost exclusively obtained from silver ore, amalgamated with mercury, and afterwards refined, melted, and cast into ingots. There was no analogy whatever between the old process of extracting silver and that discovered by Mr. Pattinson. It had long previously been found that silver, though a very beautiful metal in appearance, was almost useless, either for the manufacture of utensils or for current coin, on account of its extreme softness; articles made from pure silver being easily bent or misshapen, and coins losing their impression by wear and abrasion. But it was fortunately discovered that an addition of 10 lb. of copper to every 90 lb. of silver, so hardened and strengthened the silver as to render it eminently adapted both for the manufacture of utensils, and also for current coin. This valuable alloy of copper and silver was accepted by all European Governments as a standard alloy to be stamped as " *silver*," and it has been in universal use for many years, just as steel *alloyed* with *carburet of manganese* passes current *as steel*, the alloy having also been in public use for many years. But the silver obtained from lead pigs by Mr. Pattinson's new process, like that obtained from silver ore, was, of course, too soft to be used in that state. Now, if some speculative patentee had, on the first announcement to the world of Mr. Pattinson's great discovery, rushed to the Patent Office to claim the sole right to put 10 per cent. of copper into silver obtained by Pattinson's process, under the plea that this silver had been produced by a new method, it is self-evident that the claim could not here be substantiated. To admit it would have been simply to destroy all future great inventions; the whole idea is too absurd to require further argument.

On September 22nd, 1856, Mr. Mushet took out yet another patent, claiming the employment of one of nature's compounds : a compound which steel - makers have used for the production of steel as far back as the history of steel-making extends, and which consists of iron found in the mine associated, or combined, with manganese and oxygen. Such ore, when smelted, produces a pig iron which contains iron, carbon, manganese, silicon, and generally phosphorus, sulphur, and other matters in small quantities, in combination with the iron. In his third patent Mr. Mushet did not mention my name, or designate any patent of mine, as the invention which he proposed to improve by the use of spiegeleisen ; and again the Crown and the public were told that, for the purposes of his invention, "the iron may be purified by the action of air in the manner invented by Joseph Gilbert Martien," as will be seen by the following quotation, reproduced from a printed copy of Mushet's specification, published by the Commissioners of Patents :—

The iron may be purified by the action of air, in the manner invented by Joseph Gilbert Martien, or in any other convenient manner. The triple compound or material which I prefer to use is pig or cast iron made from spathose ore, such ore and the pig or cast iron made from it containing a proportion of manganese, as well as the iron and carbon of which cast iron is usually composed.

If Mr. Mushet had taken the trouble to examine my early patents for the manufacture of steel, he would have found that the re-carburation of converted metal by the addition thereto of molten pig iron, was perfectly well understood, and had been patented by me more than a year prior to the date of either of his three manganese patents. Mr. Mushet also appears to have entirely overlooked my description of the several modes of making alloys in my process, as set forth in my patent, dated May 13th, 1856, sixteen weeks prior to the date of either of his three patents. This description was not given for the purpose of claiming any such alloys, but, on the contrary, its object was to disclaim the right to make alloys in my converter of any metals previously used in the trade to form an alloy with steel, and by such disclaimer and publication to prevent anyone from obstructing me in the free use of all such well-known alloys. In order to show what I really did say in my patent, I give a copy of the paragraph from my specification.

When employing *fluid metal for alloying with malleable iron or steel*, I pour it through an opening in the converting vessel, so that it may fall direct into the fluid mass below; but when employing metal in a solid form, I put it into the upper chamber through the door *g*, and allow it to acquire a high temperature, after which it may be pushed with a rod, through the opening *d*, into fluid iron or steel; and when using salts or oxides of metals for the purpose of producing an alloy or mixture with the iron or steel, I prefer to introduce such salts or *oxides in the form of powder at the tuyéres*, or to put them into the vessel previous to running in the fluid metal. I would observe, that I am aware that zinc, copper, silver, *and other metals have* before been combined with iron and steel otherwise manufactured, I therefore make no general claim thereto.

This paragraph clearly points out how such alloys are to be made, and I mention as examples, silver alloys, once used and greatly esteemed as "silver steel"; also alloys of zinc, patented as a detergent to carry off phosphorus from steel; I also mention copper as used in stereo metal for the manufacture of guns in Austria, and other metals *heretofore used in steel-making*. Surely, after I had thus published and disclaimed the use of any alloys previously used, no one could obtain a valid patent for alloying steel in my process with metals used to alloy steel then in common use.

The result of my early experiments in re-carburising confirmed the view I had taken from the first, viz., that it was best to stop the process as soon as steel of the proper quality was arrived at, for the continuation of the blowing process until malleable iron was obtained, had the disadvantage of consuming from 2 to 3 per cent. more iron than when steel was made; and, what was still worse, the metal got very much overcharged with oxygen, causing violent ebullition in the mould. I had an idea that this occluded oxygen could be got rid of without any addition to the metal. I had noticed that when superoxydised molten malleable iron came in contact with the cold-iron mould, it boiled and threw off large quantities of gas, as its temperature was reduced, the action being similar to that which takes place in the cooling of large masses of molten silver, which sputter and make a sort of little volcanic mound on the top of the ingot, owing to the spontaneous disengagement of occluded oxygen. In the case of steel, this throwing off of carbonic acid, or carbonic oxide, gas was a source of great unsoundness in ingots, and appeared to be a very important subject for investigation. I consequently had a small apparatus constructed, with a view of seeing how far this gaseous matter could be prevented from escaping in the form of bubbles by being surrounded with a dense atmosphere, to suppress ebullition; and also how far it could be removed by considerably lowering the pressure of the surrounding atmosphere, thus favouring ebullition and the removal of the gas from the metal.

I may here mention, incidentally, that these experiments were the starting-point of my patents for casting under gaseous pressure, and also under the pressure of an hydraulic plunger, acting direct on the fluid metal. Under this latter patent, I granted a license to Sir Joseph Whitworth to make his compressed steel. The experimental apparatus for removing gas *in vacuo* just referred to, was simply a short cylindrical vessel, on to which a conical cover was fitted; the flanges which formed the junction between the two were accurately surfaced, and formed an air-tight joint. At the top of the apparatus a small circular piece of plate glass was inserted, through which the eye could, by means of the light emitted by the incandescent metal, see distinctly whatever was going on inside the chamber.

This apparatus is shown in section in Fig. 76, page 270. Having converted some pig iron into highly-carburised steel by means of a fire-clay blow pipe, a crucible about half filled with this steel was put into the chamber. The pipe and stop-cock shown on one side of it were made to communicate with an exhaust pump, or with an exhausted vessel, the effect of which was at first to cause a few bubbles to rise to the surface of the metal; but only a comparatively gentle ebullition was produced, however high a vacuum was attained. If mild steel, however, was so treated a much more violent ebullition took place; and if a 20-lb.

crucible containing about 10 lb. only of wholly decarburised pig iron was put into the chamber, and a high vacuum was produced, the ebullition set up by the rapid escape of gas caused the steel to boil over the top of the crucible, and occupy the lower part of the chamber, as shown in the engraving.

Many experiments were made with this simple apparatus, and they convinced me at the time that it was far preferable to blow the metal only to the condition of steel, using the recarburising process to as small an extent as possible. Thus it happened that in my early patent

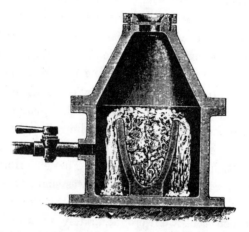

FIG. 76. EXPERIMENTAL APPARATUS FOR EXPOSING MOLTEN STEEL
TO THE ACTION OF A VACUUM

of October 17th, 1855, I described the recarburising process in the words which I reproduce from my printed specification, which dates more than one year prior to Mr. Mushet's patents.

During the decarbonizing process, the state of the metal may be tested by dipping out a sample with a small ladle, as practised in refining copper; if too much carbon is retained, the pipe G may be again introduced for a short time, or a small quantity of scrap iron may be put into it; but if too much carbon has been driven off, an addition may be made of some melted iron from the finery or cupola furnace: a little experience will, however, enable the workman to regulate his process so as to produce the different qualities of steel which he may require.

This quotation shows that, from the earliest date, I fully understood and appreciated the facility which molten carburet of iron gave for regulating the state of carburation of the converted metal; and if I used any kind of manganese pig iron for converting into steel, as I had a perfect right to do, the addition of some of this molten iron "from the cupola furnace" to my converted metal, would of necessity involve the recarburising, by the use of a "triple compound of iron, carbon, and manganese."

Now the particular manganese pig iron, called in Styria spiegeleisen, the use of which Mr. Mushet claimed by his patent, may in round numbers be fairly stated to consist of 4 per cent. carbon, 8 per cent. manganese, 2 per cent. of some half a dozen other elements, and 86 per cent. of iron. These proportions are by no means well adapted for the deoxydation of mild steel, and it is impossible to use such a metal when soft decarburised iron is desired, as steel, and not malleable iron, would be produced.

I have before stated, that in my earliest experiments the quantity of oxygen taken up by the metal was but small, if the process was stopped when the desired quality of steel was arrived at. But if I continued the blowing process until soft iron was produced I had a double disadvantage: I burnt and destroyed—as I have already stated—from 2 to 3 per cent. more of the iron than was lost when making steel, and I immensely increased the quantity of oxygen absorbed. It was this fact that induced me to persevere in decarburising only to the extent necessary to make steel of the precise quality desired; and where this system has been pursued in Sweden and in Austria, it has proved commercially a great success.

It will at once be seen how ill-adapted are the proportions of carbon, manganese, and iron, in spiegeleisen, because enough of the per cent. of manganese present cannot be put into the converter to deoxydise the malleable iron, without introducing at the same time so much of the 4 per cent. of carbon present as would make the whole of the malleable iron treated, into cast steel. For this reason the very soft or mild quality of steel required for ship and boiler-plates should be recarburised with an alloy of something like the following proportions: 60 per cent. of manganese, 4 of carbon, and 36 of iron.

Now, if Mr. Mushet had invented a new triple compound of iron, carbon, and manganese, in somewhat about the proportions indicated, and had shown a cheap and ready way of producing it on a commercial scale, he would have been entitled to a patent for his mode of producing such an alloy, and also for the use of such an artificial compound in any other process to which it might be applicable. But it was not new to improve steel by alloying it with manganese : a method long before known to, and daily practised by, hundreds of workmen in the steel trade.

This patent of Mr. Mushet, claiming the sole use of manganiferous pig iron, had simply the effect of calling the attention of steel-makers to a makeshift alloy, and thus diverted for some years my attention, and doubtless that of many other persons, from the pursuit of a ready means of producing such an alloy of manganese as would be better suited for the purposes for which spiegeleisen had been employed. All the difficulties in making boiler and ships' plates of the degree of mildness necessary to ensure their safety under the severe strains to which they are subjected, arose from the excess of carbon and the deficiency of manganese in the natural alloy spiegeleisen.

I may here state that, very soon after commencing the manufacture of steel at my Sheffield works, this difficulty about mild steel plates was strongly felt when using British coke-made iron. I attained complete success with Swedish charcoal iron, and thus could make tool steel and gun steel as good as, or better than, any in the market. On these steels there was a large profit, and the cost of the material was not important. But when the steel had to be sold in competition with iron plates, it was necessary to use cheaper pig iron, and it was with this iron that the difficulties arose. However, I found that another of Nature's compounds, wholly differing from spathose ore, or white carbonate of iron, from which spiegeleisen is obtained, existed in large quantities in New Jersey, in the United States. The mineral referred to is a ferriferous oxide of zinc, and on its discovery it was given the name "Franklinite," in honour of Dr. Franklin. When the zinc is driven off, in the form of vapour, there results an alloy of iron and manganese, usually containing from 11 per cent. to $11\frac{1}{2}$ per cent. of manganese, which is far better adapted for the deoxydation of mild steel than spiegeleisen, containing only 8 per cent. of that metal. Consequently, "Franklinite" was much used at my works in Sheffield, pending my introduction of ferro-manganese into the trade. This, unfortunately, from a variety of circumstances, was delayed until 1862, when I induced a Glasgow firm to go into the manufacture of ferro-manganese, both for our own use at Sheffield, and for the benefit of my licensees. The subjoined extract will show how valuable this ferro-manganese was, more especially for plate-making, and how much the Bessemer mild steel plates of that early date suffered in reputation by the undue introduction of carbon into the metal from the use of spiegeleisen, so rich in carbon, and so poor in manganese. I quote one of the highest living[*] authorities, a gentleman who enjoys both an American and a European reputation as an iron and steel manufacturer and metallurgist. I refer to Mr. Abram S. Hewitt, the United States Commissioner to the Universal Exposition at Paris in 1867, who, in his able report to the American Government, commented on the Bessemer process and its application to the manufacture of plates as follows :—

MANUFACTURE OF BESSEMER PLATES.

The application of the Bessemer process to the production of plates either for boilers or for ships, girders, etc., is one of the most important that could be made. Nevertheless the amount of metal used for this purpose in England falls much below that employed for other purposes. This is due to a certain amount of distrust of steel plate, doubt as to its reliability under varying strains of tension and compression, its capability of being punched and sheared without injury to itself, and of its action under the influence of heat and water as in the fire-box of a boiler. In other countries, as for example Austria, as will be shown when we come to speak of the manufacture as carried on in that country, this has not been the case, and large quantities of plates have been produced and successfully applied to a variety of uses.

The secret of the distrust in regard to Bessemer plates in England is that in nearly all cases the percentage of carbon contained in the metal has been too large. The spiegeleisen used in England is not particularly rich in manganese—seldom exceeding nine per cent. of that element, while it generally contains from four to four and a half per cent. of carbon. It is difficult, therefore, with such materials to deoxygenate the metal sufficiently without

---

[*] Living, 1896; died, January 18th, 1903, in his 81st year.

N N

introducing also a considerable percentage of carbon. About 0.4 per cent. of the latter is as large an amount as is proper for plates which are to resist severe strains, and though a greater proportion adds materially to the tensile strength of the metal when measured simply by a direct pull, it renders it also much harder and more liable to crack under the treatment to which it is exposed in the ordinary methods of construction. The difficulty in the way of producing good soft plates for boilers or other uses appeared at one time to have been satisfactorily overcome by the substitution of ferro-manganese in the place of the ordinary spiegeleisen. The manufacture of this substance was commenced by a firm in Glasgow as a branch of another business in which they were engaged, and plates made with it as a deoxygenator gave most excellent results. Unfortunately, however, the firm who had undertaken the manufacture shortly afterward became insolvent, and the patentee of the process has not as yet re-established the manufacture (which requires a considerable expenditure for suitable furnaces) elsewhere in England. Had the use of this substance continued for a longer time, so as to make the excellence of the steel produced with it fully appreciated by the public, there would have been a demand for plates urgent enough to have immediately secured the re-establishment of the manufacture.

This unbiassed judgment of the United States Commissioner amply endorses my views on the subject, and shows how much my process suffered by the adoption of a rough-and-ready mode of supplying a want, which scientific inquiry into the relative proportion of the elements present in spiegeleisen would have at once condemned.

Before dismissing Mr. Hewitt's report, it will be interesting to briefly notice what he had to say to his Government as to the carrying out of the Bessemer process both in Sweden and in Austria.

Under the head of Sweden, Mr. Hewitt made the following remarks :—

SWEDEN.

An examination of the specimens of Bessemer steel from Sweden in the Exposition shows us that the metal there produced is of a far superior character to that made in England, and naturally leads to inquiry as to the cause of the difference, and whether we may hope to attain the same success in the United States. First, we observe coils of wire of all sizes, down to the very finest, such as No. 47, or even smaller. This they have not been able regularly to produce in England. In the next place we notice a good display of fine cutlery, and the writer is informed by a competent authority that this metal answers so well for this purpose that it is now used almost to the exclusion of any other. This statement is corroborated by the fact that in the miscellaneous classes of the Swedish department, where cutlery occurs not as an exhibition of steel, but merely as a display of workmanship by other parties in the same manner as other articles of merchandise, cases of razors are exhibited with the mark of the kind of steel of which they are made stamped or etched upon them as usual, and these are all " Bessemer," but from a variety of different works, viz. :—Högbo, Carlsdal, Österby and Söderfors. The ore used in Sweden for producing iron for the Bessemer process is exclusively magnetic, and of a very pure quality. An analysis of a mixture of those used for the iron employed at the Fagersta works before roasting gives the following composition :—

Carb. acid	8.00
Silicium	17.35
Alumina	0.95
Lime.	6.50
Magnesia	4.35
Protoxide of manganese	3.35
Magnetic oxide.	32.15
Peroxide of iron	27.40
	100.05
Phosphoric acid.	0.03

All the pig made from this mixture of ores, the exhibitors state, will give a steel without the use of spiegeleisen, which is not at all red-short.

The analysis of gray iron from the same works, used for the Bessemer process, is given as follows :—

Carbon combined	1.012
Graphite	3.527
Silicium	0.854
Manganese	1.919
Phosphorus	0.031
Sulphur	0.010

The analysis of mottled pig (*la fonte truité*), consisting of two-thirds gray and one-third white, is—

Carbon combined	2.138
Graphite	2.733
Silicium	0.641
Manganese	2.926
Phosphorus	0.026
Sulphur	0.015

Of each of these it is stated that the steel produced without the employment of spiegeleisen is not at all red-short (*cassant à chaud*). The most noticeable feature in the composition of these irons is the large percentage of manganese which they contain, together with the extremely minute proportion of sulphur.

In the process of conversion, from motives of economy, a fixed form of vessel is employed, instead of one mounted on trunnions, as in England and elsewhere. The tuyères, about nineteen in number, are placed horizontally just above the bottom of the vessel, and are inclined a little from a radial direction so as to give a rotary motion to the mass of molten metal.

Here we see that fine cutlery was exhibited in 1867 with the name "Bessemer steel" conspicuously stamped upon it as a mark of superiority. Wire of the finest numbers had been produced of superior quality, etc.; the crude metal was run direct from the blast furnace and blown to steel in a fixed converter; no spiegeleisen or re-carburation was needed. This was precisely my original mode of operating, as described in my Cheltenham paper.

Again, Mr. Abram S. Hewitt, in his report, gives an interesting account of the manufacture of Bessemer steel as represented by exhibits in the Austrian Department of the Paris Exposition of 1867, and from this account I give the following quotation :—

AUSTRIA.

The conditions under which Bessemer metal is produced in Austria are in many respects similar to those existing in Sweden. The iron employed is smelted with charcoal, is nearly free from sulphur and phosphorus, and contains a large percentage of manganese. There are differences in the manner of conducting the process, but these important conditions insure the production of a metal of similar excellence to the Swedish, and, like this, much superior to the ordinary metal produced in England.

The principal works in Austria are at Neuberg, in the province of Styria, and are carried on by the government. The iron is obtained from spathic ores smelted in two furnaces 43 feet high, and yielding from 100 to 150 tons per week. The iron produced is found by analysis to contain 3.46 per cent. of manganese, and, as in Sweden, it is used for recarbonizing in the place of the usual spiegeleisen. Originally a fixed vessel was erected at these works similar to those used in Sweden, but this has been superseded by a pair of three-ton vessels of the ordinary construction. Fixed or Swedish vessels are, however, still in use at other Austrian works. The metal is run directly from the blast furnaces into the converters.

Here we have a full confirmation of the successful working of the original fixed vessels in Austria, the metal being used direct from the blast furnace. In those cases where it was recarburised, this was not done with spiegeleisen, but by using the same metal as that used for conversion, as described in my patent of 1855. If my invention had gone no further than this, and I had never introduced any of the mechanical improvements, which together constitute an entirely new system of steel manufacture, the accomplishments of such results as Mr. Hewitt saw and described would have been by itself a new departure in steel-making, and would have profoundly altered the condition of the crucible steel trade of this and other countries.   Also, the facts recorded show how far the Bessemer converter and the Sheffield crucible are in one essential feature in perfect accord, viz., the Sheffield crucible process can make excellent cutlery steel from Swedish charcoal pig iron without the use of manganese in any form.

But the Sheffield crucible process cannot make good steel from British iron smelted with mineral fuel without the employment of manganese in the steel pot.   Nor can the Bessemer converter make good steel from British iron smelted with mineral fuel without the employment of manganese in its converter.

Nothing can more clearly show that the application of manganese to Bessemer steel was not a discovery or novel invention, for with what kind of iron it was necessary to use manganese, and with what kind of iron it was not required, was perfectly well known to Sheffield steel-makers many years before Mr. Mushet claimed the use of it.

The perfect success that was obtained from the very first working of my process*, both in Sweden and in Austria, excited the greatest

---

* Referring to the development of the Bessemer process in Europe, Mr. Abram S. Hewitt said, in his Report on the 1867 Exhibition :—"It will be interesting to those who are watching the advancement of the new process, to know that it is already rapidly extending itself over Europe.   The enterprising firm of Daniel Elfstrand and Co., of Edsken, who were the pioneers in Sweden, have now made several hundred tons of excellent steel by the Bessemer process.   Another large works has since started in their immediate neighbourhood, and two other Companies are making arrangements, to use the process. The authorities in Sweden have most fully investigated the whole process and have pronounced it perfect.   The large steel circular saw-plate exhibited was made by Mr. Göranson, of Gefle, in Sweden; the ingot being cast direct from the fluid metal, within fifteen minutes of its leaving the blast furnace.   In France, the process has been for some time carried on, by the old established firm of James Jackson and Son, at their steel works, near Bordeaux.   This firm was about to go extensively into the manufacture of puddled steel, and indeed had already got a puddling furnace erected and in active operation, when their attention was directed to the Bessemer process.   The apparatus for this was put up at their works last year, and they are now greatly extending their field of operations, by putting up more powerful apparatus at their blast furnaces in the Landes.   There are also in course of erection four other blast-furnaces in the South of France, for the express purpose of carrying out the new process.   The long and well-earned reputation of the firm of James Jackson and Son is, in itself, a guarantee of the excellent quality of the steel produced by this process.   The French samples of bar steel exhibited, were manufactured by this firm. Belgium is not much behind her neighbours in the race, as the process is being put into

interest in those countries. My first licensee in Sweden, Mr. Goransen, of Gefle, came over to England as soon as the printed notice in the press of my Cheltenham paper had reached him. He was a man possessed of great energy as well as practical knowledge ; he saw the converting process at my experimental works in London, and he erected a fixed vessel like the one he saw. In this he used the molten iron direct from his blast furnace, and converted it into steel without recar-burising ; in fact, he kept strictly to the mode of operating described in my Cheltenham paper. In a very short time he had his steel works in operation, and sent over some ingots to show me what splendid steel he was making. One of these ingots was rolled in Sheffield into a circular saw - plate, $\frac{3}{16}$ in. thick and 5 ft. in diameter. So great was the interest excited in Sweden by the successful production of high-class steel by the Bessemer process, that Prince Oscar took a journey of over 200 miles to see it in operation at the works of Mr. Goransen, and the impression made on the Prince's mind was so favourable that it resulted in my being made an honorary member of the Iron Board of Sweden, in recognition of the value of my invention : a compliment which I shall ever highly esteem.

The circumstances attending the introduction of my process into Austria were very different, but were equally satisfactory.

I had no Austrian patent, and therefore did not take any steps to introduce my process into that country. The principal iron works are at Neuberg, in Styria, and belong to the Government. The intelligent managers of those works early applied to me for information regarding my steel process, and, as I had no patent, they desired to know under what terms I would supply all such plans as would enable them to put it in operation. I offered them detailed drawings of all the apparatus, a written description of the process, and a trial of their pig iron at the Sheffield Works, in the presence of one of their own employés, for which I asked a fee of £1000. This offer was at once accepted, and the

operation at Liége. While in Sardinia preparations are making to carry it into effect, Russia has sent to London an Engineer and a Professor of chemistry to report on the process, and Professor Müller of Vienna, and M. Dumas and others from Paris, have visited Sweden, to inspect and report on the working of the new system in that country."

agreement thus entered into was carried out to our mutual satisfaction; in due time, the works at Neuberg were got into active operation, and were entirely successful. In fact, with their splendid pig iron, it would have been difficult to have made a failure. Prince Demidoff inspected the works, and gave such a favourable report to the Emperor that His Majesty conferred on me the honour of "Knight Commander of the Order of His Imperial Majesty Francis Joseph," which, with the scarlet collar and gold and enamelled cross of the Order, was presented to me by His Excellency the Austrian Ambassador in London. This decoration I highly prize, and I have worn it on many public occasions.*

* The following extract from *Men and Women of Our Time* (Routledge and Co.), summarises the many distinctions conferred on Sir Henry Bessemer :—

"The first honorary recognition of the importance of the Bessemer process in this country was made by the Institution of Civil Engineers about 1858, when that body awarded Mr. Bessemer the Gold Telford Medal, for a paper read by him before them on the subject. A knowledge of the new process soon spread to Sweden, Germany, Austria, France and America, and the inventor has received from these countries many honours and marks of distinction. In the early days of the invention, Prince Oscar of Sweden travelled many miles to witness the process in operation, and, as a mark of his approval, made the inventor a member of the Iron Board of Sweden. In Austria, the honour of the Knight Commander of the Order of his Imperial Majesty Francis Joseph was presented to him by the Emperor, together with the gold and enamelled cross and ribbon of the Order. The Emperor Napoleon desired to present him with the Grand Cross of the Legion of Honour, but the British Government would not allow him to accept it. The Emperor in person presented him with a superb gold medal instead. He also received the Albert Gold Medal, which was awarded by the Council of the Society of Arts, presented to him by the Prince of Wales at Marlborough House. The King of Wurtemburg also presented to the inventor a handsome gold medal, accompanied by a complimentary testimonial. His Majesty the King of the Belgians, who has always taken a deep interest in the Bessemer process, has on several occasions honoured the inventor by personally visiting him at his residence on Denmark Hill. The Freedom of the City of Hamburg was also presented to him in due form. He was also made a member of the Royal Academy of Trade in Berlin, and a Member of the Society for the Encouragement of National Industry of Paris; and in England he was made a member of the Royal Society of British Architects, and a member of the University College, London, a member of the Society of Mechanical Engineers of England and America. He succeeded the late Duke of Devonshire as President of the Iron and Steel Institute of Great Britain, and during his presidency he instituted the Bessemer Gold Medal, which has since been awarded annually for the most important improvement in the iron and steel manufacture made during the year. He also instituted the Bessemer Bronze Medal and five-guinea prize of books, annually presented to the most successful student at the Royal School of Mines at South Kensington. The

In the latter part of 1856 and the commencement of 1857, I steadily pursued my experiments, with a view to improve the quality of the steel I was making, and to get rid of red-shortness. I sought for information on this point in old books and encyclopædias, where very little information could be gained. I also re-perused such metallurgical works as I possessed, and had already skimmed over too lightly, and in one of them I found some most valuable information, which I at once saw was applicable to my case. It related to an invention that had been introduced into the Sheffield steel trade, about sixteen years previously, by means of which iron of inferior quality was made to produce excellent steel, and to receive the property of welding. The article referred to was written by my old and esteemed friend, Dr. Andrew Ure, and appeared in a supplement to the third edition of his *Dictionary of Arts, Manufactures, and Mines*, published by Longmans and Co. in 1846.

It has many times been remarked that some of the most important events which shape and control our lives or fortunes, arise from fortuitous circumstances which apparently have no possible connection with the

Institution of Civil Engineers awarded him a splendid Gold Cup, being the Howard Quinquennial Prize. He was also presented with the Freedom of the Cutlers' Company of London, and the Freedom of the Turners' Company; and, at a specially-convened meeting at the Guildhall, on May 13th, 1880, Sir Henry Bessemer was presented with the Freedom of the City of London, beautifully illuminated, and contained in a massive gold casket, "in recognition of his valuable discoveries, which have so largely benefited the iron industry of this country, and his scientific attainments, which are so well known and appreciated throughout the world;" the same evening he was entertained at a banquet given in his honour, at the Mansion House, by the then Lord Mayor, Sir Francis Wyatt Truscott. But it may be truly said that in no part of the world has the Bessemer process been developed to the extent and with the energy that has marked its progress in America. In several different parts of the United States, where nature has richly endowed them with those aids to civilisation, coal and iron, manufacturing cities have been established, to which, by common consent, they have given the name of Bessemer. Thus we have the rapidly-increasing and important City of Bessemer, Gogebec County, Michigan; the City of Bessemer, chief town of the County of Bessemer, Alabama, with its Mayor and Corporation, its street tramways and electric lighting, and its large manufacturing works, public schools, and numerous churches. There is also the City of Bessemer, Lawrence County, Pennsylvania, the seat of the great Edgar Thompson Steel Works, the largest in America. There is also the City of Bessemer, Botetourt County, Virginia; the City of Bessemer, Natrona County, Wyoming; and the City of Bessemer, Gaston County, North Carolina."

events they have in reality brought about. My readers will remember that in the early part of this volume (page 13) I gave an account of my acquaintance with Dr. Ure, and related how I had shown him some medallions which I had coated with a thin deposit of copper from its acid solution. I told of the great interest Dr. Ure had taken in my discovery, and how, in November, 1846, he published a supplement to his work, in which he gave an account of my invention under the article " Electro-Metallurgy." Hence, I naturally purchased a copy of this, to me, most interesting volume. It was an article on the manufacture of steel, contained in this supplement, which first enlightened me on the subject of manganese and Heath's invention; this culminated in the production of ferro-manganese.

I read this account of Heath's invention with deep interest, and at the same time I scored a line under a few of the sentences which very forcibly struck me ; in order that my readers may see precisely the kind of information this article furnished, I have had the whole page photographed, and I reproduce it in Fig. 77, Plate XXXIV.

On reading this well-authenticated account of Heath's invention, I at once saw that red-shortness would be cured by its use, for I had found that my red-short steel crumbled away under the hammer if raised to a welding heat. Here, in the book of my old friend, Dr. Ure, was ample proof that inferior brands of iron could be made into weldable cast steel simply by alloying them with 1 per cent. of carburet of manganese. This fortunate discovery of what had already been practised for years came like a revelation to me ; and as this patent of Heath's had long expired, and his invention had become public property, I at once investigated the whole subject, commencing with inquiries into the law proceedings referred to by Dr. Ure, where I gained much additional information. In the reports of " Noted Cases on Letters Patent for Inventions," by Thos. Webster, barrister-at-law, published in 1855, I found the complete specification of Heath's patent, and also much evidence given in the Exchequer Court, in the case of "Heath v. Unwin," Hilary Term, 1844, by experts who had studied the subject both theoretically and practically. From these reports I subjoin the following extract :—

o o

Evidence was given on behalf of the plaintiff by manufacturers of steel, and of long experience in the trade, to the effect that cast steel suitable for the manufacture of cutlery, before the introduction of the plaintiff's process, could only be made from high-priced foreign iron, that the use of carburet of manganese in the manufacture of welding cast steel was new at the date of the plaintiff's patent; that the introduction of the plaintiff's invention caused a revolution in the trade; that the plaintiff had, after long investigation and experiments, discovered that when black oxide of manganese was combined in such proportions with carbonaceous matter as to form a carburet, it enabled the manufacturer to produce a welding cast steel suitable for the manufacture of cutlery from low-priced British iron, which had never been done before, and which reduced the price of the steel from about 70*l*. to about 35*l*. per ton.

Here was the remedy I was in search of, clearly pointed out; experienced Sheffield steel-makers had testified on oath that the use of carburet of manganese, added to the cast steel, enabled the latter to produce welding cast steel suitable for the manufacture of cutlery from low-priced British iron, which had never before been done. No sooner had I ascertained these facts than I commenced experiments on the production of Heath's carburet of manganese in crucibles, using the air-furnace which I had many years previously successfully employed to produce all the various alloys of metal required in my bronze-powder manufactory at Baxter House.

I well remember how much trouble I had with the first few experiments, in which I used charcoal and black oxide of manganese, the charcoal, ground to a very fine powder, being much in excess of the quantity actually required. This was a great mistake, as the reduced oxide remained in minute metallic particles, intermixed with the overdose of charcoal powder. This mistake was afterwards remedied, coarse granular charcoal in suitable proportion being used. I have never publicly referred to these early experiments, simply because I was unaware that I had, or could show, any evidence of the fact; and, as is my rule in all such cases, I preferred to remain absolutely silent, not only in reference to these early experiments to produce carburet of manganese, but also as to my initiation of the manufacture of alloys of iron rich in manganese, which are now so well known under the name of ferro-manganese. But a purely accidental circumstance has, within the last few years, furnished me with such conclusive evidence of the fact as to make me no longer hesitate to show how far I was instrumental in the production of that valuable alloy, ferro-manganese.

In searching through the contents of an old box I had brought to Denmark Hill from Queen Street Place on my retirement from business, I came upon six old pocket-memorandum books, in which I, from time to time, had recorded many experiments on alloys, mechanical contrivances, suggestions for new patents, etc. In one of these old books, bearing on its flyleaf the date January 8th, 1852, written forty-five years ago by my deceased partner Longsdon, I found several memoranda relating to my first attempt to make Heath's carburet of manganese, which were the direct outcome of the information I had obtained from Dr. Ure's book. These researches were made about a month before any one of Mr. Mushet's patents was published or could possibly be known to the world. It will be seen that these memoranda were roughly made on the spur of the moment, and were simply for my own guidance, or to prevent ideas and experiments from being forgotten.

I give a facsimile of some of them in Fig. 78, page 284.

It will be remembered by many members of the Iron and Steel Institute that it was in one of these old memorandum books that I came upon my notes relative to the manufacture of what were designated "Meteoric Guns," to be made by alloying malleable iron or steel with 3 per cent. of nickel; a photograph of these notes was communicated by me to the Institute, and published in their *Journal*, Vol. 18. Had it not been for this accidental discovery of memoranda made at the time, and the existence of which had been entirely forgotten, I should never have reverted to this subject, since the mere adoption of Heath's process could in no way add to whatever credit I may be entitled to for the discovery and development of the Bessemer process.

These old records of experiments will serve to show the difficulties that one meets with from the most trivial circumstances. The fact was that my air-furnace, which was designed for making bronze alloys, was deficient in temperature when treating such a refractory ore as oxide of manganese, and produced only a few buttons of reduced metal.

I had found, in making alloys of copper and tungsten for bronze
powders that the mineral wolfram was most difficult to bring to

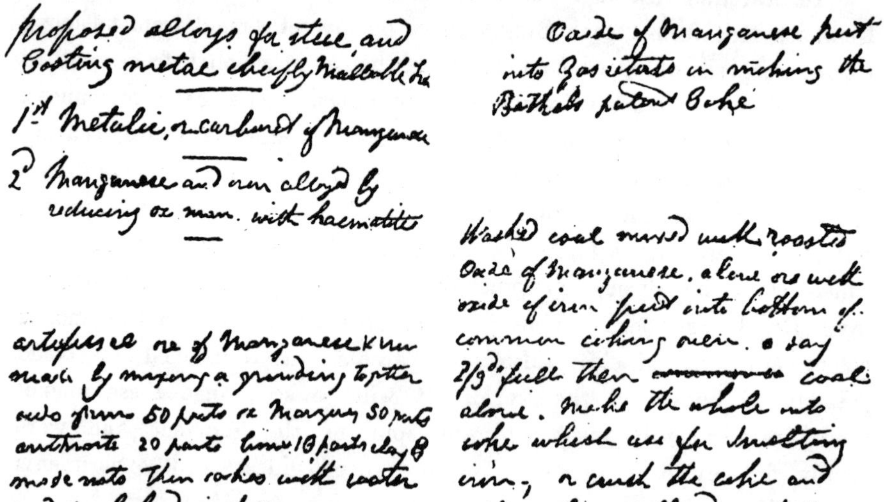

FIG. 78. FACSIMILE REPRODUCTION FROM BESSEMER'S NOTE-BOOK

the metallic state, but was reduced easily if crushed and mixed with
oxide of copper, or with refuse "copper-bronze," that is, a fine powder
with pure copper. Thus copper, alloyed with tungsten, was readily
obtained. This fact of the union of metals in the act of simultaneous

reduction from their oxides, of which I had some practical experience, at once suggested to me that the difficulty in reducing oxide of manganese would be removed, by combining it in the form of powder with oxide of iron, which is so easily fused, and then reducing the two metals simultaneously. I clearly saw, at the same time, that this system of alloying the manganese with iron would prevent the spontaneous decomposition of pure metallic manganese when exposed to ordinary atmospheric influence, as the manganese would be protected by the iron present. This mode of producing an alloy of iron and manganese, in almost any assignable proportions, appeared to me to be such an important step in advance as to render all further experiments in making Heath's pure carburet of manganese quite unnecessary; these ideas were at once jotted down in my pocket-book, and simply embody the first rough views taken of this important manufacture. The memoranda referred to are photographically reproduced in Fig. 79, Plate XXXV.

With reference to Bethel's patent coke, I may mention that this coke is made by the destructive distillation of coal-tar in closed retorts, which leaves a porous hard coke which is almost pure carbon. This process would have been excellently adapted for the reduction of oxide of manganese on a large scale, and such a system of coke-making in a retort would have been far less expensive than Heath's crucible process. What I wanted to obtain, however, was the substance I had designated "artificial ore of manganese and iron." Such artificial ore could be smelted like other iron ores, and thus offered all the prospective advantages of quantity and cheapness. This particular scheme I never lost sight of until it culminated in the production of ferro-manganese at Glasgow. Since my invention was kept in abeyance, so far as steel-making from British iron was concerned, I was desirous of making a series of experiments on all the rich alloys of iron and manganese. I, therefore, had my furnace enlarged and the draught improved. I then applied to Messrs. Bird and Company, of London, who were agents for the Workington Hematite Iron Company, to obtain for me some of their pure hematite ore for my experiments. There was some delay in getting this ore, and in the meantime both Mr. Martien's and Mr. Mushet's patents were published. Then, for the first time, I realised that an obstacle had been created, which might prevent my using manganese in my process in any and every form in which that metal was known, or had previously been in public use. Nevertheless, I felt not the slightest hesitation in making use of spiegeleisen, or any other manganesian pig-irons, which were covered by my prior patents. I was, however, unfortunately diverted for the time from the pursuit of the richer alloys of manganese which would have prevented all those troubles met with in producing steel of sufficient mildness for plates, so deeply engrossed did I become in the introduction of my process to the trade, and in keeping watch against the many attempts to encroach on my rights. Coupled with these there was constant and laborious work at the drawing-board in making the original drawings for my own further improvements, and in the development of the many mechanical devices necessary to the commercial use of my invention on a large scale. With all these imperative calls on my time, something had to go to the wall, and the rich manganesian alloys were for the time crowded out. In this busy year—that is, from September 1856 to November 1857—I had taken out eleven new patents. I had settled the mechanical details of each one, and had personally made the whole of the drawings for the eleven specifications. Every day had its new labours, and every day the need for these rich alloys of manganese became more evident.

About this time I had a long conversation on this subject with Mr. William Galloway, one of the partners in our Sheffield firm, and we seriously thought of putting up a blast furnace for making rich manganesian pig-iron. Mr. Galloway had some land at Runcorn, on the Mersey, which he suggested should be utilised for this purpose as a private speculation of our own. I made many inquiries about manganese mines at the "Mining Record" Office, and got a good deal of useful information from Mr. Robert Hunt, the indefatigable head of that most valuable institution. My inquiries and numerous visits on the subject awakened a deep interest in Mr. Hunt, and before the summer was over it was arranged that I should accompany him in his usual annual visit to the principal tin and other mines in

Cornwall. I much needed this little holiday, and Mr. Hunt drove me nearly all round the county of Cornwall in an open phaeton, a journey full of deep interest to me. My friend—for so I am proud to call him— was a positive living encyclopædia, and neither the longest journey, nor the lonely parlour of the village inn, was ever dreary with such an agreeable companion. We visited some of the manganese mines, which were not very promising, being situated in localities far removed from shipping ports, to which their output must have been transported by horse and cart over bad roads.

While Mr. Hunt pursued his professional duties, I made a short halt at Penzance, and rambled over the enormous granite rocks leading down to Land's End. At some works in the district I found a pair of dwarf serpentine columns of great beauty, which I purchased as a memento of this most interesting journey. They are at present (1896) in good company, for between them stands a massive pedestal, 4 ft. high, made of Algerian onyx, forming the base of a large Parisian clock, with a life-sized bronze figure holding a revolving pendulum. The serpentine columns support busts of Enid and Prince Geraint from the "Idylls of the King," sculptured in white Carrara marble. This group stands on one side of the entrance-hall of my residence (see Fig. 80, Plate XXXVI.).

On my return to London a plain, business-like review of all the circumstances connected with the supply of manganese ore from Cornwall was unsatisfactory. My old friend Galloway was getting on in years, and not over-anxious to embark in new undertakings, while the pursuit of my own business, and the spread of the process throughout Europe, engrossed my whole attention. Thus time rolled on; we made shift with Franklinite, which was 40 per cent. richer in manganese than spiegeleisen, but it was not all we could desire. A little later, it occurred to me that oxide of manganese was a waste product in the manufacture of chlorine and bleaching powder, and I knew that the firm of Tennant and Co., of St. Rollox, Glasgow, were most extensive manufacturers of this article. At that time Mr. Rowan, of Glasgow, was making Bessemer steel under a license from me, and I wrote to him saying that I was coming down to Glasgow, and hoped that he would be able to get me an introduction to Messrs. Tennant. In reply, Mr. Rowan invited me to come to his house and stay a week. I did so, and, in talking over the matter, he said: "I know a Mr. Henderson, who is a good chemist, and is carrying out a scheme of his own at the works of Messrs. Tennant and Co., where he is operating on iron pyrites, and one of his waste products is pure iron in the form of powder. I will, if you wish it, ask him to come and dine with us to-morrow."

The next evening I explained to Mr. Henderson how I proposed to manufacture an artificial metallic ore, consisting of iron and manganese, by combining hematite, or white carbonate of iron, with oxide of manganese, in equal proportions. These materials were to be held together with clay, or with clay and lime, to form a fluid cinder, either with or without the addition of carbonaceous matter. I proposed to mix these materials in a common brickmaker's pug-mill, to dry the mixture in moderate-sized lumps, and to convert this artificial ore into the metallic state in an ordinary blast furnace. I told Mr. Henderson that I wanted some large firm to take up the manufacture, as I had no time to attend to it, and did not wish to make such manufacture a source of profit. All I wanted was to be supplied with a manganesian alloy of iron, of not less than 50 per cent. of manganese, for my own use and that of my licensees, who would most assuredly become large purchasers. Mr. Henderson was very anxious to take the matter in hand, but he feared to encounter the large cost of erecting a complete blast furnace plant. He said that he had no doubt he could produce the alloy in a less expensive furnace, and was willing to risk the cost and trouble of doing so. I, on my part, gave up the idea of pressing it upon Messrs. Tennant, as I originally intended, and left the whole matter in Mr. Henderson's hands. The result of this was that he took out a patent for manufacturing these rich manganese alloys in a reverberatory gas-furnace, and so far succeeded as to produce alloys containing from 20 to 25 per cent. of manganese, with which he supplied our Sheffield firm until his works were, unfortunately, closed, owing to the insolvency of the iron-founder on whose premises his furnace was erected.

Thus was inaugurated the manufacture of ferro-manganese, the production of which I had followed up as closely as my many engagements permitted, from the very first inception of the idea, dating from the reading of a chapter on steel in Dr. Ure's *Dictionary of Arts and Manufactures;* followed by the perusal of Heath's patents, and the evidence of the Sheffield steel-manufacturers given in one of Heath's law suits, as published in *Webster's Law Reports.* I never lost sight of the object, so successfully arrived at, which would have been attained long before had not the inferior alloy, spiegeleisen, been an article of commerce at once procurable ; this delayed the production of an alloy specially suitable for the purpose. But, valuable as this ferro-manganese really was, neither that, nor spiegeleisen, could make good steel from the ordinary quality of pig iron used for the manufacture of iron bars, nor from the hematite iron as then made, since the hematite pig iron, like all other British pig, was greatly contaminated with phosphorus, owing to the use of puddler's tap cinder to flux the hematite ore in the blast-furnace, and thus obtain a fluid cinder. It was not until I had, with the assistance of my own chemist, prescribed new furnace charges, omitting tap cinder and substituting shale, and thus producing Bessemer pig, that any British coke-made iron could be converted by my process into good steel. The universal presence of phosphorus was the primary barrier which stopped my way; and when this difficulty was removed, by the absence of tap cinder from the hematite furnaces, we could obtain pig iron which was as free from phosphorus as the puddled bar iron used in Sheffield for conversion into steel ; and with this Bessemer pig good steel could readily be made by my process, when there was used in conjunction with it the well-known remedy for red-shortness, carburet of manganese.

In the meantime, our Sheffield works had commenced commercial operations, and we made no secret that we used spiegeleisen for recarburising the converted metal. We patiently waited for the injunction in Chancery that was to stop its use. But neither Mr. Mushet nor others took any steps to enforce their patent rights. Another year or two passed quietly by, and our steel works at Sheffield, and those of our licensees, were daily increasing the quantity of Bessemer steel placed upon the market. No attempt was made to prevent us using

P P

manganese; but, nevertheless, for some months the air was filled with vague reports of legal proceedings. A "round-robin" had, it was said, been filled up with subscribers to the extent of £10,000, and even high legal luminaries and eminent engineers and experts in Great George Street were supposed to be definitely retained. These rumours were very vague; nevertheless, they cropped up in various different quarters over a period of many months. I personally took very little heed of them, feeling absolutely secure in my patent claims; no doubt a careful search through a thousand old iron patents might unearth a few vague expressions to which legal ingenuity, under the new light thrown upon the subject by me, might give an outward appearance of similarity with my invention; but I had always remembered that my claim was "to force atmospheric air beneath the surface of crude molten iron *until it was thereby rendered malleable,* and had acquired other properties common to cast steel, while still retaining the fluid state." This I felt absolutely certain no man but myself had patented, and so I slept soundly in spite of rumour, which, however, I did not doubt had some foundation.

For a period of more than two and a-half years (1857-60) after the date of Mr. Mushet's three manganese patents, I had no intimation of any kind that either I, or my licensees, were infringing any of these patents. But about three or four months prior to the date when a further £100 stamp was required to be impressed on them, to prevent their forfeiture, I received a letter from a Mr. Clare, of Birmingham, calling himself Mr. Mushet's agent for the sale of steel, and requesting an interview with me and my partner at my office in London on the following morning. On his arrival, he explained the object of his visit; it was simply to say that Mr. Mushet was prepared to grant me a license to use his manganese patents for a nominal sum; he merely wanted his rights acknowledged. I then told Mr. Clare that we considered that Mr. Mushet had acquired no rights under either of his three manganese patents, and that we entirely repudiated them. I also told him that we were anxious to meet any claims legally preferred; that we were prepared, on any day to be mutually arranged, to receive Mr. Mushet and his solicitors and witnesses at the Sheffield Works; that we would allow them to see the crude iron converted and re-carburised with spiegeleisen, made into an ingot and forged into a bar, and that I would personally take that bar to one of my customers and sell it to him in their presence; and then the prosecution of our firm for infringement would be a very simple matter. This offer resulted in Mr. Clare's retirement from my office, and after that interview we never heard from him, or from Mr. Mushet, on the subject.

It will be within the memory of my readers that when we had got into full swing with the new process at Sheffield, and had been successful not only in making high-class tool steel from Swedish charcoal pig iron, but also mild steel for constructive purposes from Bessemer pig, I read a paper at the Institution of Civil Engineers, on which occasion many beautiful samples of steel were exhibited, made by my process in France, in Sweden, and at Sheffield. At the reading of this paper Mr. Thomas Brown, of whom I have frequently spoken, was present.

Referring to my process, Mr. Brown said that he had been sanguine of its success, and had spent £7000 in endeavouring to carry it out; but he did not say that he had no license from me to make this secret use of my invention. The annexed extract from the *Proceedings* of the Institution of Civil Engineers furnishes a report of his remarks :—

Mr. T. Brown said he had taken great interest in this process, when it was first brought forward, after the meeting of the British Association, at Cheltenham. He had been sanguine of its success, even in opposition to the opinion of others, who had no faith in it from the commencement; and he had spent £7,000 in endeavouring to carry it out. It appeared to be thought that the quality of the iron ore had an important influence upon the success of the operation. Now, he had succeeded in making samples, equal perhaps to those exhibited, from spathose ores from the mines of the Ebbw Vale Company, in the Brendon Hills, Somersetshire, with a mixture of Pontypool iron. But the difficulty he experienced—amounting, indeed, to an impracticability—was in finding a completely refractory material for the furnace. He was astonished at the price which had been stated as that at which the article could be produced. He thought a very simple calculation was sufficient to disprove it; for the iron and the material, without manipulation, made up the amount; in fact, the article in its first state, supposing Indian pig-iron to be used, cost £6 10s. per ton. He did not wish to say anything which could be looked upon as discouraging, because he had originally been one of the warmest supporters of the invention; but he believed Mr. Bessemer was now falling into the same error as to cost as he had done at Cheltenham. With regard to waste, under the most favourable circumstances, there was a loss in the manufacture of nearly 40 per cent. of metal; and on one occasion his agent informed him that the whole of the metal was consumed, and that nothing but cinder remained.

In 1862 I thought I had reason to fear the advent of a rival process brought forward by Mr. George Parry, of the Ebbw Vale Iron Works, whose name figures in a patent for the manufacture of iron and steel, bearing date November 18th, 1861.

Before making any further reference to this patent, I would remind those of my readers who are not practically acquainted with the details of my steel process, that it consists in decarburising iron which contains too much carbon to constitute steel, and in some cases this process of decarburisation is carried through every grade of steel until the carbon element is wholly removed, and soft malleable iron is the material arrived at. Now, in describing this operation in my patent, I made use of the well-known and ordinary terms by which iron in its various states of combination with carbon is commercially known; thus, I claimed to force air into and beneath the surface of molten crude iron (that is, molten iron as it leaves the blast furnace), or re-melted pig or cast iron (that is, re-melted, broken or useless castings). If, instead of using these trade terms, I had said that I claimed forcing air beneath the surface of carburet of iron, this would, in scientific language, not only have included these three ordinary qualities of iron, but it would have embraced any and every compound of iron and carbon from which I desired to eliminate the latter, and which was, in fact, the real object, meaning, and intention of my invention.

It must be remembered that my royalty of two pounds per ton on all ingots of iron or steel made by my process was holding out a great premium for the production of a carburet of iron for conversion into steel, which, from the nature of its manufacture, might so far differ from ordinary crude or pig iron as to remove it from the actual trade class of iron which I claimed to convert; such iron, even if it cost £1 per ton more than commercial pig iron, would avoid my royalty of £2, and save the patentee £1 per ton. The ostensible object of this patent of Mr. George Parry for the manufacture of iron and steel was to produce a superior quality of steel by the employment of malleable scrap iron in lieu of pig, or crude iron ; for this purpose the scrap iron was melted with coke in a small blast furnace, from which it was run into a converter similar to mine, and blown with air forced upward through it by tuyères, the orifices of which were beneath the surface of the metal ; all this was a pure and simple copy of my decarburising process. But the malleable iron scrap could not be fused when distributed and mixed up with lumps of coke in the blast furnace, without its absorbing about two per cent. of carbon, and thus producing white iron or forge pig ; it would also absorb some sulphur from the coke, and would contain that amount of phosphorus which is always present in ordinary British bar - iron, and which is an inadmissible quantity in cast steel. The metal thus produced would, in fact, be crude iron, although the various impurities present might differ in proportion from those in ordinary blast furnace iron. Such iron would, further, be deficient in that necessary heat-producing element, silicon, which is always present in considerable quantity in all pig-iron suitable for the converting process ; and this, combined with the deficiency of carbon, would form an absolute barrier to its conversion into fluid mild steel, as the necessary heat could not be produced from such a quality of carburet of iron. This process, as might have been expected, proved unsuccessful.

One more incident referring to my relations with Mr. Mushet remains to be chronicled before I close this Chapter. In December, 1866, one of my clerks announced the visit of a young lady, who did not send in her name, but wished to see me personally. She was asked into my private office, and, on my going to her, she gave the name of Mushet. She told me that the gravest misfortune had overtaken her father, and that without immediate pecuniary help their home would be taken from them. She said : " They tell me you use my father's invention, and are indebted to him for your success." I said : " I use what your father had no right to claim ; and if he had the legal position you seem to suppose, he could stop my business by an injunction to-morrow, and get many thousands of pounds' compensation for my infringement of his rights. The only result which followed from your father taking out his patents was that they pointed out to me some rights which I already possessed, but of which I was not availing myself. Thus he did me some service, and even for this unintentional service I cannot live in a state of indebtedness; so please let me know what sum will render your home secure, and I will give it you." She then handed me a paper setting forth the legal claim against him; I at once took out my cheque-book and drew for the amount, viz., £377 14s. 10d., and handed it to her. She thanked me in a faltering voice as I bade her good afternoon.

On joining my partner after this interview with Miss Mushet, I explained to him what had occurred; he listened to me with surprise, and with more impatience than I had ever seen him evince. He thought that what I had done was most unfortunate and imprudent, since from Miss Mushet's words it was evident that the idea was abroad that I had in some way taken advantage of her father. He feared lest my cheque should be considered evidence of my indebtedness. I was much distressed to find my friend Longsdon so much annoyed, for a more conscientious and just man I never knew; he was, however, somewhat reassured when I told him that I considered it a purely personal matter, and had, of course, drawn the cheque on my private bankers. He said he was glad it could never appear as an act of the firm, though he thought it would be long before I should hear the last of it.

Events proved that he was right, for not many months elapsed (about 1867) before a friend—I believe a relation of Mr. Mushet— wrote asking me to make Mushet a small allowance. I objected to do this at first, but afterwards yielded, though I did not then care to give my reasons for doing so. There was a strong desire on my part to make him my debtor rather than the reverse, and the payment had other advantages: the press at that time was violently attacking my patent, and there was the chance that if any of my licensees were thus induced to resist my claims all the rest might follow the example, and these large monthly payments might cease for such a period as the contest in the law courts might last. The annoyance, if nothing else, would have been very great, and I had neither time nor patience to wage a paper war from year's end to year's end with unscrupulous writers. In the hope that an allowance to Mr. Mushet might have the effect of restraining these attacks on me, I offered to pay him £300 a year, aiming at abating an intolerable nuisance which I had no other means of preventing. While we were paying over £3000 per annum in the form of income tax, the £300 was but a small additional tax on my resources, so I allowed it to drag on until Mr. Mushet's decease, in 1891, having thus paid him over £7000. So, naturally, ends this part of the history of my invention, as far as Mr. Mushet is concerned.

CHAPTER XIX

EBBW VALE

IN the preceding Chapter I have referred to Mr. George Parry, who was furnace manager at the Ebbw Vale Works in 1857. In that year he applied for a patent having for its object the decarburation of crude iron, by blowing forcibly down upon it in a closed chamber without fuel, instead of blowing up through it, as in my process; this patent, however, was not completed. In 1861, as already stated, Mr. Parry took another patent for making carburet of iron in a small blast furnace, the iron so produced containing some portion of all the ordinary constituents of pig iron, but differing in their proportions; in consequence of this difference it was proposed to convert this iron into steel by blowing air up through the fluid in a closed vessel, and to make it into ingots precisely in the manner directed in my patents. I think it was quite natural that efforts at competition on these and other lines should be made persistently; my process was advancing with rapid strides in every State in Europe, and immense profits were being realised in this country by the proprietors of ironworks who had taken licenses under my patents; in fact, thousands of tons of Bessemer steel rails had been sold at £18 to £20 per ton. Some two or three years had glided away after the date of Mr. Parry's second patent, which had been quite forgotten by me. I had at this time (about 1864) occasion to go to Birmingham on business, and had left Euston at 9 P.M. I was quietly reading my newspaper in the snug corner of a first-class compartment, containing only two other occupants beside myself. These were two young gentlemen, who appeared much elated at some success, or contemplated success—it might be a race, a Stock Exchange bargain, or any other matter of ordinary interest. Being quite young men they were naturally very enthusiastic, and somewhat loud in their conversation, which rather disturbed my reading. After some remarks by one of them, the other exclaimed, in a very loud tone, "I wonder what the devil Bessemer will say?" There could be no mistake as to this plain reference to me, since, with the exception of the members of my family, I alone answered to that name. It then occurred to me for the first time that all this excited language and jubilation had some reference to me; I had not the remotest idea as to what had previously been said, or to what it referred. By this time we had reached Watford, and as the train went on I kept my paper before me, but could not prevent my attention being directed to the lively sallies of these young men. Little by little, I became conscious that the exciting cause of this boisterous hilarity was some new joint-stock company that was to be floated in two or three days. It might be a gas company, a brewery, or anything else, for up to this point I had no indication of its nature, and only wondered why they should question as to how Bessemer would receive the news. But one at a time words were dropped that startled me not a little, and riveted my attention to their conversation, which was very much veiled, as though the scheme, whatever it might be, were to be kept a profound secret at present from the outer world. But here and there some casual word or two was dropped, about mines and works, and a journey up from Wales, and what David Chadwick had said about all the shares being taken up in two days for certain. Thus I soon began to grasp the meaning of the fragments I had heard, and to fit these disjointed sentences together; but there was no absolute certainty that I had guessed the true meaning.

We had by this time arrived at Leighton, and my fellow-travellers got out, as I supposed, to take some refreshment, but the train went on without them, and I was left alone to think over this curious incident. Then I remembered that Mr. Joseph Robinson, the manager of the Ebbw Vale Company's London offices, lived at Leighton. These young men might probably be his sons; and this formed another startling confirmation of the theory I had arrived at, viz., that the Ebbw Vale Iron Works were going, in a few days, to be formed into a joint-stock company, to take over the works and mines and the other property of the present owners, and that Mr. David Chadwick, whose name I distinctly heard, was the financial

QQ

agent employed to form the company. I was not long in realising all that this meant to me, and I saw that it was necessary to take immediate steps to protect myself. Hence I became very impatient to arrive at the next station, which was Blisworth, and there I got out. It was now about 11 P.M., and the next up train was nearly due. I had by this time worked myself into a considerable state of excitement, and paced the station platform so rapidly as to attract the attention of the station-master, who asked me if anything were wrong, or if he could do anything for me. I said, " No ; I have heard some news on my way down which renders my immediate return to London advisable." The up train soon arrived, and conveyed me back to Euston. I took a cab to Denmark Hill, where I arrived about 2 A.M., and somewhat alarmed my wife by my return home at such an unseemly hour. Sleep did not come readily that night, my mind was too much disturbed ; but in the quiet hours of the early morning I calmly reviewed the whole situation, and rehearsed every detail of the plan of campaign. Then I got a couple of hours' sleep, and by the time breakfast was over I felt sufficiently refreshed, and fully nerved, to carry out the plan which, after renewed consideration, I had determined to follow. I now fully realised the disadvantageous position I should be placed in if this company, with a couple of millions capital, was formed and I was left to fight them single-handed. Even now, after the lapse of so many years, this marvellous revelation, coming as it did at the precise moment necessary to be effective, seems more like an act of eternal justice than one of the ordinary affairs of life. I was startled by it at the time, and, momentous as were the interests involved, I was not unnerved, but, on the contrary, felt greatly encouraged ; for though not possessed of that very great physical courage natural to more robust men, I have ever stood firmer in the face ot a great, an appalling danger, than when encountering some of the smaller risks we all have to run at times.

On the morning following my unexpected return to London, I paid a visit to Mr. David Chadwick, at 11 A.M. ; I said I had called to discuss an important question in relation to the great iron and steel company that was to be formed to purchase and take over the Ebbw Vale

Ironworks and Mines. He started with surprise, but I had so directly assumed the fact that he made no effort to conceal it. I said : " I wish to call your attention to some facts with which you are probably wholly unacquainted, but which most nearly concern your personal interest, as well as that of myself and of your Ebbw Vale clients." I then told him, as briefly as I could, of the attempts that had been made to destroy the value of my invention by cornering manganese, and thus to force me to sell my patents for less money than they were worth. I also referred to Mr. George Parry's patents, neither of which could be worked without directly infringing mine ; therefore that the proposed company could not manufacture cheap cast steel without a license from me, and, what was of still greater importance to him and to them, was the fact that the New Ebbw Vale Steel and Iron Company could not even be formed at all without my consent and permission.

Mr. Chadwick, not unnaturally, doubted this confident expression, and said : " That's got to be proved." I said : " You must excuse my plain speaking, and allow me to call a spade a spade ; I have but to express what is my determination, unless my terms of surrender are accepted. Do not suppose me weak enough to calculate on gaining a single point by mere bluff ; I know, by reputation, that you are a very unlikely person to be led away by such means. I also know, on the other hand, that you might readily enough in your own mind come to this conclusion : ' Well, let Bessemer do what he likes in law ; it will take him some months, but we shall have got our capital in a few days, and shall be in good fighting trim, with £2,000,000 to back us, and can thus afford to laugh at any threat from him.' Now this is just the very thing I have set myself to frustrate. I can fight the question now with £100, and obtain a victory in two or three days, but if I once let you get your capital, it might cost me £10,000, and a couple of years' struggle in the Law Courts ; so you see I must choose this very day to fire the first shot, unless your clients make an immediate and unconditional surrender ; or unless you hold out a flag of truce for two days to enable you to communicate with your clients."

" Now there are two ways of carrying on such a war. If I were bent on fighting, I should mask my batteries, and so fall upon you unawares, you thinking that my armament was very small ; but I have no desire to fight unless I am driven to do so, in which case I should know how to defend myself. There is a great disadvantage in some cases in allowing your enemy to underrate your strength and to rush headlong into war, hence it is my policy just now to show you how completely I have you in my power. What I want, and must have, is the giving up by the Company of all obstructive patents in their possession, and the immediate taking out of a license from me to use my patents instead."

"If this is refused, what is my inevitable course ? I go from here direct to my solicitor, who can readily, in two hours, make a formal written application for an injunction in the Court of Chancery to restrain the company owning these patents, or any new company formed for that purpose, from using them. Meanwhile, I get a thousand blue and red posters printed, announcing the fact that I, Henry Bessemer, have applied for four separate injunctions in the Court of Chancery to restrain the Ebbw Vale Companies using certain patents for making steel, which they are in possession of ; and, further, that I have abso- lutely refused to give a license to the present—or any future—Ebbw Vale Company to use any of my patent processes for the manufacture of cast steel. These facts I can legally publish ; I could, before the day was out, cover every hoarding in the City with these staring placards, and before the members of the Stock Exchange arrive at their offices to-morrow morning, I could have fifty cabs perambulating Cornhill and the principal City thoroughfares with similar placards posted on them, as practised at election times, and distributing hand- bills by the thousand ; if you are of opinion that under these conditions you can get £2,000,000 capital subscribed for a New Ebbw Vale Steel Company, you may try and do so.

"On the other hand your clients, if this altered state of things is communicated to them in a quiet, businesslike way by their own financial agent, will never be mad enough to lose such a chance of realising so vast a sum in ready cash for their old works and plant.

"Iron-making, as far as rails are concerned, is played out. The company must make steel or shut up the works, and they have already put it off too long. My process has rendered large buildings filled with long rows of puddling furnaces of little value; and weak old-fashioned rolling-mills, that would do for iron, must all be replaced by stronger and more modern mills for rolling steel. Your clients must be fully aware of these facts, and they will never risk their present chance of selling the works for the mere pleasure of opposing me. I know this as well as they do, and there lies my source of power. Whereas their unconditional surrender would make everything smooth, their own best interests would be secured, you would get your commission for the formation of the company, and I should get my royalty for all the steel they make. Such is the brief outline of the steps I am bound to take if my offer is rejected."

"What, then, do you propose that I should do?" said Mr. Chadwick.

"Simply this. Go and see your clients, show them clearly their altered position, and absolutely refrain from taking one single step in advance until I have been brought face to face with the owners of the property, or their fully-authorised delegates; and if you pledge yourself to this course of action, I will, on my part, remain absolutely quiescent; but, please remember, that a single word in the public press will bring me into full activity."

Mr. Chadwick was much too keen a man of business not to recognise to its fullest extent the imminent peril in which the prosperity of the new company was involved, and said: "I will at once see my clients on the subject, and will wholly abstain from any further steps for the formation of this company until they have consented, or refused, to discuss the matter with you. But I have little doubt that they will come up to London, probably the day after to-morrow." Thus far we were mutually pledged, and at parting, I suggested that it would be far more agreeable to all parties concerned if they would meet me with a plain "Yes" or "No" to my demands, and so avoid a discussion that might easily terminate in many unpleasant words. I was the more anxious to do this, as every member of the then Ebbw Vale Company was wholly unknown to me, even by name, except Mr. Abraham Darby and Mr. Joseph Robinson; and, although very plain speaking had been necessary in the case of Mr. Chadwick, in order to fully impress him with the gravity of the crisis, it was most desirable that the vendors should be put in possession of these facts in a quiet businesslike manner through their own financial agent, and be thus able to calmly review their position from this new standpoint, make up their minds what course they intended to pursue before seeing me, and thereby avoid any heated discussions on the subject.

On the second day after this interview with Mr. Chadwick, I met by appointment at his offices, Mr. Abraham Darby, who was, I believe, the chief proprietor of the Ebbw Vale Iron Works ; his partner, Mr. Joseph Robinson, was also present. We met on a friendly business footing ; my terms as given to Mr. Chadwick had been accepted, and we had merely to discuss the few details that were necessary. They laid great stress on the large sums of money their patents and their experiments had cost them, setting it down, if I remember correctly, at £40,000. Then this difficulty arose : Mr. George Parry's patent was not in their hands, and £5,000 must be paid to give them an absolute control over it. This I undertook to pay, and on their arranging to go largely into the manufacture of Bessemer steel, I agreed to deduct £25,000 from their first royalties, in lieu of paying money for the purchase of all their patents. After this deduction was made, they were to pay me the same royalties as I charged to other licensees on all the steel they produced.

Thus the two great objects I had in view were accomplished. The signing of my deed of license took the sting out of my opponents, for it contained what lawyers call an " estoppel clause," in which they, under their hands and seals, acknowledged the perfect validity of all my patents : " That they were new and useful," and " were sufficiently described in my specifications," and that " they were all duly specified within the time prescribed by law." This clause deprived them of the possibility of attacking my patents, or refusing to pay the royalties agreed upon in their deed of license.

It was also important that I should get the assignments of all their patents. Not that these patents were in themselves worth the paper they were written on, but so long as they existed and were the property of some other persons, they were fighting material, and could be utilised to keep me in the Law Courts possibly for a couple of years. This might have cost me an amount of money immensely greater than the loss I should sustain by the Ebbw Vale Company's not paying me a royalty on their first year's production of steel; which was, in fact, only the loss of what never would have been mine if I had let them go on their own way unopposed. Under these conditions I withdrew all opposition to the formation of the new steel company, and after a not very long interval I began to receive from the Ebbw Vale Company large sums quarterly in the form of royalty. I cannot, at this distant period, find all the returns of the sums they paid me, but I am under the impression that I received from them altogether in royalties between £50,000 and £60,000; added to this they had given up all the patents which had been held for years suspended over me.

Thus happily was removed the last barrier to the quiet commercial progress of my invention throughout Europe and America—an invention which from its infancy has steadily grown in extent and importance, until the production of Bessemer steel has reached an annual amount of not less than 10,500,000 tons, equal to an average production of 33,500 tons in every working day of the year, and having a commercial daily value of a quarter of a million sterling.

CHAPTER XX

THE BESSEMER SALOON STEAM-SHIP

FEW persons have suffered more severely than I have from sea-sickness, and on a return voyage from Calais to Dover in the year 1868, the illness commencing at sea continued with great severity during my journey by rail to London, and for twelve hours after my arrival there. My doctor saw with apprehension the state I was in. He remained with me throughout the whole night, and eventually found it necessary to administer small doses of prussic acid, which gradually produced the desired effect, and I slowly recovered from this severe attack. My attention thus became forcibly directed to the causes of this painful malady, which I, in common with most other persons, attributed to the diaphragm being subjected to the sudden motions of the ship. Hence, as a natural sequence, its cure appeared only to require that some mechanical means should be devised whereby that part of the ship occupied by passengers should be so far isolated as to prevent it from partaking of the general rolling and pitching motions. In this way I entered, almost without knowing it, into an investigation of the subject; and gradually, as my ideas were developed, I determined to make a model vessel, small enough to be placed on a table, and to which the usual pitching motion of a ship was imparted by clockwork.

On this model was arranged a suspended cabin, supported on separate axes, placed at right angles to each other. I obtained a patent in December, 1869, for this invention, which is represented in two sectional engravings, Figs. 81 and 82, on Plate XXXVII. The cabin, shown in the illustrations, is circular in form, with a hemispherical ceiling or roof, whose centre coincides with the axis of suspension. Seats are arranged all around its circumference, with a gallery above, provided also with

seats, while the circular floor is large enough to serve as a promenade.
A heavy counterbalance weight is suspended vertically below the floor
of the cabin, to retain it in a horizontal plane. In Fig. 81, the cabin is
shown in the position it would naturally assume when the ship is in
dock, and in Fig. 82 in the position it would maintain when the ship is
rolling, that is, with its floor quite horizontal. Immediately beneath
the large pendulous mass which controls the cabin is shown a concave
iron surface, turned quite smooth, and fixed to the ship. This surface
is made with a curve, the centre of which coincides with that of the
axis of the cabin, and the pendulous mass has a heavy cylindrical weight
within it, which is shod with wood. This can be let down so as to
come lightly in contact with the concave dish or surface, or be pressed
down upon it by a screw, if desired, thus acting as a friction brake to
prevent the cabin from acquiring a swinging motion, or when required,
to lock it fast to the ship. There were many other details planned,
which need not be now entered into, as the description I have already
given will serve to show what were the crude ideas presented to my
mind in the early stages of the investigation of this subject. All this
was hurried on in a short time, and I felt determined to put the general
scheme to the test of actual experiment at sea, trusting to remove
defects in the details as experience showed them to be necessary. I
therefore planned a small steamer suitable for carrying this circular
saloon, and entered into a contract with Messrs. Maudslay, Sons and Field
to build it for me for £2,975. This sum was further augmented
by slight alterations of the original plan, bringing up the net cost
of the vessel to £3,061, which was duly paid on the delivery of the
ship to me at Greenwich.

While this small steamer was being built, I continued to study the
subject more deeply, and in doing so, I felt some serious misgivings
as to the motions of translation of certain parts of the ship, depending
on the distance of such parts from the centre about which the vessel
rolled and pitched, and which would tend to set up an oscillating motion
of the cabin. I saw it was necessary to place the axis of the cabin as
near as possible to the point about which the vessel pitched and rolled,
and then the question of absolute personal control of the cabin by a

R R

steersman arose in my mind. This gradually shaped itself into a necessity, if perfect quietness in the cabin was to be ensured. These improvements were of vital importance, and I could not hide from myself the fact that the small steamship which was then being built for me by Messrs. Maudslay, Sons and Field could not be so altered as to give these ideas a fair trial. I therefore abandoned all intention of fitting up my suspended saloon in it, and I eventually sold it in an unfinished state for what it would fetch, so I lost about £2,000 by the first move. I was not, however, discouraged, but on the contrary I felt more confidence than ever in the success of the plans that time and study had so far developed.

The fact that I could not venture out to sea to try my experiments was a great drawback to me, and to meet this difficulty I determined to make a large working model, to try the mechanical motions and other details of my plans, on land. For this purpose I constructed the central part of a fair-sized vessel, omitting the bows and stern portions, which, as will be hereafter shown, had nothing to do with the trials to be made.

This model had 20 ft. beam and was 20 ft. long ; that is, it represented a slice cut, as it were, out of the central part of a vessel as large as the Thames above-bridge passenger steam-boats. It was fitted into a square opening or pit, formed in the ground to such a depth as to represent its natural immersion had it been placed in water, the level of the land surrounding it consequently representing the level of the water in which the model was assumed to be floating. This structure was erected in a meadow at the rear of my residence at Denmark Hill, and was supported on axes in the line of the keel ; it was made to roll by a steam engine actuating a crankshaft and connecting-rod, so arranged as to give a gentle motion to the whole fabric, which weighed several tons. The angle of roll was 15 deg. on each side of a horizontal line ; that is, a complete roll of 30 deg. In the central part, and on a level with the deck of this model ship, was a small saloon 12 ft. by 14 ft. inside, with seats along each side of it, and a row of small windows above them. This cabin was large enough to conveniently accommodate a dozen persons at a time. The ceiling was flat, and its upper surface formed a little promenade deck, with a light iron hand-rail all round it. In the centre of the cabin a small sunk space, surrounded by a railing, permitted the steersman to stand with his head and shoulders a foot or two above the floor, and before him was a spirit-level placed in position at right angles to the axis on which the model ship was made to roll. The steersman had a small double handle immediately in front of him, very like the steering bar of an ordinary bicycle; this handle actuated an equilibrium valve, so easy of motion that a mere child could work it. The valve admitted water under pressure to one side of a piston, and allowed its escape from the other side, thus silently and quietly controlling power capable of holding in absolute check any amount of force tending to put the floor out of a true horizontal plane. If twelve or fourteen persons walked suddenly over in a body from one side of the cabin to the other, it made no perceptible difference, for the steersman had only to watch the spirit-level, and by gently moving the handle keep the bubble permanently in the centre, and thus insure absolute steadiness. If the steersman took his hands off the steering handle, the cabin immediately partook of the motion of the model, which was fully equal to the roll of a small ship in a heavy sea. This sudden transition from absolute quiet to a most unpleasant roll generally resulted in loud shouts of "Stop her!" from the persons seated in the cabin : an order which, after well shaking up the passengers, the steersman always attended to. He applied his hand once more to the lever, when absolute quietness was restored, to the relief of all. As the mechanical demonstration of my scheme, the effect was perfect. This experiment was witnessed and the result admitted by some of the first engineers and scientists of this country, many of whom will recognise in the two illustrations, Figs. 83 and 84, on Plates XXXVIII. and XXXIX.), a correct representation of the apparatus they did me the honour to inspect at my house in 1869.

To facilitate entering and leaving the cabin at all times, notwithstanding the continued rolling motion of the model, half a dozen steps led to a small fixed staging supported by posts driven into the ground. Between this platform and the moving hull were two stout circular steel rods, working in sockets at each end, horizontally parallel to each other. A number of flat oak bars, having a small round hole near each end, were slipped on to these steel bars, a rubber washer between adjacent bars keeping them a short distance apart, and the whole forming a sort of *grille* extending from the fixed stage to the moving hull, and gradually partaking of the slope of the latter; thus, any person could walk with perfect ease from the fixed to the moving part, or *vice versâ*.

My idea of an improved Channel service became generally known, and I had the satisfaction of seeing that I was not alone in my opinions as to its ultimate results. My plans were submitted to the judgment of practical men of the highest mechanical ability. All was said that could be said theoretically on the subject, *pro* and *con.*, and the time for action had now arrived. I therefore laid my plans before the well-known financial agents, Messrs. Chadwick, Adamson, and Co., who undertook the formation of a limited joint stock company, to run steamships between England and France, provided with saloons steadied by the hydraulic apparatus secured under my patents. The prospectus was issued, and in due course the company was registered, with a nominal capital of £250,000, the amount actually subscribed being much below that required even to build the first ship: a fact to which I objected, but which I was assured, was an everyday occurrence. My original intention as patentee was to grant licenses to shipbuilders and passenger steam companies to use my invention, charging a small extra sum for all passengers booking for the saloon. But the company just formed insisted on having the entire monopoly of the ships running between the English and French ports, thus absorbing a great part of the value of the patent, and shutting it up until after their first ship came into use. In order to meet this sweeping demand, I consented to take 10 per cent. on the cost of the ships, which was to be paid concurrently with the remittances to the shipbuilders, and I further conceded to the company a share of my half-crown per head royalty. I thus received no cash payment for this share or participation in my patent, although I had already spent considerably over £5,000 in the construction of a steam-ship and other models, trials and patents, etc. Notwithstanding this, I was among the first subscribers to the company's capital, and as soon as the shares were ready for issue, I applied for £10,000 in ordinary shares, which on allotment I paid for in cash, the best evidence I could give of my entire confidence in the Bessemer Saloon Ship Company.

At the time when the company was formed, I was much pressed to become its chairman, but I declined to do so, or even to take the position of director, because I had not only a great interest in the Saloon Ship Company, but I had other interests as a patentee, which might possibly come in conflict with those of the company. I felt well assured that no man can serve two masters, and I emphatically declined to place myself in so false a position. At the same time, I also declined to make myself the servant of the company in any way; but as they desired my advice and opinion on matters connected with the saloon and its machinery, I accepted the office of Consulting Engineer without fees.

Mr. (afterwards Sir) E. J. Reed, who then held an important position in Earle's Shipbuilding Company at Hull, was appointed Naval Constructor to the Bessemer Saloon Ship Company, and we also had on our Board of Directors, Admiral Sir Spencer Robinson, who was an influential Director of Earle's Shipbuilding Company. It was understood that the ship in all its details should be designed by Mr. Reed, subject to such modifications as the necessities of the saloon imposed, and which were few and simple, although they undoubtedly introduced important structural difficulties.

First, I decided that the saloon, as far as hydraulic control was concerned, should move on axes parallel with the line of the keel, and that pitching in the short sea of the channel should be reduced, as far as possible, by the great length of the ship. It occurred to me that the bows of the ship would not be lifted so high in meeting a high wave, or mound, of water, in front of her, if she had a low freeboard. The forecastle would then receive part of the weight of the mound of water, and not be floated upward to the same extent as if constructed with high bows, which might be surrounded by a heavy rising wave. This was simply a landsman's view of the conditions to be met, in which, however, Mr. Reed concurred, and designed his ship with a low freeboard at both ends, as she was intended to run in and out of the harbours without turning round.

Secondly, to reduce the amount both of pitching and rolling of the saloon, I required a space equal to 70 ft. in length and 30 ft. in breadth

in the centre of the ship for the reception of the saloon, which was
to extend so low down as to bring its turning axis as near as possible
on the line about which the centre would roll. These conditions were
provided for by Mr. Reed, and it only remained for me to design
the saloon and the governing machinery required, all the drawings and
plans for which occupied many months of close application. Here it may
be desirable to refer generally to the means employed for governing the
motions of the saloon; for this purpose I have given an engraved copy
from one of my old drawings (Fig. 85, Plate XL.), showing a cross-section
through the centre.

Two large A-shaped frames, shown partly by dotted lines, were
securely bolted to the main framing of the ship; these frames were several
feet apart, and were held together by stretcher-bars, which passed through
curved slots in the webs of a horizontal pair of large " working beams."
There were strong angle-brackets formed on the upper side of these beams,
which supported the axis about which the saloon moved, similar axes
being provided at or near each end of the saloon floor coinciding in position
with the central axis, as shown on the engraving, thus firmly supporting
the weight of the saloon by strong axes and carrying frames at three
points in its length. It will be seen that at each end of the large working
beams, and coupled in the space between them, were two hydraulic
cylinders hanging vertically from massive girders connected with the main
deck frames, so that any movement or oscillation of the working beams
permitted these hydraulic cylinders and their piston rods to oscillate
slightly, and follow the radial motion of their beam ends. A suitable
set of hydraulic force pumps, driven by a separate steam engine, was
so arranged as to furnish a constant supply of water under any required
uniform pressure.

The "steersman," or controller, was provided with a handle controlling
a set of delicately-balanced equilibrium valves, forming a connection
between the water in the air vessel, or pressure chamber, of the force
pumps and the vertical hydraulic cylinders, which were always kept
full of water on both sides of their pistons by means of a loaded valve
at the discharge end of the exhaust pipe, but with a very much greater
pressure on one side of each piston than on the other. Things being thus

arranged, it will be readily understood that if the ship were in harbour and at rest, the steersman by moving his handle so as to admit water under great pressure into the lower part of the left-hand cylinder, would expel the water which was above the piston through the loaded valve. At the same time the left-hand end of the beam would be forced downwards; and the valve would have admitted water under great pressure on the upper side of the piston contained in the right-hand cylinder, thus forcing or lifting up the right-hand end of the working beams, and so on. It will be seen that if the floor of the saloon could be thus made to oscillate on its axis by means of the hydraulic cylinders when the ship was in dock, the reverse would take place when the ship was rolling at sea; that is while the ship rolled, the use of the hydraulic cylinders would enable the floor of the saloon to remain horizontal. The distance through which a roll takes place, and the time occupied in performing the roll, constantly vary; but by means of equilibrium valves under personal control, this variation could be easily provided for. The spirit-level directly under the eye of the steersman instantaneously indicated to him any movement of the floor from a true horizontal plane, by the travel of the bubble from the centre towards one end. A slight turn of the handle by the steersman prevented further movement. All he had to do was to keep the bubble in the centre of the gauge; and it was found in the working model erected at Denmark Hill that, when going as fast as ten complete rolls per minute, and rolling through an angle of 30 deg., a position of the floor not deviating more than 1 in. or 2 in. from the horizontal was maintained with ease, and with absolute freedom from jerks, a result which the *vis inertiæ* of the heavy mass forming this large saloon would tend to still more favourably secure. The larger the flywheel attached to irregularly-moving machinery the more perfectly are these irregularities controlled by it; and it must always be borne in mind that all oscillating motions in nature commence very slowly and acquire a maximum velocity, gradually becoming less rapid, until motion absolutely ceases in that direction. Then the infinitely slow reaction in the opposite direction takes place, and goes on until a maximum velocity is again arrived at. Let anyone for a minute or two watch the beautiful motions of the pendulum of a common clock; there is no jerk, it does not travel through its whole range at a uniform speed and then start back in the same way, but, like the oscillations of all heavy bodies, obeys those laws which bring the control of oscillations in such bodies within the sphere of applied mechanics.

To prove the confidence felt by my colleagues in the certain success of my scheme, I cannot do better than reproduce here a letter from that eminent authority, Sir E. J. Reed ; this letter was published in *The Times* on November 26th, 1872.

To the Editor of "The Times."

Sir,

The discussion upon Channel steamers has proceeded so far and taken such a form in your columns that it seems proper for me, as the designer of the vessels which are to carry Mr. Bessemer's saloon, to submit the following observations upon the subject. I should have said nothing about the Dicey project had not one of the directors of the Dicey Company made it necessary for the proposers of the Bessemer vessels to defend their work ; and even now I shall offer but a very few words upon it, as the able letter of Colonel Strange, which you published on Saturday, contains nearly all that it is necessary to say.

I believe the "Dicey" ship to be wrong for the following reasons :—*First*, where one of the primary objects is to secure small draught of water, and therefore lightness of structure, the plan in question renders a very unusual weight of hull necessary, because it gives the ship four sides instead of two, and introduces a heavy superstructure for the purpose of yoking the two half ships together ; *secondly*, unless this superstructure is extremely well designed and very strongly built, it will not keep the two half ships effectually together in a heavy storm ; and their separation would be fatal to both ; *thirdly*, there is great reason to suppose that two half ships of equal size and large proportions, placed 30 ft. apart, and yoked together, however propelled, would be circumstanced very unfavourably for high speed, because of the interference with each other of the waves of displacement in retreating from the inner bows ; *fourthly*, there is also much reason, and some experience, to suppose that such a vessel, propelled by an interior wheel, would be under very great additional disadvantages as regards the obtaining of extreme speed—such vessels have, in fact, failed from want of speed ; and, *fifthly*, the Dicey ship, being made (by the separation of the twin portions) of very unusual breadth from stem to stern, is peculiarly unadapted for entering the narrow harbours of Calais and Folkestone in bad weather. I will only further add that the experiment which Admiral Elliot promises with a twin Dicey steamer no bigger altogether than a "Citizen" boat will throw little or no light upon any one of the above questions ; and the very fact that such a vessel is being prepared for the purpose of proving to the public that the Dicey ship is right, when great size and speed are to be realized, strongly inclines me to believe that its advocates have neither considered nor understood the real difficulties that will oppose their success and frustrate their good intentions. In seeking to reduce rolling they have looked past other equally important conditions. I know Admiral Elliot has attempted to silence the objections of mere ship-builders and engineers like myself by telling us that it is as a sailor that he contradicts us, and that it is in the name of sailors that he speaks; but I do not consider that a sailor is any better entitled than other persons to pronounce dogmatically upon such questions as these, nor do I believe that Admiral Elliot has authority to speak in the name of the naval profession in this matter.

I now come to the Bessemer ship, and will state as briefly as I can what she is to be, and why she has been made so.

The present discomforts of the Channel passage are almost wholly due to the smallness of the present steamers, which have been kept of small dimensions and light draught to enable them to frequent the French harbours of Calais and Boulogne; and being so small, they knock about terribly in rough weather. I will not say that these vessels are the best that can be produced of their dimensions, but some of them are well designed, and no improvements without increase of size would make them even approximately fit for the Channel passenger service. The first thing to be done, therefore, is to build much larger vessels,— that is to say, vessels of much greater length and breadth, for the draught of water must not be substantially increased. The limit of length has hitherto been fixed by the breadth of the harbours, because the vessels, which must of necessity run in bow first, have had to be turned round into the opposite direction, with bow seaward, before starting again.

We must first, therefore, dispense with this necessity of turning the vessels round within the harbours, and the only way to do this is to make them capable of steaming equally well in either direction. Now, this, although not by any means so easy a thing to do as many suppose, is nevertheless quite practicable. Both ends of such a vessel can be made quite efficient as a bow, and equally efficient as a stern, provided the necessary steps are taken. This has frequently been attempted, only to result in failure; it has less frequently been done successfully. Those who underrate the difficulties fail; those who truly estimate them and take the necessary pains to meet them succeed. They occur in the hull, in the rudder, in the steering gear, in the locking apparatus, in the engines, and in the paddle-wheels, and we believe that in the Bessemer ship we have well considered and carefully met them all, and thus have secured the power of leaving harbour without turning round. We have consequently escaped from the limit of length hitherto imposed, and have been made free to go to larger dimensions. This is the first important step.

The dimensions we have adopted are: length, 350 ft.; breadth at deck beam, 40 ft.; outside breadth across paddle-boxes, 65 ft.; draught of water, $7\frac{1}{2}$ ft. On these dimensions we have been able to provide for the Bessemer Saloon (the extra weight of which, with all its appliances, is, in fact, not great), and for engines and boilers which will deliver more than 4000 horse-power. At this stage the Bessemer Saloon claims primary consideration, and we have allotted to it the central part of the ship for 70 ft. in length. This splendid saloon, and all connected with it, has been so well described already in your columns that I will not add a word respecting it, except to say that if the mechanical difficulties of working it were far greater than they really are, the mechanical genius of Mr. Bessemer would be fully equal to their mastery. My chief duty is to make the ship thoroughly capable of sustaining the saloon, and of giving ample support to its bearings. This duty has required, of course, novel and well-considered structural arrangements; but more difficult things have been done in our ironclads, and I need not, therefore, dwell upon it. The saloon being in the centre, we had to place the engines and boilers in some other position. They have been placed in duplicate portions

S S

immediately before and abaft the saloon, the vessel consequently having two sets of paddle-wheels. I anticipate some disadvantage in point of speed from this arrangement, and have accordingly provided somewhat more steam power than would otherwise have been needed; not much, however, because the loss will not, in my opinion, be more than a small fraction. As a compensation we have the great advantages of avoiding the risks that attend the use of very large forgings in paddle engines, and of securing the ship against total disablement by engine accidents. The importance of this latter advantage to the owners of such a vessel is great.

I now come to the low freeboard at the extremities. This feature was suggested during the progress of the design by Mr. Bessemer, who considered that it would promote the longitudinal steadiness of the vessel, or, in other words, reduce pitching. Now, had the vessel been intended for ocean purposes, I should have altogether dissented from this proposal, had Mr. Bessemer made it, as he probably would not in that case have done. I feel as strongly as Admiral Elliot can possibly do that a low freeboard at the bow of a fast ocean steamer, or indeed of any ocean steamer, is utterly wrong. In the case of the "Devastation" I incurred much odium because I insisted on giving her a forecastle, and I carefully predicted that even with the forecastle she would be deeply deluged forward by Atlantic seas I have seen the Holyhead packets, which have additional rather than reduced freeboard forward, steam down the long slope of a great wave in the Irish Sea until one-fourth of their length disappeared from view under the succeeding wave. I am no advocate, therefore, for low bows in heavy seas; but the case of the steamer to run between Dover and Calais is a very different one. There the waves, even in the worst weather, are comparatively short—so short as to present an altogether different set of conditions. The pitching of a well-designed ship, 350 ft. long, could there never be great, and the problem that Mr. Bessemer and I had to solve was, not to reduce extreme pitching motions, but to make pitching motions, already necessarily small, still smaller. For this purpose I believe the low freeboard will prove advantageous, or, to say the least, innocuous; and if we should be mistaken on this point the low freeboard can easily be got rid of by prolonging the upper deck and the sides to the extremities—an inexpensive addition. I do not, however, believe this addition will prove desirable, and I hope it will not, because we have gained another very great advantage indeed by adopting the low freeboard at the ends. That advantage is this: although the ship is 350 ft. long in the water, she is only 250 ft. long above the water, where she is exposed to the wind; so that not only shall we escape the risk which other long ships will be exposed to of being blown across the harbour entrance in a gale, but we shall positively be better off than smaller vessels in this respect, because, while we shall have a comparatively small surface exposed to the wind, we shall have a greatly-lengthened surface immersed in the water to resist the leeway resulting from the wind's action. This is a great advantage which the Bessemer ship will possess, and which no other competing vessel that I know of does possess.

I will not seek to further trespass upon your space by dwelling upon other features of the Bessemer vessel. For my part, I do not put her forward as a perfect remedy for sea-sickness in all cases, although I think she will be found a sufficient remedy in the Straits of Dover. Her advantages seem to me to be that she will be large enough herself to escape all but very small movements as regards lifting bodily and pitching. The moderate pitching which she would otherwise experience will be diminished by the low ends, and what remains of it will scarcely be felt at all in the centre saloon. The rolling of the ship, which is the only remaining movement of importance, will be perfectly neutralised by Mr. Bessemer's hydraulic arrangements

In other respects the ship will be fast, capacious, well furnished, and well ventilated. I am, therefore, of opinion that, although she may not fulfil every random prophecy that has been printed respecting her, she will thoroughly fulfil the object which the travelling public desire— namely, that of enabling us to cross to and from the Continent with health, decency, and comfort, instead of being subjected, as we now are in bad weather, to conditions which violate all these, and are in every respect disgraceful to the age we live in.

I have the honour to be, Sir,

Your obedient Servant,

E. J. REED.

The general appearance of the Bessemer saloon steam-ship is clearly shown in Fig. 86, Plate XLI. Her low freeboard at each end is distinctly seen, and the position of her boilers and engines fore and aft of the long saloon, which is to a great extent hidden by a line of deck cabins extending from one pair of paddle wheels to the other.

Separate tenders for the construction of the ship, and for the engine and boilers, were issued, and that for the ship by Earle's Shipbuilding Company, of Hull, was accepted. The contract for the engines and boilers was given to Messrs. John Penn and Company, of Greenwich. Knowing, as I did, what a light and compact class of engine this firm turned out, I was satisfied that we should be sure of admirable design and splendid workmanship. Here, unfortunately, began the first of a series of antagonisms naturally arising from powerful dual interests. Mr. Reed and Sir Spencer Robinson expressed the opinion that it would be difficult to tow the saloon ship from Hull to Greenwich to be engined, but they did not suggest that it would be easy to send Mr. Penn's engines in parts to Hull, as steam-engines are sent all over the world. Finally, the tender of Mr. J. Penn was, by his consent, given up, and the construction of the engines and boilers was handed over to Earle's Shipbuilding Company.

Unfortunately, financial difficulties and misapprehensions occurred at this early stage; some of the latter were so erroneous that I find it impossible to pass over this subject (as I should have wished to do) in silence, but will content myself by simply stating facts which the Company's books, the list of shares issued, and my vouchers for payment, render absolutely indisputable. Several months after the formation of the Company, the amount of cash in the bank was getting very low, and I subscribed first £3,000, and then £2,000 more in the purchase of shares, thus bringing up the amount of my ordinary shares of the Company to £15,000. Very soon after this, and for the same reason, I took a further sum of £5,000 in Debenture Bonds, raising my investment to £20,000.

In the interim, I had received from the Company £3,000, being 10 per cent. on the early payments to the shipbuilders (Earle's Shipbuilding Company, Hull); but for more than a year afterwards, the Saloon Ship Company, although they found money to pay Earle's, could not do so to fulfil their engagements with me. At last, I consented to take £6,000 more in Debentures, in lieu of the cash then owing to me on the 10 per cent. account. The Company were still short of funds, and as none of the large capitalists connected with it would take any more of the Debentures, I had again to put my hand into my pocket for another £5,000, for which I accepted Debentures at par, bringing up my investment to £31,000. This money was soon absorbed, and tradesmen who had done work on the ship, or had supplied goods to the Company, could not be paid, and they were becoming clamorous for their money. I naturally felt much annoyed to find this state of things going on in a concern with which my name was so intimately connected, and, in spite of my knowledge of the embarassed state of the Company, I offered to lend them £3,000 for a week or ten days, as money was expected within that time. They accepted my offer, but handed me a bill at three months' date for the amount; and having waited that time I was requested not to present it, as there were no funds provided to meet it. I accordingly held it over, but firmly determined not to allow my sympathy with the objects of the Company to draw me into further risks.

But very soon after this prudent resolve there came a worse pinch than ever. The boat was lying in the Millwall Docks; the London, Chatham, and Dover Railway Company wanted to run it for the holidays; but it must be insured before it could safely be sent round to Dover. There was the further difficulty that the debentures could not be legally issued, for one of the conditions attached to them was that the boat should be insured for £100,000, and the premium on this insurance was no less than £7,000. There appeared to be no means of raising this sum : all the shares were fully paid up, debentures could not be issued, for no one could be induced to take them. I knew the Company was deeply in debt, and wholly without the means of paying, and that, therefore, they could give me no sound security for my further advances. But, on the other hand, a collapse was imminent, and to prevent this catastrophe I lent the £7,000 to cover the insurance and get the boat round to Dover, thus bringing up my total investment and advances to the Company to £41,000. This sum of £7,000 was borrowed from me under promise of a special resolution of the Board, stating that the two sums of £3,000 and £7,000 should be repaid as soon as £10,000, which had been promised to be placed to the credit of the Company on the security of £20,000 in Debentures, had been received. In due course this £10,000 was placed to the credit of the Company, and a cheque was drawn for me—not, however, for the £10,000 owing me, but for £7,000 only. I pressed hard for the £3,000 in cash, which by a special resolution of the Board I had a right to, and which was in their possession ; but I failed to get the money, and after a time I was glad to take £3,000 in Debentures, in lieu of the money lent them, although I knew at the time that these Debentures were of very questionable value. I, therefore, held in the Company £15,000 in Ordinary Shares and £19,000 in Debentures.

I have shown that by Agreement and Deed of License I was only to give my advice and opinion, but being above all things desirous for the success of the enterprise, I took upon myself an immense amount of practical detail. I had been some years without doing any actual work at the drawing-board ; my staff of assistants was scattered, and I feared to entrust so important a matter as the arrangement of all the details of the saloon machinery to strangers. I consequently went to the drawing-board myself, working long hours for many weeks together. At my then time of life, and with the effects of my former efforts still hanging about me, this work proved too much. I suffered constantly from severe headache and want of sleep, and at last my health broke down so completely that my friends became alarmed, and I consulted

Dr. Jenner, who ordered me at once to leave home and all business matters for some months, enjoying perfect quiet and repose. I, however, held on, and got such further professional assistance as was necessary to finish the work before I left London. I also went to the cost of many photographs and two large coloured drawings of the interior of the saloon, by means of which the Directors were enabled to see precisely what was intended to be done. Everyone seemed to pride himself on the beautiful saloon, and not a word was raised about the expense of it; each of the contracts for oak carving, cartoon paintings, and gilt decorations passed the Board with full approval.

I may mention that, during my study of the best means of governing the saloon, I proposed to employ a gyroscope, driven at a very high speed by a steam turbine on the same axis. There was enough doubt about so novel a contrivance to prevent me from feeling quite justified in advising the Company to go to the expense of trying it; and, on the other hand, I believed it would be a splendid success if it acted at all in the way proposed, and seeing that the cost of the apparatus would not exceed £500, I volunteered to go to this expense, and had a beautiful instrument made on a large scale; I also went to the further expense of taking out patents, at home and abroad, so as to secure its use to the Bessemer Saloon Ship Company. But when the latter fell into liquidation, this beautiful instrument, which was chiefly constructed of gun-metal, was sold as old metal; only the fly-wheel remained to give an idea of its size.

The Bessemer Saloon Ship Company were most fortunate in finding in Mr. Forbes and Capt. Godbold, of the London, Chatham and Dover Railway Company, gentlemen who not only thoroughly appreciated the advantages to the Channel service promised by the Saloon Steamboat, but who also, with enlightened liberality, seldom equalled in a public body, gave the most valuable help on all occasions: lending the services of their experienced Commodore, Captain Pittock, and in a dozen other ways affording the most generous assistance.

Among other things they organised a trial trip of the Saloon boat, to take place on the 8th May, 1875, having invited the most influential officers and others connected with their Company and with the Channel service. Now, what was the result of this elaborately-organised trial trip, from which so much had been expected ? On a beautiful calm day, in broad daylight, at a carefully-chosen time of the tide, and with all the skill of the best Channel navigator, the ship dashed into the pier at Calais for the second time out of three attempts to enter the harbour, doing damage for which the authorities claimed £2,800 (a sum greatly in excess of the injury done); and for this an undertaking had to be given before the vessel was allowed to depart. On this run, with every effort, she did not steam faster than the small boats, although from the huge columns of smoke issuing from the funnels it was evident to all on board that she was consuming coal at a furious rate. The fact that the boat did not answer her helm was sworn to by Captain Pittock before the Consul at Calais.

That this mishap did in no way arise from any failure of the saloon itself, or from any inefficiency in the machinery used to control it, I have the testimony of Mr. E. J. Reed (the Company's Naval Architect), as well as that of Admiral Sir Spencer Robinson. The facts of the case were given by these gentlemen in the most clear and emphatic terms.

During the week immediately following the catastrophe at Calais I remained in Paris, and on my return to England I had placed before me, by the Secretary of the Saloon Ship Company, a letter for my approval, which, as endorsed thereon, was intended to be sent to all the London daily papers. The letter was written by no less an authority than Mr. E. J. Reed, and it had also been approved by Admiral Sir Spencer Robinson, who had, at his own discretion, made some additions to the latter part of the letter, leaving intact, and without one word of alteration, all the part of it having reference to the saloon and its machinery, which I therefore quote as being purely independent evidence, written without my knowledge or suggestion, and intended to convey to the world, through the medium of the public press, the simple facts of the case. This letter was to appear in the papers as if written by the Secretary under the authority of the Board, and to be signed by him; but as Captain Davis (one of the Directors) did not agree with some of the statements made in the latter part of the letter in reference to the steering powers of the boat, the publication of it was postponed, and it was never sent to the press.

Here follows a correct copy of so much of the letter in question as refers to the Saloon and its machinery :—

THE BESSEMER.

I am instructed by the Board of this Company to request your kind insertion of the following remarks upon a subject which appears to be of sufficient public interest to justify this request.

The facts that the Bessemer is not yet running between England and France, and that on two occasions the pier of Calais has been injured by her in entering, have led some persons to state that the vessel has failed, and that the object which the Company had in view cannot be accomplished. That this is a hastily-drawn inference will appear from the following.

The Bessemer was built primarily for the purpose of showing that the rolling motion of a passenger steamer might be neutralised in a saloon supported upon axles and controlled by hydraulic power. It is well understood that this was a great experiment, and all reasonable persons expected that the totally novel machinery required for keeping the saloon at comparative rest, however successful in principle, would require some experience, and probably some minor modifications, in order to put it successfully to work.

Now up to this present moment Mr. Bessemer and his representatives have been able to make but extremely few trials, and there does not appear to be the slightest ground for alleging that he will fail in his object. He has amply proved the sufficiency of his machinery for applying to the Saloon all the power that is requisite for the purpose. The only changes which he has yet found desirable have been of a minor kind, and connected only with the valves and levers. These improvements have not yet been properly tried, for it is not an easy thing, particularly at this season of the year, to find suitable opportunities for working the cabin at sea, and for making such adjustments as experiment only can indicate. Any supposition of failure, therefore, with regard to Mr. Bessemer's plans, is altogether premature and without proper foundation.

After this second collision with the Calais pier, nothing was done to test the powers of the hydraulic machinery; not a single thing was done or alteration made, not even a screw was undone or touched, so that the saloon and its hydraulic governing machinery still remains an untried mechanical problem. And it is important that it should be understood how it happened that the machinery connected with the saloon was prevented from being completed by a similar accident, or collision, with the Calais pier about three weeks prior to the fatal smashing of the pier on the public trial on May 8th, 1875. The simple facts are these :—

Immediately after this public trial-trip had been decided upon, the Saloon Ship Company thought it prudent to have a rehearsal, and it was arranged that Captain Pittock, the able Commander of the Chatham and Dover steamboats, should run the boat into Calais harbour at mid-day, and return at once to Dover. Matters being thus arranged, Captain Pittock started from Dover for this private trial-trip about the middle of April; and, notwithstanding his long experience in daily navigating the Channel for twenty years, in daylight and in darkness, in calm and in storm, yet on a bright Spring morning, with a gentle breeze, he failed to steer safely into Calais Harbour, which he knew so well, and where at all states of the tide, and in all weathers, he had steered his Channel ships thousands of times without a mishap of any kind. On this rehearsal trial he was unable to keep the Bessemer ship off the pier, which she crashed into, not with her bows but with her paddle-wheels, doing much damage to the pier, but still more damage to one paddle-wheel and adjacent parts of the ship. He was, however, able to back out of the harbour that he had partially entered, and by the aid of the other pair of paddle-wheels to crawl back again into Dover Harbour, thus deranging the whole programme, and altering all that had been decided to be done during the three weeks pending the great demonstration advertised to be made on the 8th May, and which could not be put off.

The saloon machinery was nearly completed, but the whole of its working parts had never once been put together, and the trial referred to in the letter written by Mr. Reed had reference only to the testing of joints and connections, steam pumps, etc. ; no trial whatever up to the present hour has ever been made with the complete apparatus, which, in fact, was never finished. The interval of about three weeks between the middle of April and May 8th would have enabled me to complete my work, and also to get a first rehearsal of the saloon with its machinery absolutely finished, prior to the public use of it on the 8th May, had it not been for the smashing of the paddle-wheel. But the first thing to be done after the accident was to render the ship itself capable of performing the advertised voyage, and with this object every available man was put on the repairs of the disabled paddle-wheel, and the other parts of the vessel injured in its collision with the pier.

There was scarce time, by working night and day, to get the ship again in good order for the 8th of May. It was impossible that time could be allowed me to have a trial-trip and a proper rehearsal of the Saloon machinery, and I did not feel justified in subjecting our visitors to the first trial of so novel an invention, with a steersman absolutely without practice. Seeing this was to be the case, I employed the few hands that could be spared in riveting some plates and stays to the underside of the saloon, and securing their opposite ends to the main ribs and bottom of the ship, thus making the saloon, for the time being, a part and parcel of the ship itself, like any other fixed cabin, and quite safe for persons to go into it, or crowd upon its upper open deck, as they did on the journey to Calais.

Thus, owing to the want of control of the rudder, this first smash of the Calais pier destroyed the only opportunity I ever had of trying the action of the saloon with all its mechanical arrangements complete. With the rigidly-fixed saloon the invited company started for Calais; everyone was charmed with it, the proportions being so unlike the cabin of a Channel boat. It formed a room 70 ft. long by 30 ft. wide, with a ceiling 20 ft. from the floor; its beautiful morocco-covered seats; its fine carved-oak divisions and spiral columns; its gilt, moulded panels, with hand-painted cartoons; its groined ceiling, tastefully decorated, gave an idea of luxury to the future Channel passage which all seemed to appreciate.

I have given an illustration of the interior of the saloon (see Fig. 87, Plate XLII.) in section, taken from a large water-colour painting, closely following all the details of the structure; but it requires a very fertile imagination, when looking at this small black-and-white illustration, to fill in the exquisite oak carving and arabesques in its numerous panels, its bold cartoon filling each space between the spiral oak columns, with the beautiful colouring intermixed, with just enough gilding to convert the decorations into one harmonious whole, pleasing to the eye but not distracting to the senses: a room which did infinite credit to its able and truly artistic decorators, Messrs. B. Simpson and Son. Everyone on board on that fatal 8th of May roamed over the various small cabins connected with the saloon, and ascended to the upper deck. They all had gone over the ship, and commented, according to their different tastes and ideas, on the many novelties in this new structure; and in the interim we had arrived—very slowly, it must be admitted—at the entrance of Calais Harbour. I, knowing what had occurred on a previous occasion, held my breath while the veteran Captain Pittock gave his orders to the man at the helm. But the ship did not obey him, and crash she went along the pier side, knocking down the huge timbers like so many ninepins!

I knew what it all meant to me. That five minutes had made me a poorer man by £34,000; it had deprived me of one of the greatest triumphs of a long professional life, and had wrought the loss of the dearly-cherished hope that buoyed me up and helped to carry me through my personal labours. I had fondly hoped to remove for ever from thousands yet unborn the bitter pangs of the Channel passage, and thus by intercourse, and a greater appreciation of each other, to strengthen the bonds of mutual respect and esteem between two great nations, while it still left us the silver streak for our political protection. All this had gone for ever.

It will be readily understood that this second catastrophe at Calais finally determined the fate of the Bessemer Saloon Steamboat Company, which had thus become hopelessly discredited; its financial position was equally bad, and there only remained the formal act of winding up the Company, from which I withdrew myself, much disappointed.

Had this unfortunate ship been able to steam rapidly and steer safely, all might still have been saved, for Captain Godbold, the Foreign Traffic Manager of the London, Chatham, and Dover Railway Company, distinctly stated to me that had the Bessemer been capable of steering safely into Calais Harbour at the promised speed, his Company would have run her regularly on this station, even if her saloon had always been kept locked fast to the boat, on terms that would have yielded a handsome profit to the Saloon Company, and thus have afforded ample opportunities between her trips across the Channel for practising and perfecting the controlling apparatus, and training two or three men to this new occupation.

I have already explained the conditions under which I granted a monopoly of a portion of my patent rights to the Saloon Company, and for which I was to receive 10 per cent. on the cost of the ships built by them. Now, if there is any force in a sealed contract, deliberately entered into by men of business, I had a clear right, both moral and legal, to the advantages secured by that contract. I state this distinctly, so that there may be no mistake, and that no one shall be able to say that my determination never to apply to my own private use one shilling of the money so obtained from the Saloon Company, was because I had any doubts of my moral and legal rights thereto. It was purely because I would not put into my own pocket one shilling earned by my invention, while there were tradesmen and manufacturers who had done work and supplied material to the Saloon Company, but who remained unpaid their just debts.

As before stated, I had received £3000 in cash, and had been obliged to take £6000 in debentures in lieu of that amount in cash. These I was fortunate to sell for £3000, making my gross receipts from the Company £6000. I held only three-fourths of the saloon patents, and I therefore handed over £1500 to my friends who held the remainder, thus reducing the amount personally received by me to £4500.

I then requested my solicitor to write to each of the creditors of the Bessemer Saloon Steamboat Company, fixing a day, viz., Wednesday, June 23rd, 1875—when, on applying at the offices of Messrs. Watkin, Baker, Bayllis, and Baker, of 11, Sackville Street, their accounts would be paid in full. It appeared that there were twenty-one creditors, whose united claims on the Company amounted to the sum of £3328 18s. 9d., which was duly paid to them. I am, at the time of writing, still in possession of all the receipts given by these creditors.

The payment of these twenty-one accounts left me with a balance of £1172 out of the monies I had received from the Saloon Company, and which sum I had proposed to hand over to some public charity, but one of the unfortunate shareholders suggested that "Charity begins at home," and I therefore handed over this balance to the liquidators. All these details would for ever have been buried in oblivion but for the fact that I had, through no fault of my own, been identified with the affairs of this bankrupt company; and I consequently feel bound to vindicate my character, and to show that I had, time after time, helped with a liberal hand to extricate the Company from its financial difficulties by taking further shares. I desire also to show that I had not benefited by my connection with the Company to the extent of a single shilling, either for my arduous personal services or for the sale of a portion of my patent rights to them. And, further, that the collapse of the Company was not caused by any failure of my invention, which remains to this hour an untried mechanical problem, in which I have still the most perfect confidence. Indeed, nothing has happened to lessen or destroy the confidence with which I had followed it up from its first inception to the time of the Calais smash; and, even when all seemed lost, I could not resist one more attempt to save the Company and the unfortunate shareholders. We were not bound to that particular ship, and if an opportunity could only be obtained to show that the Saloon when finished would do what was expected of it, all might yet be well. But who was to lead this forlorn hope? I, if anyone. But how dare I run such a ship on my own responsibility? I was not mad enough for this; but the ship was worth so little that the liquidators might be induced to risk taking her to sea, after the completion of the Saloon machinery had been effected, and I was willing to risk yet another £1000 to get this done and the device properly tried at sea. For this purpose I proposed to place the sum of £1000 in the London Joint Stock Bank in the names of Mr. J. O. Chadwick, one of the liquidators, and Captain Henry Davis, a director of the Saloon Company. In order that the fund thus provided should be applied in a manner that would be satisfactory to the liquidators, I proposed to form a committee of three competent engineers—viz., Mr. John Beckwith, manager of Messrs. Galloway and Sons, who made all the hydraulic apparatus for the saloon; Mr. Robert Charles May, of Great George Street, Westminster, an eminent civil engineer, and myself. These three persons were to decide by a majority on all the steps to be taken, and to draw cheques on this £1000 for the payment of fitting up, completing, and working experimentally the hydraulic apparatus at sea. I insisted, however, that I, personally, should not be held responsible for any damage the Saloon ship might do to herself, or to other vessels she might collide with or run into, etc. This offer, if accepted, would in all human probability have saved the whole property of the Company from wreck by proving the success of the saloon machinery, but it was refused by the liquidators, who thus gave the final *coup* to this most unfortunate undertaking.

In writing an account of the more salient incidents of my professional career, it was impossible for me to omit the story of the Saloon Steam-ship, about which the general public very naturally came to the conclusion that my system of controlling the motion of the saloon by hydraulic power had proved an entire failure, and that the collapse of the Saloon Steam-boat Company had consequently ensued. Nothing could be more absolutely untrue, but a simple denial of that fact on my part would have had no weight against the fact that the Company had collapsed. I was, therefore, obliged to choose between two alternatives ; I must either for ever remain under the stigma of this supposed failure, or I must combat that erroneous impression by placing unreservedly the leading facts of the case before the public, and thus bring home, even to the untechnical reader, evidence that no fair-minded person can hesitate to accept. It must be borne in mind that I had personally expended over model ships, patents, and experiments, some £5000 prior to the formation of the Saloon Company. Now this Company ought certainly to have been a source of gain or profit to me, or, at any rate, have recouped my initial outlay; but, on the contrary, I had to prop it up, taking an undue amount in shares, and ultimately losing £34,000 on them. This may be said only to show great confidence in the invention on my part, but when all was over, and the Company was in liquidation, while still smarting under my pecuniary losses, and still more so over my loss of professional reputation, by the supposed failure of the untried Saloon, what could have induced me to place £1000 in the bank to give my invention a trial, if it had already proved a failure? This is, at least, a point upon which the common-sense of unscientific people will enable everyone to form a sound opinion, and accept the fact that my hydraulic controlling apparatus was never completed, was never tested at sea, and consequently *never failed.*

CHAPTER XXI

[The Bessemer Autobiography terminates with the preceding page. The obvious intention to continue it, practically to completion, was never carried out; for although to within a few months of his death Sir Henry was busily occupied in collecting notes of an active though retired period, the narrative to be evolved from these notes was not commenced. The alternative was, therefore, to present an unfinished story, or to complete it with the assistance of his eldest son, Mr. Henry Bessemer. The latter alternative being considered the more desirable, and Mr. Bessemer having kindly offered his collaboration, the following Chapter has been added to this book.—Ed.]

THE unfortunate destruction of my father's copious notes relating to those years of his life after he had retired from active business, but not from usefulness, has made my task a difficult one, because I have to rely on memory, aided by some memoranda and letters ; and because I was not at that time in constant touch with my father, as he resided in London and I in a rather distant part of the country.

I have read the pages of this Autobiography with much care, and with a critical desire that no paragraph shall go forth to the world that can in any way reflect on my father's memory or do an injustice to those who were apparently hostile to him, and I find no statement which goes beyond the limit of accuracy. To be sure, all the matters referred to are ancient history. Most of the actors, too, in what was a vivid drama, are dead and, perhaps, forgotten, but they were very real to my father when he wrote his story—nearly as real as when he was the chief actor thirty years before. Perhaps I may be prejudiced, but it seems to me that, old as it is, my father's story must always remain full of interest, not only because it records in detail the early history of the greatest invention of modern times, but also because on almost every page there is a lesson to the young inventor—a revelation of the secret of success. From the commencement to the end of his long and most honourable career, my father never failed to put in practice his motto, "Onward Ever!"

I find a few matters on which I should like to touch before I turn to the more difficult task I have undertaken—a sketch of my father's later years of leisure and retirement. Some of these matters are of small importance, but they possess an interest from the fact that several of his casual inventions, dating many years back, have in more recent times been re-invented, and taken their place among the everyday necessities of our lives. I am not claiming for my father any special merit in this: almost every true inventor anticipates the wants of the public (or some of them) before that public even knows of the requirement, and solves the problem years before the need for it is realised. To give two or three illustrations.

In 1846 he obtained a patent in which, among other matters, was described a method for making an elastic communication between the ends of railway carriages, so that the whole train could be continuous from end to end. The device consisted in stretching leather or other material over collapsing frames, after the fashion of the hood of a landau; the ends of all the carriages being fitted with such hoods, they could be brought together and secured so as to make a covered connection between the vehicles. This was certainly an anticipation of the vestibule train.

That my father was a prolific inventor is evident from his Autobiography, and in one place he makes special reference to the large number of patents for inventions which he secured. I have been at some pains to make as complete a list as possible of these patents and applications for patents, and this list I subjoin, arranged chronologically. It will be noticed that the patents chiefly refer to four main subjects: The manufacture of glass, the manufacture of iron and steel, improvements in ordnance, and the manufacture of sugar. Of these, only the patents relating to the manufacture of iron and steel bore a plentiful harvest. As already explained, no patents of importance were obtained by my father for the manufacture of bronze powder.

LIST OF PATENTS GRANTED TO HENRY BESSEMER, 1838-1883.

1838	March 8.	No. 7,585.	Casting, breaking off, and counting printing types.
1841	Jan. 6.	No. 8,777.	Checking or stopping railroad carriages.
1841	Sept. 23.	No. 9,100.	Manufacture of glass.
1843	June 15.	No. 9,775.	Manufacture of bronze and other metallic powders.
1844	Jan. 13.	No. 10,011.	Preparing paint and varnishes for fixing metallic powders or leaf.
1845	Dec. 5.	No. 10,981.	Atmospheric propulsion, and exhausting air and other fluids.
1846	July 30.	No. 11,317.	Manufacture, silvering, and coating of glass.
1846	Aug. 26.	No. 11,352.	Railway engines and carriages.
1847	July 17.	No. 11,794.	Manufacture of glass.
1848	March 22.	No. 12,101.	Manufacture of glass.
1849	Jan. 31.	No. 12,450.	Manufacture of glass.
1849	April 17.	No. 12,578.	Manufacture of cane sugar.
1849	May 15.	No. 12,611.	Manufacture of oils, varnishes, pigments and paints.
1849	June 23.	No. 12,669.	Raising and forcing water.
1849	Sept. 20.	No. 12,780.	Preparation of fuel and stoking machinery.
1850	July 22.	No. 13,183.	Figuring and ornamenting surfaces.
1850	July 31.	No. 13,202.	Manufacture and treatment of sugar.
1851	March 20.	No. 13,560.	Manufacture and refining of sugar.
1851	Nov. 19.	No. 13,819.	Ornamenting woven fabrics and leather.
1852	Feb. 24.	No. 13,988.	Manufacture of sugar.
1852	July 24.	No. 14,239.	Manufacture of sugar.
1852	Nov. 19.	No. 795.	Treatment of cane juices.
1852	Nov. 19.	No. 796.	Manufacture of sugar.
1852	Nov. 19.	No. 797.	Treatment of washed sugar.
1852	Nov. 19.	No. 799.	Concentrating saccharine fluids.
1853	June 18.	No. 1,483.	Manufacture of waterproof fabrics.
1853	July 14.	No. 1,687.	Refining and manufacturing sugar.
1853	July 15.	No. 1,689.	Manufacture of bastard sugar from molasses and scums.
1853	July 15.	No. 1,691.	Manufacture and refining of sugar.
1853	Dec. 2.	No. 2,811.	Manufacture and refining of sugar.
1853	Dec. 9.	No. 2,875.	Railway axles and brakes.
1854	Aug. 21.	No. 1,835.	Treatment of slag.
1854	Aug. 25.	No. 1,868.	Naval and military guns.
1854	Nov. 24.	No. 2,489.	Projectiles and guns.
1855	Jan. 10.	No. 66.	Manufacture of iron and steel.
1855	Jan. 10.	No. 67.	Manufacture of ordnance.
1855	June 18.	No. 1,382.	Screw propellers, cranks and propeller shafts.
1855	June 18.	No. 1,384.	Manufacture of cast steel and mixtures of steel and cast iron.
1855	June 18.	No. 1,386.	Manufacture of ordnance.

U U

1855	June	18.	No.	1,388.	Manufacture of rolls or cylinders for shaping metals, crushing ores, etc.; and calendering, glazing, embossing, printing, and pressing.
1855	June	18.	No.	1,390.	Manufacture of railway wheels.
1855	Oct.	17.	No.	2,317.	Manufacture of anchors.
1855	Oct.	17.	No.	2,319.	Manufacture of railway bars.
1855	Oct.	17.	No.	2,321.	Manufacture of cast steel.
1855	Oct.	17.	No.	2,323.	Metal beams, girders, and tension bars used in constructing buildings, viaducts, and bridges.
1855	Oct.	17.	No.	2,325.	Ordnance and projectiles.
1855	Oct.	17.	No.	2,327.	Railway wheels.
1855	Dec.	7.	No.	2,768.	Manufacture of iron.
1856	Jan.	4.	No.	44.	Manufacture of iron and steel.
1856	Feb.	12.	No.	356.	Manufacture of malleable iron and steel.
1856	March	15.	No.	630.	Manufacture of iron and steel.
1856	May	31.	No.	1,290.	Shaping, pressing, and rolling malleable iron and steel.
1856	May	31.	No.	1,292.	Manufacture of iron and steel.
1856	Aug.	19.	No.	1,938.	Manufacture of iron and steel.
1856	Aug.	25.	No.	1,981.	Manufacture of iron and steel.
1856	Nov.	4.	No.	2,585.	Manufacture of railway rails and axles.
1856	Nov.	10.	No.	2,639.	Manufacture of iron and steel.
1856	Nov.	18.	No.	2,726.	Manufacture of iron.
1857	Jan.	24.	No.	221.	Manufacture of iron and steel.
1857	Sept.	18.	No.	2,432.	Manufacture of cast steel.
1857	Nov.	5.	No.	2,808.	Treating iron ores.
1857	Nov.	6.	No.	2,819.	Manufacture of malleable iron and steel, and of railway and other bars, plates, and rods.
1857	Nov.	13.	No.	2,862.	Treating and smelting iron ores.
1857	Nov.	20.	No.	2,921.	Manufacture of iron and steel.
1858	July	30.	No.	1,724.	Cleaning pit coal.
1858	Dec.	1.	No.	2,747.	Wheels and tyres.
1859	March	16.	No.	670.	Manufacture of crank axles.
1860	March	1.	No.	578.	Apparatus for the manufacture of malleable iron and steel.
1861	Jan.	26.	No.	216.	Ordnance and projectiles.
1861	Feb.	1.	No.	275.	Manufacture of malleable iron and steel, and apparatus therefor.
1861	April	27,	No.	1,069.	Projectiles and ordnance.
1862	Jan.	8.	No.	56.	Apparatus for the manufacture of malleable iron and steel.
1863	Jan.	5.	No.	37.	Apparatus for pressing, moulding, shaping, embossing, crushing, shearing, and cutting metallic and other substances.

1863	Jan.	13.	No.	114.	Manufacture of malleable iron and steel, and furnaces and apparatus therefor.
1863	June	9.	No.	1,439.	Construction of hydraulic presses and machinery.
1863	Nov.	5.	No.	2,744.	Manufacture of railway bars.
1863	Nov.	5.	No.	2,746.	Manufacture of malleable iron and steel.
1864	Jan.	25.	No.	217.	Manufacture of projectiles.
1864	Jan.	30.	No.	265.	Manufacture of armour plate.
1865	May	1.	No.	1,208.	Manufacture of pig-iron or foundry metal, and of castings thereof.
1865	Nov.	3.	No.	2,835.	Manufacture of iron and steel, and apparatus therefor.
1867	Aug.	14.	No.	2,343.	Ordnance.
1867	Nov.	11.	No.	3,193.	Grindstones and artificial stones.
1867	Dec.	9.	No.	3,501.	Manufacture of firebricks, retorts, and crucibles.
1867	Dec.	31.	No.	3,714.	Treatment of cast iron and manufacture of malleable iron and steel.
1868	March	21.	No.	965.	Manufacture of iron and steel.
1868	March	21.	No.	967.	Manufacture of iron and steel.
1868	March	31.	No.	1,095.	Manufacture of iron and steel, heating and melting of metals.
1868	Nov.	10.	No.	3,419.	Manufacture of cast steel and homogenous malleable iron.
1869	Feb.	23.	No.	566.	Apparatus and buildings for manufacture of cast steel and malleable iron from pig iron.
1869	May	10.	No.	1,431.	Manufacture of malleable iron and steel, and furnaces therefor.
1869	May	10.	No.	1,432.	Furnaces for obtaining cast steel or homogeneous malleable iron from wrought iron or pig.
1869	May	10.	No.	1433.	Conversion of molten pig iron into homogeneous malleable iron or steel.
1869	May	10.	No.	1,434.	Treatment of pig iron and apparatus therefor.
1869	May	10.	No.	1,435.	Blast furnaces, their gaseous products, and the construction of blowing engines.
1869	Aug.	10.	No.	2,397.	Melting and casting metals.
1869	Dec.	22.	No.	3,707.	Vessels for prevention of sea-sickness.
1870	Feb.	24.	No.	553.	Vessels for prevention of sea-sickness.
1870	May	27.	No.	1,559.	Vessels for prevention of sea-sickness.
1870	May	30.	No.	1,580.	Steamships for prevention of sea-sickness.
1870	June	17.	No.	1,742.	Vessels for prevention of sea-sickness.
1870	Nov.	29.	No.	3,130.	Ordnance and ammunition.
1871	Jan.	27.	No.	223.	Marine artillery.
1871	Feb.	15.	No.	386.	Repairing and converting vessels.
1871	June	1.	No.	1,466.	Ordnance and projectiles.
1871	July	4.	No.	1,737.	Asphalte pavement.
1872	Oct.	1.	No.	2,897.	Passenger vessels.

1873	March 22.	No.	1,076.	Controlling, etc., suspended saloons; discharging marine artillery.
1874	Sept 24.	No.	3,274.	Ships' saloons, cabins, etc.
1874	Sept 28.	No.	3,319.	Supplying water.
1874	Dec. 10.	No.	4,258.	Ships' saloons, cabins, etc.
1875	Dec. 31.	No.	4,552.	Reflectors, lenses, etc.
1879	April 5.	No.	1,368.	(A. G. Bessemer and Sir H. Bessemer.) Making tinplate and blackplate.
1879	Oct. 10.	No.	4,110.	Tinplate bars or slabs.
1880	March 6.	No.	987.	(A. G. Bessemer and Sir H. Bessemer.) Making malleable iron; making castings or ingots.
1882	Oct. 30.	No.	5,171.	Loading, etc., merchandise; rolling stock of railways.
1883	Jan. 18.	No.	305.	Loading, etc., merchandise; rolling stock of railways.

In the earlier pages of the narrative, my father relates the story of a visit he paid to the works of some friends of his, Messrs. Hayward and Co., manufacturers of paints and varnishes, in London. He tells how he was struck with the time-honoured, wasteful, and imperfect process of making drying oils in an iron pot over an open fire: a crude method, always attended with uncertainty, danger, and not infrequently with a complete loss of the whole charge. We are told how he recommended a new, simple, and certain plan to replace the old primitive and dangerous method—a plan that had occurred to him as he walked through the works, and which he embodied in a sketch. The idea was put into practice by his friends, to their lasting profit, as they for years kept it a secret in the colour trade. The new plan (not described in the Autobiography) was this: instead of a small charge of two or three gallons being heated over an open fire, some fifty or sixty gallons were run into a tank, in the bottom of which was a pipe terminating in a large rose-head. Connected with this pipe was a coil that could be heated to any desired temperature, and air could be forced through this coil, escaping from the rose-head into the oil. The exact degree of heat required could be thus maintained, and the process completed with certainty and safety, without waste, and, above all, without any discoloration of the oil. This may seem but a small matter—as, indeed, it was so far as my father was concerned, for the incident passed from his mind until he was reminded of it later. But it proved a fortune to the firm, and to-day exactly the same method, carried to a further degree of oxidation, is the foundation of the vast linoleum industries throughout the world.

I do not find that my father refers (except now and then quite indirectly) to his skill as a draughtsman and designer; yet he possessed both these qualities to a remarkable degree. No doubt they were inherited, with other mechanical gifts, from his father, who, as we have seen, occupied a prominent position in the Paris Mint at an early age, and possessed a rare combination of mechanical and artistic skill. My father speaks of a knack he possessed, as a boy, of modelling in clay, which he put to use at a very early age; later, when the family had removed to London, the production of dies for embossing cardboard and metal, and especially the ornamental designs that characterised them, depended wholly on himself, and would not have been possible without natural gifts carefully cultivated. Perhaps even more interesting was his skill in the application of art to mechanical purposes, as evidenced by the engraving of his new stamp dies (see Fig. 5, Plate III.); by his designs and preparation of the deep-cut cylinders for making figured Utrecht velvet (Fig. 15, Plate VII.), and a number of other applications of art to mechanics that are only briefly referred to, or even not mentioned, in the course of his Autobiography.

In another direction his skill and assiduity as a draughtsman were remarkable. He made, with his own hand, all the drawings that accompanied his patent specifications, at least for a great number of years; and, near the close of his business career, we find that he himself prepared nearly all the drawings for the Bessemer Saloon.

In lighter vein, his designs for alterations and additions to his own residences, and those of his children, were quite remarkable. This is a matter to which I shall have occasion to again refer, later on.

The story of his great invention, the "Bessemer Process," is told in the Autobiography at much length and with characteristic vigour; but in this story my father has omitted a few noteworthy details which should not be lost. It is interesting that, in the month of June, 1859, Bessemer tool steel was first quoted in the printed price lists of the trade. The *Mining Journal* of June 4th, 1859, gives the necessary evidence on this point. It says:—

In this day's *Journal* we quote, for the first time, amongst the metallic manufactures of this country, the steel produced by the process patented by Mr. Bessemer, and we are informed that the new material can be supplied in almost any quantities. The usual price of engineers' tool steel is from £2 15s. to £3 5s. per cwt., while Mr. Bessemer offers an article, which prominent judges pronounce equal to the best, at £2 4s., his other kinds of steel being proportionately lower. As to the quality of the article, there can be little doubt, since the tests to which it has been submitted at Woolwich gave much satisfaction to the officials; and, we understand, a contract for a considerable period has been concluded with Mr. Bessemer. With a steel of equal quality a little more than two-thirds the usual price, it would appear almost impossible for success to be wanting to the seller, while the pecuniary advantage to the consumer will be at once verified; so that it is needless to commend Bessemer's steel to the consideration of our readers.

Sir Henry speaks of the early struggles and ultimate success of the Sheffield works; how great that success was may be gathered from the following passage, given in almost similar words on page 177.

Some idea may be formed of its importance as a manufacture when I state the simple fact that on the expiration of the fourteen years' term of partnership of our Sheffield firm, the works, which had been greatly increased from time to time, entirely out of revenue, were sold by private contract for exactly twenty-four times the amount of the whole subscribed capital of the firm, notwithstanding that we had divided in profits during the partnership a sum equal to fifty-seven times the capital; so that, by the mere commercial working of the process, apart from the patents, each of the partners retired, after fourteen years, from the Sheffield works with eighty-one times the amount of his subscribed capital, or an average of nearly cent per cent. every two months.

But, during the early days (from 1858 to 1861), the success was problematical and the anxiety very great. The subjoined statement shows the financial results obtained at the Sheffield works during the first ten years of its existence.

Year						£	s.	d.
1858	...	...	...	...	Loss	729	12	2
1859	...	...	...	...	,,	1,093	6	2
1860	...	...	...	...	Profit	923	2	1
1861	...	...	...	...	,,	1,475	10	2
1862	...	...	...	...	,,	3,685	18	4
1863	...	...	...	...	,,	10,968	6	3
1864	...	...	...	...	,,	11,827	0	4
1865	...	...	...	...	,,	3,949	5	11
1866	...	...	...	...	,,	18,076	18	4
1867	...	...	...	...	,,	28,622	1	8

My father omits any reference to the first steel rails put into actual service; and, curiously enough, he does not mention the historic occasion when he persuaded Mr. Ramsbottom, then the chief mechanical engineer of the London and North Western Railway, to make a trial. He has, however, described this interview in a letter :

Perhaps there was no better practical engineer in Great Britain than Mr. John Ramsbottom, of the London and North-Western Railway; and when I proposed steel rails to him, Mr. Ramsbottom, looking at me with astonishment, and almost with anger, said : "Mr. Bessemer, do you wish to see me tried for manslaughter?" That observation was the natural result of the then state of knowledge as to what could be done with steel. At that time steel was made almost exclusively for cutting purposes, and it was highly carbonized, and certainly too hard for rails. After seeing my samples, however, Mr Ramsbottom, whose mind was thoroughly open to conviction, said : "Well, let me have 10 tons of this material that I may torture it to my heart's content. . . ." A steel rail was rolled by Mr. Ramsbottom from a portion of the 10 tons mentioned, and it had been twisted cold by clamping one end in the reversing brasses of a rolling mill, and putting the other end in connection with the shaft driven by the engine, till it was twisted into two pieces. I carefully measured that sample, and I found that in a part measuring 6 ft. along the centre of the web, each of the flanges measured 8 ft. 1 in. This twisted rail was a good example of what mild steel was in those days. To show that such material, which twisted so well cold, would endure in the hot test, a 4-in. square bar was twisted hot. It was twisted till it came in two in the centre. The angles were thus made to form a sort of screw with threads $\frac{5}{8}$ in. to $\frac{1}{4}$ in. apart.

The first steel rail was laid down between two adjacent iron rails, at the Camden Goods Station of the London and North-Western Railway, on May 9th, 1862, and as the first of so many millions of tons, its history should not be forgotten. It is summarised in *Engineering*, of January 5th, 1866, as follows :—

The Bessemer rails, judging from the experience of the London and North-Western Railway Company, are cheaper than iron rails at £50, or more, per ton. There is the remarkable rail under the Chalk Farm bridge of the North-Western line, which, but a few weeks ago, when we saw it, was wearing out the seventeenth or eighteenth face of wrought-iron rails adjoining it, and which were subjected to exactly the same conditions of traffic. This comparison is an extraordinary one. A steel rail, rolled at Crewe, from an ingot cast at the Sheffield Steel Works of Messrs. Henry Bessemer and Company, was selected at random, and laid down between contiguous iron rails, at the Camden Goods Station, May 9th, 1862. In 1864, a gentleman about to proceed to Belgium upon business connected with steel rails, asked and received permission from the London and North-Western directors to copy the records of this and the contiguous rails, as filed in their office. The steel rail, when examined in September, 1864, had never been turned, top for bottom, and showed "but little signs of wear." Eight thousand goods trucks pass over this line in twenty-four hours, and it is estimated that nearly 10,000,000 trucks, or more than 20,000,000 wheels, have passed over this rail from the first. The next iron rail, contiguous to it, was laid down, quite new, on the same date, May 9th, 1862. It was found necessary to turn it in the following July; and on September 9th of the same year, a second iron rail had to be laid down in the place of the first. This required to be turned on November 6th, and on January 6th, 1863, a third iron rail had to be put down. This was turned on March 1st, and a fourth new rail put down on April 29th. This was turned top for bottom, July 3rd, and a fifth new iron rail laid down September 29th. It was turned over on December 16th, and on February 16th, 1864, a sixth new rail was put in. This was turned April 12th, and a seventh new rail put in its place, August 6th, 1864. Mr. Bessemer exhibited his rail at the last meeting of the British Association at Birmingham, after it had worn out additional and neighbouring wrought-iron rails. It was not greatly worn, and had never been turned. The Crewe station, with nearly three hundred time-table trains through it daily, besides a great amount of shunting almost always going on, was laid with rails rolled from ingots cast by Messrs. Bessemer and Company, November 9th and 10th, 1861, and a year later, on the Prince of Wales's Birthday, Mr. Bessemer's exhibition rails, 35 ft. long, rolled at Crewe, were laid in the up line, just out of the station. None of these rails have yet been turned, and we believe that Mr. Ramsbottom, and Captain Webb, of the Crewe Works, anticipate sending the 35-ft. rails in good order to the next International Exhibition, to which we are looking forward, in 1872.

In 1861, the price of Bessemer steel rails was £22 per ton, but in the following year the Metropolitan Railway paid only £17 per ton. Although my father does not say so, steel rails were a conspicuous feature in the Bessemer collection in the London Exhibition of 1862. The following description of this exhibit is quoted from an appreciative article in *The Engineer* :—

There are also some close bends of rails, one of which is deserving special notice. Mr. Ramsbottom, the able engineer of the railway works at Crewe, had this piece taken up while covered with sharp frost, and placed under the large steam-hammer, where it stood the blow necessary to double both ends together, without showing the smallest indications of fracture. . . . . There are also some extraordinary examples of the toughness of the Bessemer steel, made from British coke pig iron, among which may be enumerated two deep vessels of 1 ft. in diameter, with flattened bottoms and vertical sides. At the top edge, one of these is $\frac{5}{8}$ in. and the other $\frac{7}{8}$ in. in thickness. A 4-in. square bar has been so twisted, while hot, that its angles have approached within less than half an inch of each other, so that what was originally 1 ft. length of surface has now become 26 ft., while the central portion of the bar still preserves its original length of 1 ft. ! . . . . By the present process, although the number of operations is reduced by casting steel in large masses, its cost as compared with that of wrought iron is somewhat increased. Still, it compares favourably considering its greater strength. The present causes of the costliness of steel are principally these : Melting the metal is expensive. Such a high temperature is required that the pots for very low steel stand only one or two meltings. The subsequent heating of immense ingots (one of Krupp's, in the Great Exhibition, was 44 in. in diameter, and 8 ft. long) requires time and skill ; drawing them under ordinary hammers, not to speak of its injurious effects, is a very long operation. The careful preparation and selection of the materials add considerably to the cost. Again, the business is now monopolised by a few manufacturers. Standard qualities of low steel bring a price much more disproportionate than that of wrought iron, compared to the cost of production. Some of the processes are secret, others are covered by patents ; but the chief difficulty is, that very few establishments out of the whole number have undertaken the manufacture. Many of the large British establishments have introduced the Bessemer process. In this country several ironmasters pronounce this process a failure, and propose to stick to puddling and piling. At the same time others are doing all they can to develop this and similar improvements, but are indifferently encouraged. There is no doubt, however, that within a few years low steel will be produced at a cheap rate all over the world. The wonderful success and spread of the Bessemer process in England, France, Prussia, Belgium, Sweden, and even in India, all within three or four years, prove that great talent and capital are already concentrated on this subject, and promise the most favourable results.

And again :

Among the specimens of Bessemer metal in the Exhibition of 1862 was a 14-in. octagonal ingot, broken at one end, and turned at the other end, to show that the metal was perfectly solid. The turned end looked like forged steel. An 18-in. ingot, weighing 3136 lb., was the six thousand four hundred and tenth "direct steel" ingot made at the works of Messrs. Henry Bessemer and Co.

There were also exhibited a double-headed rail, 40 ft. long ; a 24-pounder and a 32-pounder cannon ; a 250 horse-power crank-shaft, and several tyres without welds. The specimens showing the wonderful ductility of the metal have been referred to. The Bessemer process has been adopted during the last two or three years, since its early embarrassments were overcome, with such great success, and by so many leading manufacturers in England, France, Sweden, Belgium, and other European States, that its general substitution for all processes for making either fine wrought iron or cheap low steel is now considered certain.

One of the most important chapters in the history of the Bessemer steel industry concerns its introduction and development in the United States, and although he makes many suggestive references to this subject in his Autobiography, Sir Henry does not give any details. An attempt should therefore be made to supply this deficiency, though only in a very brief and imperfect way. Alexander Lyman Holley says, in his book on Ordnance and Armour, written in 1863, that at that date, the Bessemer process was to be tried on a working scale at Troy, in the State of New York, by Messrs. Winslow, Griswold, and Holley.

x x

HENRY BESSEMER

The process was then in operation on a large scale in many places in England and on the Continent; but, almost seven years before, experiments had been carried out by Messrs. Cooper and Hewitt at their iron works in Philipsburgh, New Jersey, following the information given them in Mr. Bessemer's British Association paper of 1856. At the meeting of the British Iron and Steel Institute in America, in 1890, Mr. Abram S. Hewitt, said:

> Mr. Bessemer read his celebrated paper describing the process of producing steel without fuel, at the Cheltenham meeting of the British Association for the Advancement of Science in the summer of 1856; an imperfect report of this paper was published in the journals of the day, and attracted my notice. The theory announced seemed to be entirely sound, and the apparatus simple and effective. I gave orders at once, without further information than that derived from the published report, to erect an experimental vessel for the purpose of testing the possibility of producing steel direct from the blast furnace. In the same year in which this paper was read the experiment was tried at the furnace of Cooper and Hewitt, at Philipsburgh, in New Jersey, and the result served to show, beyond all doubt, that the invention of Mr. Bessemer was one that could be successfully reduced to practice.

This fixes the date of the first application of the Bessemer process in the United States. However, nothing on a practical scale was done until after A. L. Holley's visit to England in 1862, when, on behalf of a syndicate known as the "Bessemer Association," one of the most active members of which was Mr. Hewitt, he opened negotiations for the purchase of the Bessemer American patents. These negotiations were completed in 1864; but the opposition which Mr. Bessemer encountered up to that date, as is vividly related in his Autobiography, delayed developments in the United States, and the Bessemer Association was deterred from taking action by the widespread hostile articles in the British press. A letter written by Holley, in September, 1866, shows the actual position at that date.

> In view of the diverse statements of the English journals regarding the success of the Bessemer process in this country, and of the improvements actually developed here, I trust that some account of our practice will be interesting. The Bessemer process was first experimentally practised in this country with a 3-ton converter, at the ironworks of Mr. E. B. Ward, at Wyandotte, near Detroit, under the superintendence of Mr. L. M. Hart, who had learned the Bessemer process at the works of Messrs. Jackson, in France. . . . .
>
> Before the Wyandotte experiments were commenced, Messrs. Winslow, Griswold and Holley, of Troy, had completed an arrangement with Mr. Bessemer and his associates for the

purchase of the Bessemer patents in the United States, and had commenced the erection of a 2-ton experimental plant. This plant was started in February, 1865, and has since been in constant operation. The first ingot made had a tensile strength of 65,000 lb. per square inch in the cast state, and 121,000 lbs. when hammered to a 2 in. bar. A 2-in. bar was bent double cold. The first ingot was a fair representative of all the steel that has since been manufactured at Troy. The pig iron used was smelted with charcoal from the hematite and from the magnetic ores of the Lake Champlain, the Hudson River, and the Salisbury regions, and the Lake Superior iron, smelted either with charcoal near the mines, or with bituminous coal in the Mahoning Valley of Ohio. Some of the Pennsylvania and New Jersey anthracite irons produce steel equal in quality to that made from the English hematites of the Cumberland regions, but not equal to that made from the American charcoal irons mentioned. Some 100 tons of the best steel have been made from the Iron Mountain ores of Missouri, smelted with charcoal, and from the charcoal hematite irons of central Alabama. Sufficient experience has already been gained in the mixing of these various pigs to produce, uniformly, all grades of steel. The only irons that have failed are those reduced from surface ores, containing an excess of phosphorus, and those that have been smelted with very sulphurous coal. As far as tested, probably three-quarters of the American pigs produce first-rate Bessemer steel.

Meanwhile Messrs. Winslow, Griswold and Holley had commenced the erection of a pair of 5-ton converters, and the Wyandotte Works were producing a good quality of steel from the Lake Superior irons. The re-carboniser at both works has been the Franklinite pig iron of New Jersey, which is slightly richer in manganese than spiegeleisen. . . . At the present time the 2-ton converter at Troy is producing 10 tons of ingot (six charges) per twenty-four hours ; the 5-ton converter will be in operation next December. The Wyandotte Works are producing a smaller quantity, but a good quality of steel. Of the licensees, the Pennsylvania Steel Company, at Harrisburg, will be in operation early next year, with two 5-ton converters, a 25-in. three-high rail mill, a tyre mill, a plate mill, and a forge suited to the manufacture of all ingots under 12 tons weight. Similar works at Chester, Pennsylvania, at Cleveland, in Ohio, are partially completed, and will be running during the next year. Several other works in Pennsylvania and at the West will probably produce steel within eighteen to twenty months of the present writing (September, 1866).

The plans for the Pennsylvania Steel Company's Works were prepared by my father himself, at Sheffield, and the plant was almost wholly of English manufacture. The converters were made by Messrs. Galloway and Sons, of Manchester, and the hammers by Messrs. Thwaites and Carbutt, of Bradford.

The first charge of Bessemer metal made in the United States was run into ingots at Troy, on February 16th, 1865. The works were very small, with two 2-ton converters, but the results obtained under Holley's able management were so surprising that the Bessemer

Association required no further proof, and both practical steel-makers and capitalists were convinced. During the month of May, 1865, no less than eighty converter charges were run into ingots. This was rapidly followed by the installation of two 5-ton converters at Troy, and the construction of the Pennsylvania Steel Company's Bessemer Works at Harrisburg. During fifteen years Holley's life was spared to build up the Bessemer process in America, and to make a lasting monument for himself. In March, 1865, the two small converters at Troy made 118 tons of Bessemer steel, or at the rate of 1,400 tons in the year. In 1880 the output from two 5-ton converters was 14,000 tons. In 1868 the total output of Bessemer steel in America was 8,500 tons; the same year it was 110,000 tons in England. Eleven years later the American production had equalled that of this country, and since then it has always exceeded it.

The Report of the Twelfth Census of the United States contains an interesting account of the position of the Bessemer steel industry in 1900. In that year forty-two establishments owned Bessemer converters; of these thirty-three were active and nine idle. The number of active converters was seventy, and their daily capacity was 34,925 gross tons. The total production for the year exceeded 7,500,000 tons, and its money value was nearly £27,000,000. These astonishing amounts have been exceeded since 1900, as will be seen by the following figures for 1902; those for Germany, Great Britain, and France being also added :—

						Tons.
United States	...	...	...	...	...	9,138,363
Germany ...	...	...	...	...	...	5,229,939
Great Britain	...	...	...	...	...	1,825,779
France ...	...	...	...	...	...	1,010,000

These figures show clearly the stupendous growth of the Bessemer steel industry in the United States, from the small converters at Troy in 1865 : a development chiefly due to Holley's untiring energy and skill. The secret of his progress and success was defined by Mr. Robert W. Hunt, in a paragraph of a paper read by him before the American Institute of Mining Engineers, and called " A History of the Bessemer Manufacture in America." Mr. Hunt said :—

After building the first experimental works at Troy, Mr. Holley seems to have at once broken loose from the restraints of his foreign experience, and to have been impressed with the capabilities of the new process. The result is that, mainly through his inventions and modifications of the plant, we in America are to-day enabled to stand at the head of the world in respect of amount of product.

Referring to the modifications and improvements made in the Bessemer process by Holley, and to which the great output of the American Bessemer steel works is largely due, Mr. Hunt said further :—

He did away with the English deep pit, and raised the vessels so as to get working space under them on the ground floor; he instituted top-supported hydraulic cranes for the more expensive English counterweighted ones; he put three ingot cranes around the pit instead of two, and thereby obtained greater area of power. He changed the location of the vessels, as related to the pit and smelting-house. He modified the ladle-crane, and worked all the cranes and the vessels from a single point; he substituted cupolas for reverberatory furnaces; and last, but by no means least, introduced the intermediate or accumulative ladle, which is placed on scales, and thus insures accuracy of operation, by rendering possible the weighing of each charge of melted iron, before pouring it into the converter. These points cover the radical features of his innovations. After building such a plant, he began to meet the difficulties in manufacture, among the most serious of which was the short duration of the vessel bottoms, and the time required to cool off the vessel to a point at which it was possible for workmen to enter and make new bottoms. After many experiments, the result was the Holley vessel bottom, which, either in its form as patented, or in a modification of it as now used in all American works, has rendered possible, as much as any other one thing, the present immense production. Then he tried many forms of cupolas at Troy, adopting in the original plant a changeable bottom, or section below the tuyères; then, later, at Harrisburg, assisting Mr. S. B. Pearce, in developing the furnace to a point, which rendered the many bottoms unnecessary, chiefly by deepening the bottom and enlarging the tuyère area. Upon his rebuilding the Troy works, after their destruction by fire, Mr. Holley put in the perfected cupolas. At this time the practice was to run a cupola for a turn's melting, which had reached eight heats or forty tons of steel, and then dropping its bottom. This was already an increase of 100 per cent. over his boast about the same amount in twenty-nine hours.

By the year 1865 the Bessemer process was firmly established on the Continent. Krupp, of Essen, had installed a plant; the Bochum Works had four 3-ton converters; the Hoerde Company, near Dortmund, had two converters; a steel works in Düsseldorf was completed with two converters; the Neuberg Works, in Styria, works at Grasse and at Witkowitz, all made Bessemer steel; as also did John Cockerill, of Seraing; Petin, Gaudet and Co., at Rive de Giers; James Jackson

and Co., of St. Severin, near Bordeaux, besides many others. During
that year the production of Bessemer steel on the Continent was about
100,000 tons. This and the following few years ripened the golden
harvest for my father, his royalties from all sources reaching a very high
figure. In 1869, however, his leading patent expired; and while this
gave a great impetus to the production of Bessemer steel, his income
arising from the royalties was much diminished.

In the early 'sixties, when my father's process for the manufacture
of steel had triumphed over the many difficulties described in the
previous pages, he determined on the establishment of steel works in
or near London. I may say here that this intention was chiefly for
the benefit of my brother and myself, the idea being that we should
carry on the works under the general supervision of my father. After
careful consideration, a site of about three acres was secured on the
banks of the Thames, just below Greenwich; for this a long lease was
obtained. It was determined that the works should be only on a small
scale, comprising two $2\frac{1}{2}$-ton converters, and all the plant necessary for
the production of steel on that basis. This plant included one $2\frac{1}{2}$-ton
steam-hammer and another of smaller size; the buildings were carefully
designed, with the intention that the establishment should in all respects
be a model one.

At the time I speak of, the Thames was a very busy shipbuilding
centre, and we naturally expected to find a large number of customers
ready to our hand. But before we were able to commence operations
the shipbuilding trade had deserted the Thames for the North, and
with this great change our expected customers had also disappeared.
Under these circumstances, my father did not consider it desirable for
us to open the works, although they were fully equipped, and he decided
upon letting them. This was done after considerable delay; our first
tenants being the Steel and Ordnance Company, who, however, did
not achieve much success, and the factory became vacant after a few
years. Then we let it to Messrs. Appleby Brothers, who were general
engineers, and did not propose to become steel makers. For this reason
the whole of the steel-making plant was disposed of, and the scheme
for manufacturing Bessemer steel on the Thames was finally abandoned.

Again, after some delay, we found the place on our hands, and this time my brother and I determined upon converting the long lease into a freehold: an operation effected only with much difficulty and after prolonged negotiations. Then followed a considerable period when the works remained untenanted, but we eventually let them to a company for the manufacture of linoleum. This time there was no doubt about the success of the undertaking, and the company added to the size of the works until nearly the whole of the three acres was covered with buildings. A few years since we sold the property to this company, and thus terminated our connection with the works my father had originally built for us.

It may not be generally known that long before the episode of the Bessemer steam-ship, with its swinging saloon, my father had given much attention to the Channel crossing, a voyage which he heartily disliked. At the time I speak of, his idea was to construct a large circular vessel about 200 ft. in diameter, of a double-convex form in cross-section, and large enough to float over two or three Channel waves at a time; in the hull were to be contained the necessary propelling machinery, cabin accommodation, etc., and in the centre was to be a raised circular deck about one-third of the vessel's diameter. This scheme, so far as my father was concerned, never went beyond the stage of a general design; but Admiral Popoff, at that time a prominent Russian naval constructor, and an acquaintance of my father, was much struck with the idea, and embodied it in a vessel he built for the Russian Navy. A model of this vessel, which was called the "Popoffka," is now in the Musée de la Marine, at the Louvre, in Paris.

A very conspicuous feature in my father's character was his intense love of home and its surroundings: a sentiment which endured to the last days of his life, and never grew slack, even during the busiest and most harassed periods. He always found time to make alterations and improvements, and to decorate his own home with the natural taste that belonged to him, and which he had inherited from his father. During his later years he extended this love of domestic improvements to the houses of his children, all of which bore the impress of his individuality.

During his long life he did not shift his home frequently. From the humble beginnings, commencing with his marriage, we read in his Autobiography of his moving into Baxter House; and, after some years, when the large returns from the bronze-powder business permitted it, he has told us how he indulged his natural longings for a country life by the acquisition of a house and grounds, which he called "Charlton," at Highgate. Here he lived for a number of years, until he made a last change to Denmark Hill, where he found a charming though unpretentious house, with beautiful and extensive grounds. Settled here, the alterations and improvements that he made, both in house and gardens, occupied and amused him for a number of years. This may be a matter of very small moment to the general reader, but to me it possesses a special interest, knowing as I do how the house at Denmark Hill became an inseparable part of my father's life. For this reason I venture to give, in Fig. 88, Plate XLIII., a view of the house, this view showing on the left-hand side the conservatory which he designed and erected. An interior view of this conservatory is given in Fig. 89, Plate XLIV. If serving no other purpose, it will show that my father's capabilities as a designer and decorator were of no mean order. Another example of his talents in this direction is given in Fig. 90, Plate XLV., which shows the interior of a grotto constructed by him in his grounds; the mound within which this was built was formed by the excavation of a large lake which he had made. As will be seen from the illustration, the grotto is of an elaborately ornate character, both design and colouring having been taken from one of the courts of the Alhambra. Illusions of distance were produced by the ingenious application of large mirrors.

From the charming Autobiography of that great engineer, James Nasmyth, I take the liberty of making the following extract, and basing upon it a comparison of his career with that of my father. Mr. Nasmyth says :—

"The 'Dial of Life' [see Fig. 91] gives a brief summary of my career. It shows the brevity of life and indicates the tale that is soon told. The first part of the semicircle includes the passage from infancy to boyhood and manhood. While that period lasts, time seems to pass very slowly. We long to be men, and doing men's work. What I have called the 'Tableland of Life' is then reached. Ordinary observation shows that between thirty and fifty the full strength of body and mind is reached, and at that period we energise our faculties to the utmost.

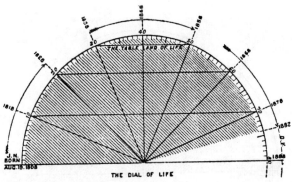

FIG. 91. "DIAL OF LIFE" FOR MR. JAMES NASMYTH.

"Those who are blessed with good health and a sound constitution may prolong the period of energy to sixty or even seventy; but Nature's laws must be obeyed, and the period of decline begins and goes on with accelerated rapidity. Then comes old age; and as we descend the semicircle towards eighty, we find that the remnant of life becomes vague and cloudy. By shading off, as I have done, the portion of the area of the diagram according to the individual age, everyone may see how much of life is consumed and how much is left—D.V.

"Here is my brief record:—

Age.	Year.	
	1808	Born, August 19th.
9	1817	Went to the High School, Edinburgh.
13	1821	Attended the School of Arts.
21	1829	Went to London to Maudslay's.
23	1831	Returned to Edinburgh to make my engineer's tools.
26	1834	Went to Manchester to begin business.
28	1836	Removed to Patricroft and built the Bridgewater Foundry.
31	1839	Invented the steam-hammer.
32	1840	Marriage.
34	1842	First visit to France and Italy.
35	1843	Visit to St. Petersburg, Stockholm, Dannemora.
37	1845	Application of the steam-hammer to pile-driving.
48	1856	Retired from business, to enjoy the rest of my life in the active pursuit of my most favourite occupations."

Y Y

It will be interesting to compare the foregoing with my father's record, which stands as follows, his "Dial of Life" being given in Fig. 92 :—

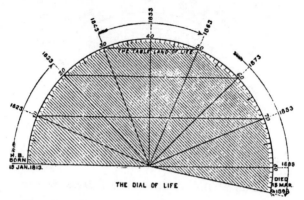

FIG. 92. "DIAL OF LIFE" FOR SIR HENRY BESSEMER

Age	Year.	
	1813	Born, 19th of January.
17	1830	Arrived in London.
20	1833	Improvements in Government stamps.
21	1834	Marriage.
30	1843	Manufacture of bronze powder.
42	1855	First patent for manufacture of iron and steel, October 15th.
43	1856	Paper read before the Cheltenham Meeting of British Association, August 24th.
45	1858	Bessemer Works started at Sheffield.
49	1862	First Bessemer steel rail laid at Camden Goods Station, May 9th.
50	1863	Bessemer steel first used in the construction of ships.
52	1865	First Bessemer Works started in America, by A. L. Holley, at Troy, U.S.A.
56	1869	Bessemer Saloon patented.
59	1872	Retired from business.
66	1879	Knighted, June 26th.
85	1898	Died, March 15th.

When describing the first announcement of his great steel invention at the British Association Cheltenham Meeting, in 1856*, my father referred to the encouraging and generous remarks made by Mr. James

* See page 162, *ante*

Nasmyth. On searching through a mass of miscellaneous papers in my possession, I came across an interesting correspondence between Nasmyth and my father, and it seems to me that the facsimile of a letter, reproduced in Fig. 93, Plate XLVI., will be read with interest; it is a characteristic letter, and shows that the generous impulse, which so encouraged my father during the discussion of his memorable paper of 1856 had remained unaltered twenty-five years later.

On turning to the Autobiography referred to in his letter by Mr. Nasmyth, as giving him so much pleasure in the preparation, I find that some references are made to this correspondence. He says :—

In 1854 I took out a patent for puddling iron by means of steam. Many of my readers may not know that cast iron is converted into malleable iron by the process called puddling. The iron, while in a molten state, is violently stirred and agitated by a stiff iron rod, having its end bent like a hoe or flattened hook, by which every portion of the molten metal is exposed to the oxygen of the air, and the supercharge of carbon which the cast iron contains is thus burnt out. When this is effectually done the iron becomes malleable and weldable.

This state of the iron is indicated by a general loss of fluidity, accompanied by a tendency to gather together in globular masses. The puddler, by his dexterous use of the rabbling-bar, puts the masses together, and, in fact, welds the new-born particles into puddle-balls of about three-quarter cwt. each. These are successively removed from the pool of the puddling furnace, and subjected to the energetic blows of the steam-hammer, which drives out all the scoriæ lurking within the spongy puddle-balls, and thus welds them into compact masses of malleable iron. When re-heated to a welding heat, they are rolled ont into flat bars or round rods, in a variety of sizes so as to be suitable for the consumer.

The manual and physical labour of the puddler is tedious, fatiguing, and unhealthy. The process of puddling occupies about an hour's violent labour, and only robust young men can stand the fatigue and violent heat. I had frequent opportunities of observing the labour and unhealthiness of the process, as well as the great loss of time required to bring it to a conclusion. It occurred to me that much of this could be avoided by employing some other means of getting rid of the superfluous carbon, and bringing the molten cast iron into a malleable condition.

The method that occurred to me was the substitution of a small steam-pipe in the place of the puddler's rabbling-bar. By having the end of this steam pipe bent downwards, so as to reach the bottom of the pool, and then to discharge a current of steam *beneath the surface of the molten cast iron*, I thought that I should by this simple means supply a most effective carbon oxidising agent, at the same time that I produced a powerful agitating action within the pool. Thus the steam would be decomposed and supply oxygen to the carbon of the cast iron, while the mechanical action of the rush of steam upwards would cause so violent a commotion throughout the pool of melted iron as to exceed the utmost efforts of the labour of the puddler. All the gases would pass up the chimney of the puddling furnace, and the puddler would not be subject to their influence.   Such was the method specified in my patent of 1854.*

My friend Thomas Lever Rushton, proprietor of the Bolton Iron Works, was so much impressed with the soundness of the principle, as well as with the great simplicity of carrying the invention into practical effect, that he urged me to secure the patent, and he soon after gave me the opportunity of trying the process at his works.   The results were most encouraging.   There was a great saving of labour and time compared with the old puddling process; and the malleable iron produced was found to be of the highest order as regarded strength, toughness, and purity.   My process was soon after adopted by several iron manufacturers, with equally favourable results.   Such, however, was the energy of the steam, that unless the workmen were most careful to regulate its force and the duration of its action, the waste of iron by undue oxidation was such as in great measure to neutralise its commercial gain as regarded the superior value of the malleable iron thus produced.

Before I had time or opportunity to remove this commercial difficulty, Mr. Bessemer had secured his patent of the 17th of October, 1855.   By this patent he employed a blast of air to do the same work as I had proposed to accomplish by means of a blast of steam, forced up beneath the surface of the molten cast iron.   He added some other improvements, with that happy fertility of invention which has always characterised him.   The results were so magnificently successful as to totally eclipse my process, and to cast it comparatively into the shade.   At the same time I may say that I was in a measure *the pioneer* of his invention; that I initiated a new system, and led up to one of the most important improvements in the manufacture of iron and steel that has ever been given to the world.

Mr. Bessemer brought the subject of his invention before the Meeting of the British Association at Cheltenham in the autumn of 1856.   There he read his Paper "On the Manufacture of Iron and Steel without Fuel."   I was present on the occasion, and listened to his statement with mingled feelings of regret and enthusiasm: of regret, because I had been so clearly superseded and excelled in my performances; and of enthusiasm—because I could not but admire and honour the genius who had given so great an invention to the mechanical world.   I immediately took the opportunity of giving my assent to the principles which he had propounded.   My words were not reported at the time, nor was Mr. Bessemer's Paper printed by the Association, perhaps because it was thought of so little importance.   But, on applying to Mr. (now Sir Henry) Bessemer, he was so kind as to give me his recollection of the words which I used on the occasion. . . .   It was thoroughly consistent with Mr Bessemer's kindly feelings towards me that, after our meeting at Cheltenham, he made me an offer of one-third share of the value of his patents.   This would have been another fortune to me.   But I had already made money enough.   I just then taking down my signboard and leaving business.   I did not need to plunge into any such tempting enterprise, and I therefore thankfully declined the offer.

I need not refer in this place to my father's reply to Mr. Nasmyth's letter; to do so would be only to repeat what he has already written

* Specification of James Nasmyth—"Employment of Steam in the Process of Puddling Iron."   May 4th, 1854; No. 1001

in the earlier pages of his Autobiography. It will be noticed that at the end of his letter, reproduced in Fig. 93, Plate XLVI., Mr. Nasmyth expresses the hope that my father was making satisfactory progress with his telescope. And this naturally leads me to say something about a pursuit which, if it had no practical and useful conclusion, at all events afforded my father a congenial occupation for many years.

With the termination of the Saloon Steam-ship episode, my father's active business career came to an end. That was in the year 1873, when he was about sixty years of age, so that nearly a quarter of a century of busy relaxation was still in store for him. Not that the collision of the "Bessemer" with the pier at Calais terminated that unfortunate incident; on the contrary, as my father has already shown, several years elapsed before the business was entirely closed. Still it occupied only a small portion of his time, and he was left free to follow congenial pursuits. During twenty-five years of his strenuous life he had accumulated what was, for that time, a large fortune, though modest enough if compared with the standard of to-day. So that his later years were not only entirely free from business cares, but also from financial anxieties.

That so active a man could remain without occupation was evidently impossible; his house and grounds, in which he took unwearying delight, had been developed and improved until they afforded him little beyond the routine occupation of management, and he naturally turned his attention to some work which should give full play to his mechanical abilities. As matters turned out, four different pursuits occupied him fully up to the year of his death. These were—the construction of an observatory and telescope; his experiments with a solar furnace; the installation of a diamond-polishing factory for the benefit of his grandson; and his Autobiography. The last named has told us the story of his active life in a characteristic manner. The diamond-polishing factory, to be referred to presently, was an assured success; as for the telescope and the solar furnace, it is as well to state at the commencement that neither proved of any practical value, although they provided for him a never-failing source of enjoyment.

My father's first ambition was to construct a refracting telescope with a 50-in. objective. Fortunately for himself, he soon realised the impossibility of achieving this ambition: not only on account of the enormous cost of so large a lens, but because no one could be found, at that time, to undertake its production with any chance of success. Therefore he was quickly led to the construction of a reflecting telescope, which did not seem to present such insurmountable difficulties. Moreover, this had the special attraction to him that he resolved to make the reflector himself—a mistake, of course, so far as the scientific outcome of his work was concerned, but a success in the main respect, namely, that of giving him congenial occupation. At first he tried a speculum metal of the same alloy as that used by Lord Rosse in his great Parsonstown reflector; that is to say, a mixture of copper and tin, in the proportions of 126.4 of the former to 58.9 parts of the latter. This alloy, however, in my father's hands did not prove satisfactory. He found it extremely brittle and difficult to cast; moreover, it was entirely unsuited for turning in the lathe, and it was my father's intention to give the reflector its proper form by mechanical cutting and grinding. His early experience of alloys naturally led him to engage in a long series of experiments with different mixtures. Ultimately, however, he determined to abandon the use of metal, and to employ instead a disc of glass, which should be trued up and polished to the correct figure, and afterwards coated with silver. Before commencing his operations, he built and equipped a workshop at Denmark Hill. Besides the necessary steam power, this workshop contained a number of ordinary tools; but the main feature was a special grinding and polishing machine, which he designed and had built from his own drawings. A sketch of this machine—somewhat imperfect, as I have made it from memory — is shown in Figs. 94 and 95. It comprised a long and rigid bed A A, with a head-stock B B, on the spindle of which are the driving pulleys C and two face-plates D E. These latter were rather more than 5 ft. in diameter, so as to take the 50 in. glass discs from which the reflectors were to be made. At the other end of the bed was a pin G, on which there was mounted a cast-iron frame F, the frame being free to swing horizontally on the pin. As will be seen from the diagrams, the size of this swinging frame was increased as it approached the head-stock, until it was large enough to enclose the latter, and to oscillate without coming into contact either with the head-stock or with the face-plate on which the glass disc was mounted. Within the large end of the oscillating frame was fixed a bar, to the end of which was secured a black diamond that formed the cutting tool. The bar could be moved backwards or forwards to vary the depth of cut; and as the frame was traversed to-and-fro, the tool described an arc of a circle which could be varied within narrow limits, so as to modify the degree of curvature, and, consequently, the length of focus

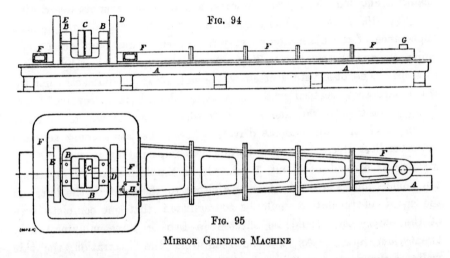

Fig. 94

Fig. 95

MIRROR GRINDING MACHINE

given to the reflector. When this apparatus was set in movement, the revolution of the face-plate, combined with the oscillating motion of the frame, gradually gave a concave surface to the disc of glass. A reverse action could also be obtained by mounting the tool-holder on the extreme end of the oscillating frame, with the diamond cutter pointing inwards, in such a way that it could operate on a disc of glass mounted on the face-plate E; in this way convex, instead of concave, surfaces could be formed. A number of both classes of discs were roughed out and polished by this machine. It was found necessary to do the finishing work at night, because during the day sudden variations in temperature occurred which altered the length of the oscillating frame F, and so changed the curvature of the surfaces produced.

The design, installation, and operation of this machine, occupied my father for a long time; it was made with great care; but the surfaces it produced lacked the accuracy and form necessary for lenses or mirrors intended for astronomical work, and it never found any useful application.

Whilst the workshop and its equipment were in progress, my father commenced the construction of the observatory in which the telescope (never to be completed) was to be mounted. It was altogether a very beautiful and ingenious piece of work, and calls for a short description.

Fig. 96, Plate XLVII., gives an excellent idea of the general appearance of the observatory; as will be seen, it was built on a slight eminence, and the circular gallery around it was approached by a flight of steps. The observatory itself was about 40 ft. in diameter, and the whole structure, vertical walls as well as domed roof, revolved very freely on a circular rail and a ring of wheels, some of which served as bearing and the remainder as driving wheels, the latter being actuated by an endless steel-wire rope; motion was imparted by a small turbine working at a head of about 70 ft. By a simple method of reversing, the dome could be caused to revolve either to the right or the left, and the speed of revolution could be so regulated, that one complete turn of the observatory could be effected in two or three minutes, or in twenty-four hours. As will be seen from the illustration, the side walls of the dome were pierced with openings for windows and a door, so that access to the interior could be always obtained from the outer gallery; the position of the sliding shutter in the cupola is clearly shown in the illustration. The telescope was mounted on trunnions, in bearings at the end of a vertical pillar resting on very solid foundations, and, of course, the observatory floor was framed solidly to the dome, and moved with it.

The telescope itself, which unfortunately was never finished, is illustrated by Figs. 97 and 98, Plates XLVIII. and XLIX. Fig. 97 shows the gallery floor, at the level of which are the trunnions carrying the telescope. In Fig. 98, Plate XLIX., which is a view on the ground floor of the observatory, it will be seen that the foundation for the telescope was surmounted by a steel ring carried on a series of short columns; on the face of this ring was a roller-path on which the central column of the telescope took its bearing, and could be turned with very little effort. The column, and with it the telescope, were turned by means of hydraulic machinery, the velocity of which could be adjusted exactly to the same rate as that of the dome, so that the turning rate of the two was identical. As will be seen by Fig. 98, the body of the telescope consisted of a very rigid open cast-iron frame, with a solid ring in the centre carrying the trunnions, and at the lower end it terminated in a ribbed and dish-shaped casting intended to receive the large concave reflector. The central band and trunnions are better shown in Fig. 97 ; on these the telescope could oscillate from a vertical to a horizontal position. This movement was effected by an ingenious arrangement, indicated in both the illustrations. A large gun-metal wheel was mounted on each trunnion, and immediately below, but not in contact, was a second wheel. On each side was an hydraulic cylinder, the plunger of which terminated in a long flat steel bar that passed between the two gun-metal wheels, already spoken of as being placed on each side of the telescope. As the plunger was run in and out, the bar moving between the two wheels gave motion to the latter, and, of course, caused the telescope to turn. The speed was controlled with the utmost delicacy by a small valve, which regulated the flow of water into the cylinders. This valve was, of course, worked by the observer. The position of the finders on the telescope is shown clearly in the illustration.

As may naturally be supposed, every detail connected with the observatory and the telescope was planned by my father, and showed throughout his characteristic ingenuity, engineering knowledge, and correct taste. The undertaking occupied him almost to the time of his death ; but unfortunately it was far from complete, and with him died the personal interest necessary for the completion of an undertaking so full of ideas.

A series of experiments which gave great occupation to my father for quite a long period grew directly out of his work with the telescope ; and though they led to no practical result, this notice would be

z z

incomplete if I did not refer to them. The object my father had in view was to ascertain to what extent concentrated solar rays could be employed in the creation of very high temperatures; he aimed, in fact, at making a solar furnace in which the most refractory material could be readily broken down. He expended a great deal of time and money in this pursuit, and the experimental furnace which he built is illustrated by the diagrams on page 355, Fig. 99 showing roughly his preliminary experimental arrangements and Fig. 100 the finished device. In Fig. 99, on a table was placed a swing mirror *a*, and at a convenient height above this was fixed a concave reflector *b*, about 12 in. in diameter; a lens *c* 6 in. in diameter was mounted above the table, the various parts being so arranged as to produce the following result : The solar rays striking the mirror at an angle, were thrown up to the concave reflector, and sent back to the lens which focussed them at the point *b*. From this experimental apparatus the solar furnace illustrated in Fig. 100 was constructed ; as will be seen, it was almost precisely identical with the earlier form. The tower-like structure was carried on wheels running on a circular rail laid in the shallow pit which formed the foundation ; the platform on which the tower was built was about 18 ft. in diameter. The tower itself was 12 ft. square and about 30 ft. in height ; the front side was open, but could be closed by sliding shutters when the apparatus was not at work ; the lower part was cut off by a large mirror placed at an angle which could be adjusted within moderate limits. Beneath this mirror was the small furnace, access to which was obtained by a door at the back of the tower. The mirror, about 12 ft. square, was carried by a strong cast-iron frame closely boarded over, so that the silvered glass was entirely supported. In the centre was cut a circular opening about 3 ft. in diameter. The cast-iron mirror frame was mounted on trunnions, so that its angle could, as already stated, be altered to suit the altitude of the sun. In the circular opening, and in the position shown in the diagram, Fig. 100, was mounted a lens 30 in. in diameter, and immediately beneath was placed the crucible already referred to; the frame on which this crucible was mounted could be raised or lowered, so as to reduce or increase the degree of heat obtained by the concentration of the solar rays. In the upper part of the tower was placed the large concave reflector, which was

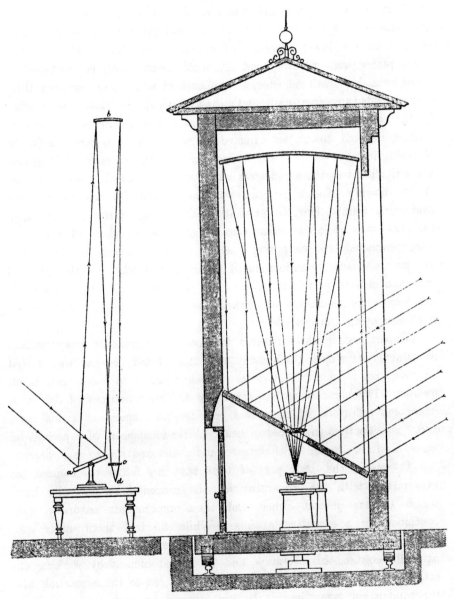

FIG. 99.   FIG. 100.

THE BESSEMER SOLAR FURNACE

about 10 ft. in diameter, and was formed of a concave cast-iron frame; into this were fitted one hundred hexagonal glass plates, each slightly concave on the lower surface, and convex at the back. The backs of these plates were silvered, and afterwards coated with copper: each of them formed a small reflector, to the back of which were secured three copper studs to receive screws, by means of which the plate was attached to the cast-iron frame. This arrangement was found necessary, because each plate had to be so adjusted as to act as a separate reflector, throwing its rays down to the 30-in. lens beneath; as may be imagined, the adjustment of these reflectors was a very long and tedious operation. As the position of each reflector was finally fixed, its face was covered over with water-colour, to prevent the focussing of the rays through the lower lens while the others were being adjusted; when all the parts were completed the paint was washed off, and the reflector was ready for use. A screen was introduced to cut off the whole of the reflected rays, and only to permit them gradually to focus on the crucible beneath the lens; this precaution was taken to prevent any great and sudden heat destroying the crucible.

Experiments made with this furnace were somewhat disappointing, as it never developed the amount of heat expected; copper was melted and zinc was vapourised, but its efficiency ought to have been much greater. The non-success was attributed to the inaccuracy of the small hexagonal reflectors, which caused considerable dispersion of the rays, and consequently a great loss in heat. After some years of experimental work, my father became disheartened, and abandoned the solar furnace.

It was in the early part of 1868 that my father commenced his experiments with the apparatus for the concentration of solar heat, which I have just described, and these experiments naturally were continued for a considerable time. While he had them under consideration, the question of the cause of the great heat of the sun naturally engaged his attention, and he busied himself by working out a theory which would account for it. He was led to the conclusion that the combustion going on in the sun attained its very great intensity largely owing to the pressure under which it took place. As the force of gravity at the sun's surface is about 27.6 times as great as it is on the surface of the earth, all the incandescent solar gases must be maintained in a state 27.6 times as dense as they would be if they formed a portion of our own atmosphere. My father was therefore struck with the idea that the great intensity of the solar heat might be simply due to the fact that combustion took place under very much higher pressure than combustion naturally does on the earth. No sooner did this solution suggest itself to him than he resolved to test it by actual experiment, or, as he expressed it, to have "a little sun of his own." He was at that time desirous of carrying out a series of researches on the fusion, vaporisation, and re-crystallisation of the more refractory metals, and of other so-called infusible substances. As the first step towards attaining this result, he constructed a small cupola furnace, so made that the products of combustion, instead of escaping freely as usual, were checked, in consequence of the mouth of the cupola being narrowed to a diameter of some 2 in. The result of this arrangement was that when air was blown into the furnace at considerable pressure the products of combustion within the furnace were raised to several pounds above the atmospheric pressure, and combustion accordingly took place under this pressure.

With this furnace my father obtained results as brilliant as the conception to which the furnace owed its origin. While the furnace was worked at pressures of from 15 to 18 lb. per square inch above the atmosphere, the temperature obtained was such that steel—and even wrought iron—might be melted more readily than cast iron in an ordinary cupola. Thus on one occasion, a piece of 2-in. square bar iron, 1 ft. long, and weighing 13 lb., was introduced cold into the furnace, and was completely fused in five and a-half minutes; while 3. cwt. of wrought-iron scrap, also introduced cold into the same furnace, was run off in a completely fused state in fifteen minutes. These results were obtained in a small model furnace, without much greater expenditure of coke than would take place in an ordinary cupola furnace melting cast iron.

There were naturally considerable mechanical difficulties to be over-come in making tight a furnace wherein such high temperatures prevailed, and one of these is worth referring to, as it illustrates the happy way in which my father could meet very serious obstacles by devices which were exceedingly simple, and at the same time eminently successful. At the place at which the joint occurred, and which, of course, was liable to be rapidly cut out by the leakage of flame and products of combustion at an intense temperature, he provided a hollow ring, which was connected to the main blast pipe, and therefore received air at a pressure of a few pounds higher than existed in the body of the furnace. The result was that whatever leakage took place was a leakage of cold air into the furnace, and this cold air completely protected the surfaces which otherwise might have been eroded by the gases.

After success had been attained in a small furnace, my father designed : first, an ordinary cupola furnace ; next, a reverberatory furnace for the melting of steel scrap ; and, later, a Bessemer converter, all of which were intended to be worked under pressure. The matter, however, was not proceeded with beyond this point. To-day the electric furnace provides still higher temperatures in a more convenient manner ; and it is not likely that combustion under pressure will ever be resorted to in practical work. Nevertheless, it was a brilliant conception, and was admirably worked out.

Whilst engaged in his experiments connected with the cutting of the glass discs to form the mirror for his telescope, my father had occasion to visit the diamond-cutting factories which existed in Clerkenwell at that time ; and in that industry he, with his ardent temperament, took immediately a keen interest. It happened that about 1884 he was anxious to establish one of his grandsons in business, and he accordingly made arrangements by which, under the name of Messrs. Ford and Wright, a diamond-cutting and polishing factory should be installed in Clerkenwell. London, two hundred years ago, was one of the chief centres of this industry ; but it gradually became supplanted by France and Holland, until the most important diamond-cutting factories in the world were established in Amsterdam, which is still the chief seat of the industry. The trade meanwhile had died out of London, and was practically re-established by the energy and ingenuity of Sir Henry Bessemer, acting on behalf of Mr. Ford, and his grandson, Mr. Wright. It is almost needless to

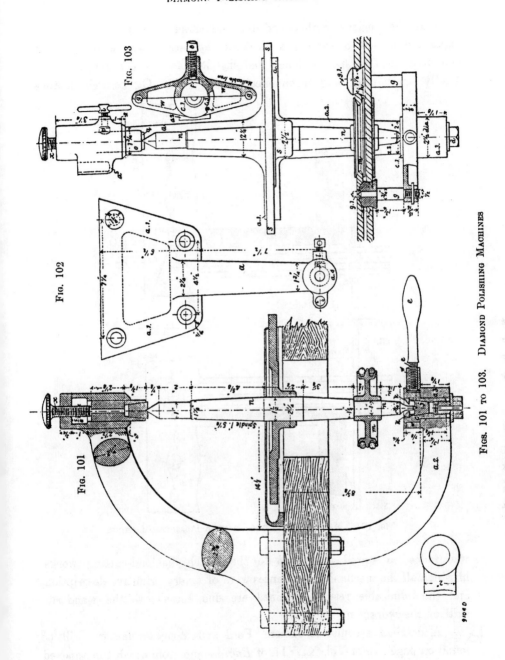

FIGS. 101 TO 103.   DIAMOND POLISHING MACHINES

say that the ancient methods of diamond-cutting and polishing still in vogue did not at all suit the ideas of my father, who could not rest contented until he had designed and installed an entirely new plant on strictly mechanical and economical lines. The Clerkenwell factory

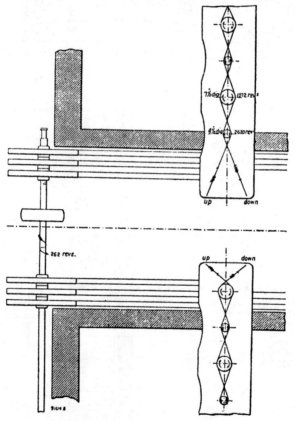

FIG. 104. METHOD OF DRIVING DIAMOND POLISHING MACHINES

was indeed a startling contrast to the Dutch diamond-cutting works, in which all the mechanical appliances were of a very primitive description, and the admirable results obtained are due entirely to the wonderful skill of the workmen.

A detailed account of Messrs. Ford and Wright's factory will be found on page 123 of Vol. XLVII. of *Engineering*, from which the annexed illustrations are taken, and a description summarised. The cutting and polishing machines were quite small, and a number of them were arranged in a row upon a bench in the workshop; the factory included several of these benches. One of the machines is illustrated in Figs. 101 to 103, page 359, while Fig. 104 is a diagram showing the method of driving a complete series. Each mill consisted in its main parts of a cast-iron bracket, above and below the bench; of a vertical spindle n, carrying above the bench a heavy disc, and below the bench a double-grooved pulley m, which received the driving cord, by which a very rapid

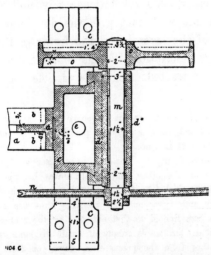

FIG. 105. METHOD OF DRIVING DIAMOND POLISHING MACHINE

rotation was transmitted to the spindle and disc. As will be seen from the illustration, Fig. 101, the top and bottom of the spindle were pointed, and ran in bearings made of *lignum vitæ*; these blocks were set in tapered metal bushes, and the position of the upper one could be adjusted by the set screw x. It will be noticed that a ring k was introduced above the lower bearing, to prevent the lubricating oil used from being thrown out; the driving disc was protected by a guard. All the mills were of the same pattern, and were driven, as explained, by endless ropes. Fig. 104 shows the method of transmission. The benches were divided by a gangway across the centre;

3 A

underneath this, in the basement, was the steam engine and transmission pulleys, which gave motion to six belts. These passed from
the main shaft on to the pulleys o, Fig. 105, these pulleys being
mounted on spindles m, carried on the brackets d, the position of
which on the vertical frame c could be adjusted by a screw. At
the other end of the pulley spindle was the rope pulley n, driving the
main cotton rope, which rose vertically to the level of the working
benches, and then was taken round the pulleys on the spindles, as shown
in Fig. 104. The speed of the engine was multiplied until that of
the pulleys was nearly 3000 revolutions; any mill could be stopped by
throwing the rope out of contact with the double-grooved pulley on the
spindle, by means of the lever e and the frame g, Figs. 101 and 103.

The following description of the manner in which these mills
were operated is extracted from the article in *Engineering* above
referred to :—

The natural angles of the stones are so sharp that if applied to the discs of the
mills they would rapidly cut them away, and practically ruin them. These angles are,
therefore, abraded in the first instance by hand. Two diamonds are mounted on sticks or
holders, and the operator, taking one in each hand, uses an angle of one gem to cut
off or reduce an angle of the other, and in this way he gradually removes them all,
carefully catching the dust which falls, for subsequent use. Originally the stone was practically shaped in this way, and it was only the polishing that was done on the mill, for the
supply of diamond dust was limited to that obtained by the mutual attrition of the gems.
But now there is obtainable an ample supply of small diamonds which are worthless for
decorative purposes, either from the presence of flaws, or from the poorness of their
colour. They are placed in a steel mortar with a tight-fitting cover, and are gradually ground
into a fine powder, which is used upon the mills, and serves to do most of the work which
was formerly effected in the way just described. The whole process is now called polishing,
the two processes of grinding and finishing being simultaneous.

After the natural angles of the stone have been removed, it is mounted in a ball
of lead about the size of a large walnut or small apple. The metal is heated till it
reaches the plastic stage, when the jewel is pressed into it, leaving visible the particular
surface which it is desired to grind. The metal is very easily manipulated by those who
have the skill, and can be "wiped" to one side or the other so as to vary the position
of the stone and give it greater or less prominence. At the opposite side of the lead ball
to the diamond there is a stalk of brass wire, and once the metal is set, this stalk offers
the only means of adjustment by being bent to one side or the other. It is perfectly
marvellous how thirty-two facets can be cut on a diamond no larger than a hemp-seed with
such means as these. The cutter is truly a handicraftsman, for he depends entirely on the
senses of sight and touch, and has no apparatus to aid him in making his minute divisions.

When the diamond has been fixed in its bed the stalk is clamped in a holder, which consists of a heavy bar with a hole to receive the stalk at one end, and two feet at the other end. Practically, the lead ball forms a third foot to the bar, which is laid down with two feet resting on the bench and the third on the disc of the mill. The disc is moistened from time to time with olive oil containing diamond dust, and runs at a speed of about a mile a minute. The small particles of diamond become rubbed into its face, which is of soft cast iron, and thus produce an abrading surface acting continually against the diamond contained in the lead holder. The keen edges of the dust, aided by the speed, and the weight of the lead ball, gradually wear away the stone, which is removed for inspection every few minutes. When the workman considers that the cutting has proceeded as far as is necessary, the lead is softened, and the gem is released, ready to be again set in another position. Thus by successive stages the cutting proceeds until the jewel finally assumes the proper form.

This diamond factory was extremely successful, and remained in active operation for several years, until circumstances which have nothing to do with this story, rendered it desirable for the partnership to be dissolved, and the works closed.

I may refer here to April, 1880, when the Freedom of the Company of Turners was conferred on Sir Henry Bessemer, "in recognition of his valuable discoveries, which have so largely benefited the iron industries of the country, and his scientific attainments, which are so well known and appreciated throughout the world." In the course of his speech, my father, in expressing his thanks for this distinction, said :—

Under the process which I had the honour of inaugurating, we dispense with every one of the intermediate processes formerly employed. We have no smelting of pig iron, we have no making of balls, we have no rolling of bars, we have no shearing of bars, we have no piling up, we have no heating furnaces.

You will readily understand why, with a process so rapid, and so entirely devoid of the use of expensive fuel, and of all those varied skilled manipulations which were necessary at every stage in the old process, the cost of manufacture is so exceedingly small as it is found to be. . . . At the time when my invention was introduced into Sheffield the entire make of steel was 51,000 tons a year; last year we made 830,000 tons of Bessemer steel, being sixteen times what was before the entire output of the whole produce of the whole country. It is anticipated that on the Continent of Europe this year's make will reach in all 3,000,000 tons. The value of these 3,000,000 tons altogether may be taken at £10 per ton, or £30,000,000 sterling; and if that metal had been made by the old process which I have described, it would have been impossible to have brought it into the market under £50 per ton, or £150,000,000 sterling.

A matter of much importance, not referred to in the Autobiography, is Sir Henry Bessemer's close connection with the Iron and Steel Institute, of which he was one of the founders in 1868, and the President in 1871 to 1873. He only contributed two Papers to the Institute. The first of these was read in 1886, on "Some Earlier Forms of Bessemer Converters"; the second was read in 1891, and is published in the "Transactions" under the title of "The Manufacture of Continuous Sheets of Malleable Iron or Steel direct from the Fluid Metal."

In 1873 Sir Henry Bessemer presented to the Institute a sum to be invested for the purchase each year of a gold medal, to be awarded under the following conditions :—

The awards are to be (1) to the inventor or introducer of any important or remarkable invention, employed in the manufacture of iron or steel; (2) for a paper read before the Institute, and having special merit and importance in connection with the iron and steel manufacture; (3) for a contribution to the "Journal" of the Institute, being an original investigation bearing on the iron and steel manufacture, and capable of being productive of valuable and practical results. The Council may, in their discretion, award the medal in any case not coming strictly under the foregoing definition, should they consider that the iron and steel trades have been, or may be, substantially benefited by the person to whom such an award is to be made.

In 1890, my uncle, the late Mr. W. D. Allen, to whom frequent reference is made in the foregoing pages, was the recipient of this medal, and on that occasion Sir Henry Bessemer addressed to Mr. Allen the following letter, which is reproduced here, not only on account of its intrinsic interest, but because it contains a just appreciation of Mr. Allen, my father's brother-in-law, and partner during so many years :—

There was a Council meeting this morning of the Iron and Steel Institute, and, among other business, we had to decide the question of the award of the Bessemer medal. I addressed the meeting, and said I had made it a rule not to throw any weight into this question, but preferred that my fellow councilmen should take the initiative, but at the same time observing that this standing aloof might be carried too far, and a great injustice done ; and, under these circumstances, I said that I felt in duty bound to name a gentleman, to whom the introduction and the successful carrying out of the Bessemer steel process was very greatly indebted ; and that I was the more able to bear testimony in his behalf, because, although once intimately associated with him in business, I had for the last dozen years ceased to have any pecuniary interest whatever in the works referred to.

I said that Mr. W. Allen, of the Sheffield Bessemer steel works, assisted me in the very first experiments I ever made, and became thoroughly initiated in all facts that related to the process; that he assisted in the building and laying-out of our Sheffield works, and had the entire management of the process as well as of the business, and in that capacity realised almost fabulous profits from an extremely small capital. Further, that in aid of the introduction and dissemination of the "art and mystery" he had done a great deal, all the early makers having derived from him that stock of knowledge with which they commenced their respective businesses; and further, that Mr. Allen had introduced many important improvements in the detail of manufacture.

I also remarked that it had been frequently said that Bessemer steel was very good for rails, but not for a higher class of goods. Now Mr. Allen had conclusively proved the contrary of this assertion; he had never made a rail, but had gone in for the better class of material now so largely used in the Sheffield manufactures. He produced a high class of Bessemer steel, which was fully appreciated by the Sheffield trade, and he consequently was able to realise most remunerative prices: in proof of which I might mention the fact that a few weeks ago Messrs. Bessemer and Co. (which is mainly Mr. Allen and his son) declared a dividend of 25 per cent. per annum on a capital of £90,000; carried £23,000 forward; wrote off £5,000 depreciation; and spent out of revenue £11,000 in new erections. Such a result, in the face of the great competition in Bessemer steel is, I take it, a strong proof of the excellence of the material which Mr. Allen has acquired the art of making.

Mr. Windsor Richards spoke of the valuable information he had received from Mr. Allen, Mr. Snelus made a similar statement, and Mr. Ellis confirmed the fact of your success as a manufacturer of high-grade Bessemer; the question was then passed, and you were unanimously awarded the Bessemer medal, which is to be presented to you at the May meeting. This award has given me a great deal of satisfaction. . . . . I do not know if you are aware that I have been engaged in designing and superintending the execution of a very handsome diploma, framed and glazed, to be presented to all who have been previously awarded the medal: a dozen of these will be sent out to-morrow. . . . . It was thought that the medal itself can be rarely shown, and that this large and beautifully got-up design might be hung up in a library or the principal office of the medallist, where the fact would show itself to all who came, whereas the medal itself was generally locked up for safety.

I cannot better close this brief reference to my father's long connection with the Iron and Steel Institute than by repeating the resolution of sympathy passed by the Council on the occasion of his death. "The Council of the Iron and Steel Institute desire to express their sincere sympathy with the relatives of Sir Henry Bessemer in their bereavement; and, recognising his great services to the Institute as one of its founders, as its President, and the generous donor of the Bessemer gold medal, and for thirteen years as trustee of its funds, deeply deplore his loss."

In connection with the early history of steel rails referred to elsewhere I found among my father's papers a very interesting letter from Mr. F. W. Webb, the chief mechanical engineer of the London and North-Western Railway. This letter is as follows :—

London and North-Western Railway,
Locomotive Department, Crewe,
26th April, 1897.

DEAR SIR HENRY,

Referring to your last communication with reference to steel specimens.

I enclose you herewith rough hand-sketch showing the size of the piece of wheel; also the length of the old piece of rail with the inscription, which is stamped on it, together with two short pieces of bent and twisted rails. I shall be glad to know whether these will meet your requirements for exhibition.

Yours faithfully,

F. W. WEBB.

Sir Henry Bessemer,
165, Denmark Hill, Surrey.

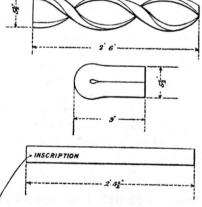

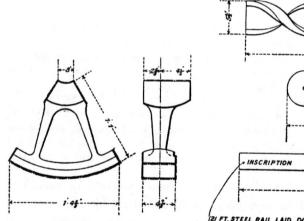

SEGMENT OF WHEEL; BRIGHT PARTS EDGED THUS:—

21 FT. STEEL RAIL LAID DOWN AT CREWE STATION 1863, TURNED 1866, TAKEN UP 1875, ESTIMATED TONNAGE 72 MILLIONS, GREATEST WEAR OF TABLES 0·85 INCH. LOSS OF WEIGHT 20 LBS. PER YARD.

FIG. 106

The sketch to which this letter refers has been reproduced in Fig. 106.

Shortly after Sir Henry's death, Mr. Webb presented the Iron and Steel Institute with a piece of this rail ; the presentation being accompanied by the following letter addressed to the Secretary ;—

May 4th, 1898.

DEAR SIR,

As I promised when I was at the last Council meeting, I am sending you to-day, by passenger train, a piece of one of the earliest Bessemer steel rails that were put down on this line, and I hope you will consider it of sufficient interest to be preserved with the other early specimens of Bessemer steel at the Institute. You will observe that I have had the following particulars stamped on the piece of rail :—"Bessemer steel rail, laid down at Crewe Station, 1863, turned 1866, taken up 1875 ; estimated tonnage 72,000,000 ; greatest wear of tables, 0.85 in. ; loss of weight, 20 lb. per yard. Presented by the London and North-Western Railway, per F. W. Webb, April 18, '98." I shall be glad if you will kindly acknowledge receipt on its arrival.

To the Secretary, Iron and Steel Institute.

On the occasion when this letter was read, Mr. E. Riley said that he had a piece of the first Bessemer rail which was ever rolled ; it was rolled at Dowlais. The rail broke in rolling. It was made from Blaenavon pig iron, at the Bessemer steel works at St. Pancras, London, and was rolled at Dowlais in 1856.

It was a source of constant pride and gratification to my father that several towns in the United States were named after him. The most important of these is the City of Bessemer, in Pennsylvania, in which are situated the Edgar Thomson steel works, founded in 1870, acquired by Mr. Carnegie after the death of Mr. Edgar Thomson, and now forming the most important unit in the gigantic steel trust of America. This City of Bessemer is situated on the Monongahela River, a few miles from Pittsburg.

In the State of Alabama there is another Bessemer, which has been raised to the dignity of a city. It was established in 1886, and is situated in the northern part of the State, in the centre of the southern Bessemer steel industry. The population of this city is over 6,000.

Bessemer, in the State of Virginia, near Clifton Ford, on the Chesapeake and Ohio Railway, is also in the southern section of the Bessemer steel industry. I am unaware to what extent this city has developed, but a few years since it promised to become a very important manufacturing town.

A fourth City of Bessemer is in the State of Michigan. This place also has, I believe, developed into considerable prosperity, as it is situated in the heart of a vast ore-producing district.

In all there are no fewer than thirteen Bessemers in the United States; the following list has been compiled officially from the United States Census of 1900.

Town.	State.	County.	Population in 1900. U. S. Census.	In Iron-Mining District?
Bessemer	Alabama ...	Jefferson ...	6,358	Yes
Bessemer Junction	Alabama ...	Jefferson ...	(a)	Yes
Bessemer (Station, Pueblo) ...	Colorado ...	Pueblo ...	(b)	Yes
Bessemer Junction ...	Colorado ...	Pueblo ...	(a)	Yes
Bessemer	Michigan ...	Gogebic ...	3,911	Yes
Bessemer Junction	Michigan ...	Gogebic ...	(a)	Yes
Bessemer	N. Y.	Tomkins ...	(a)	No
Bessemer City	N. C.	Gaston ...	1,100	No
Bessemer	Pennsylvania ...	Allegheny ...	(a)	No
Bessemer	Pennsylvania ...	Lawrence ...	(a)	No
Bessemer	Texas	Llano ...	(a)	No
Bessemer	Vancouver ...	Botetourt ...	(a)	Yes
Bessemer	Wyoming ...	Natrona ...	(a)	No

(a) Not separately returned by the Census of 1900.

(b) Bessemer City (population 3,317 as returned by Census of 1890) annexed to Pueblo since 1890.

My father possessed a special charm of conversation, and an unusual facility for explaining difficult subjects in the most graphic manner. Those who enjoyed his friendship will always remember this peculiar gift. Striking examples of it are to be found in his remarks at various scientific meetings where he took part in discussions; and on the rare occasions when he wrote letters to the newspapers, chiefly to *The Times*. As illustrations of what I mean, I reproduce here three letters which I think are of much interest. The first of these was published in *The Times* in January, 1878, under the title "A Billion Dissected."

SIR,

It would be curious to know how many of your readers have brought fully home to their inner consciousness the real significance of that little word "billion" which we have seen of late so glibly used in your columns. There are, indeed, few intellects that can fairly grasp it and digest it as a whole; and there are, doubtless, many thousands who cannot appreciate its true worth, even when reduced to fragments for more easy assimilation. Its arithmetical symbol is simple and without much pretension; there are no large figures—just a modest 1 followed by a dozen cyphers, and that is all.

Let us briefly take a glance at it as a measure of time, distance, and weight. As a measure of time, I would take one second as the unit, and carry myself in thought through the lapse of ages back to the first day of the Year 1 of our era, remembering that in all those years we have 365 days, and in every day just 86,400 seconds of time. Hence, in returning in thought back again to this year of grace 1878, one might have supposed that a billion of seconds had long since elapsed; but this is not so. We have not even passed one-sixteenth of that number in all these long eventful years, for it takes just 31,687 years, 17 days, 22 hours, 45 minutes, and 5 seconds to constitute a billion of seconds of time.

It is no easy matter to bring under the cognizance of the human eye a billion objects of any kind. Let us try in imagination to arrange this number for inspection, and for this purpose I would select a sovereign as a familiar object. Let us put one on the ground and pile upon it as many as will reach 20 ft. in height; then let us place numbers of similar columns in close contact, forming a straight line, and making a sort of wall 20 ft. high, showing only the thin edges of the coin. Imagine two such walls running parallel to each other and forming, as it were, a long street. We must then keep on extending these walls for miles —nay, hundreds of miles, and still we shall be far short of the required number. And it is not until we have extended our imaginary street to a distance of 2386½ miles that we shall have presented for inspection our one billion of coins.

Or, in lieu of this arrangement, we may place them flat upon the ground, forming one continuous line like a long golden chain, with every link in close contact. But to do this we must pass over land and sea, mountain and valley, desert and plain, crossing the Equator, and returning around the southern hemisphere through the trackless ocean, retrace our way again across the Equator, then still on and on, until we again arrive at our starting-point; and when we have thus passed a golden chain round the huge bulk of the earth, we shall be but at the beginning of our task. We must drag this imaginary chain no less than 763 times round the globe. If we can further imagine all those rows of links laid closely side by side and every one in contact with its neighbour, we shall have formed a golden band around the globe just 52 ft. 6 in. wide; and this will represent our one billion of coins. Such a chain, if laid in a straight line, would reach a fraction over 18,328,445 miles, the weight of which, if estimated at ¼ oz. each sovereign, would be 6,975,447 tons, and would require for their transport no less than 2325 ships, each with a full cargo of 3000 tons. Even then there would be a residue of 447 tons, representing 64,081,920 sovereigns.

For a measure of height let us take a much smaller unit as our measuring rod. The thin sheets of paper on which these lines are printed, if laid out flat and firmly pressed together as in a well-bound book, would represent a measure of about 1-333rd of an inch in thickness. Let us see how high a dense pile formed by a billion of these thin paper leaves would reach. We must, in imagination, pile them vertically upward, by degrees reaching to the height of our tallest spires; and passing these, the pile must still grow higher, topping the Alps and Andes and the highest peaks of the Himalayas, and shooting up from thence through the fleecy clouds, pass beyond the confines of our attenuated atmosphere, and leap

3 B

up into the blue ether with which the universe is filled, standing proudly up far beyond the reach of all terrestrial things; still pile on your thousands and millions of thin leaves, for we are only beginning to rear the mighty mass. Add millions on millions of sheets, and thousands of miles on these, and still the number will lack its due amount. Let us pause to look at the neat ploughed edges of the book before us. See how closely lie those thin flakes of paper, how many there are in the mere width of a span, and then turn our eyes in imagination upwards to our mighty column of accumulated sheets. It now contains its appointed number, and our one billion of sheets of *The Times*, superimposed upon each other and pressed into a compact mass, has reached an altitude of 47,348 miles !

Those who have taken the trouble to follow me thus far will, I think, agree with me that a billion is a fearful thing, and that few can appreciate its real value. As for quadrillions and trillions, they are simply words: mere words wholly incapable of adequately impressing themselves on the human intellect.

I remain, your obedient servant,

Denmark Hill, January 3, 1878.

HENRY BESSEMER.

The second was also a letter that appeared in *The Times* on April the 18th, 1882, and was called "Easter and the Coal Question."

SIR,

The Easter holidays have come round once more, and our boys, with their bright, beaming faces, full of mirth and cheerfulness, have been flocking home from school to dear old smoky London, all unmindful of its murky atmosphere, and intent only on the many wondrous sights they hope to see. I had just filled some loose sheets with calculations which I had been making, with a view to afford some amusement to my grandsons on their return, when, looking up from my task, I noticed a stream of small, bright objects flitting by. The sharp east wind was breaking up the large seed-pods on the great Occidental plane tree near my study window, and its taper seeds, with their beautiful little gold-coloured parachutes, were being wafted far away, falling into little chinks and unknown out-of-the-way places. Some, resting on the bare earth, may perchance be seized by some blind worm, and made to close the door of its lowly habitation, and, germinating there, may, in after-years, when all who now live have passed away, spread its huge arms, and afford a grateful shelter to those who are to come after us. Just so the broad sheet you daily publish conveys to every civilised part of the world the thoughts and sentiments of those who lead and form public opinion, while it never fails to give the latest expression of science, literature, and art. Much of all this may, like the flying plane tree seeds, fall on unproductive soil; yet who shall say in that ceaseless stream of intelligence how many a sympathetic chord of the human heart may be touched, or how many thoughts and sentiments so imbibed may germinate, and, gaining strength with years, may change the whole current of a life, and form the statesman, the scientist, or the man of letters? Thus musing, it occurred to me that the statistical results I had arrived at might, perhaps, interest some other boys than those for whom they were intended, and if thought worthy of a place in *The Times* might inspire a more than passing interest in an otherwise most uninviting subject.

Every one of late must have had his thoughts more or less turned to the prevention of smoke in large cities, and also to the exhibition of the electric light now in progress at the Crystal Palace, for every form and modification of which we are still dependent on that vast storehouse of Nature—our beds of coal, the economic use of which is of such vast importance to our national progress, and to the maintenance and spread of civilisation throughout the world, that no one can afford to remain indifferent to it. It is only when the mind can fairly grasp the magnitude of our coal consumption that the importance of its economy can be fully realised. The statistics of the coal trade show that during the year 1881 the quantity of coal raised in Great Britain was no less than 154,184,300 tons.

When the eye passes over these nine figures it does not leave on the mind a very vivid picture of the reality—it does not say much for the twelve months of incessant toil of the 495,000 men who are employed in this vast industry; hence I have endeavoured in a pictorial form to convey to the mind's eye of my young friends something like the true meaning of those figures; for mere magnitude to the youthful mind has always an absorbing interest, and the gigantic works of the ancients fortunately supply us with a ready means of comparison with our own. Let us take as an example the great pyramid of Gizeh, a work of human labour which has excited the admiration of the world for thousands of years. Though in itself inaccessible to my young friends, we fortunately have its base clearly marked out in the metropolis.

When Inigo Jones laid out the plans of Lincoln's Inn Fields he placed the houses on opposite sides of the square just as far from each other as to enclose a space between them of precisely the same dimensions as the base of the great pyramid. Measuring up to the front walls of the houses this space is just equal to 11 acres and 4 poles. Now, if my young friends will imagine St. Paul's Cathedral to be placed in the centre of this square space, and having a flagstaff of 95 ft. in height standing up above the top of the cross, we shall have attained an altitude of 499 ft., which is precisely equal to that of the great pyramid. Further, let us imagine that four ropes are made to extend from the top of this flagstaff, each one terminating at one of the four corners of the square and touching the front walls of the houses. We shall then have a perfect outline of the pyramid of exactly the same size as the original. The whole space enclosed within these diagonal ropes is equal to 79,881,417 cubic feet, and if occupied by one solid mass of coal it would weigh 2,781,581 tons—a mass less than 1-55th part of the coal raised last year in Great Britain. In fact, the coal trade could supply such a mass as this every week, and at the end of the year have more than nine millions of tons to spare.

Higher up the Nile Thebes presents us with another example of what may be accomplished by human labour. The great temple of Rameses at Karnak, with its hundred columns of 12 ft. in diameter, and over 100 ft. in height, cannot fail to deeply impress the imagination of all who, in their mind's eye, can realise this magnificent colonnade. It may be interesting to ascertain what size of column and what extent of colonnade we could construct with the coal we laboriously sculpture from its solid bed in every year.

Let us imagine a plain cylindrical column of 50 ft. in diameter, and of 500 ft. in height, our one year's production of coal would suffice to make no less than 4511 of these gigantic columns, which, if placed only at their own diameter apart, would form a colonnade which would extend in a straight line to a distance of no less than 85 miles and 750 yards

—in fact, we dig in every working day throughout the year a little more than enough to form fourteen of these tall and massive columns, which, if placed upon each other, would reach an altitude of 7,000 ft.

But there is yet another great work of antiquity which our boys will not fail to remember as offering itself for comparison ; they have all heard of the Great Wall of China, which was erected more than 2,000 years ago to exclude the Tartars from the Chinese Empire. This great wall extends to a distance of 1,400 miles, and is 20 ft. in height, and 24 ft. in thickness, and hence contains no less than 3,548,160,000 cubic feet of solid matter. Now, our last year's production of coal was 4,427,586,820 cubic feet, and is sufficient in bulk to build a wall round London of 200 miles in length, 100 ft. high, and 41 ft. 11 in. in thickness; a mass not only equal to the whole cubic contents of the Great Wall of China, but sufficient to add another 346 miles to its length.

These imaginary coal structures can scarcely fail to impress the mind of youth with the enormous consumption of coal ; and when they are told that in many of its applications the useful effect obtained is not one-fifth of its theoretic capabilities, they will be enabled to form some idea of the vast importance of the economic problem which calls so loudly for solution.   They must not, however, fall into the too-common error of supposing that the electric light by superseding gas is to do away with the use of coal in the production of light, or that dynamo-electric machines will largely replace the steam engine and boiler.

A visit to the Crystal Palace, which has for the time being become a great school of applied science, will set them right on this point.   There they will find that coal, our willing slave, still lends its powerful aid in propelling those machines by which we manufacture artificial lightning ; and there also, in its mere infancy, they will see something of the colossal power that is destined to effect such vast changes, and to carry forward by another grand leap the ever-increasing dominion of mind over matter.

Let every boy now home from school be taken to see this grand exhibition before it closes, and while still on the tablets of the brain there are left some few blank pages, let these marvels of applied science inscribe an indelible record, which, perchance, in after years, may profitably be drawn upon and improved; and in due course they may find their own names inscribed among those who, following the paths of science, have become the benefactors of mankind.

Although coal is still our great agent in the production of motive power, it must not be forgotten that Sir William Thomson has clearly shown that by the use of dynamo-electric machines, worked by the Falls of Niagara, motive power could be generated to an almost unlimited extent, and that no less than 26,250 horse-power so obtained could be conveyed to a distance of 300 miles by means of a single copper wire of $\frac{1}{2}$ in. in diameter, with a loss in transmission of not more than 20 per cent., and hence delivering at the opposite end of the wire 21,000 horse-power.*

What a magnificent vista of legitimate mercantile enterprise this simple fact opens up

---

* Since the above was written, experiments on a large scale have shown that the loss of power in transmission is much greater than stated, and also that the size of the copper wire was very much underestimated, but these facts do not materially lessen the advantages of this mode of supplying power and light to London direct from the coalfield.—H. B.

for our own country! Why should we not at once connect London with one of our nearest coalfields by means of a copper rod of 1 in. in diameter and capable of transmitting 84,000 horse-power to London, and thus practically bring up the coal by wire instead of by rail?

Let us now see what is the equivalent in coal of this amount of motive power. Assuming that each horse-power can be generated by the consumption of 3 lb. of coal per hour, and that the engines work six and a-half days per week, we should require an annual consumption of coal equal to 1,012,600 tons to produce such a result.

Now all this coal would in the case assumed be burned at the pit's mouth, at a cost of 6s. per ton for large and 2s. per ton for small coal—that is, at less than one-fourth the cost of coal in London. This would immensely reduce the cost of the electric light, and of the motive power now used in London for such a vast variety of purposes, and at the same time save us from the enormous volumes of smoke and foul gases which this million of tons of coal would make if burned in our midst. A 1 in. diameter copper rod would cost about 533*l.* per mile, and if laid to a colliery 120 miles away, the interest at 5 per cent. on its first cost would be less than 1d. per ton on the coal practically conveyed by it direct into the house of the consumer.

I am, Sir, your obedient servant,

HENRY BESSEMER.

Denmark Hill, April 17, 1882.

The third reprint is from the *Engineering Review*, of July the 20th, 1894, and is entitled "A Brief Statistical Sketch of the Bessemer Steel Industry : Past and Present."

It is an old man's privilege to look back upon the past and compare it with the present. It is no less his privilege to do so when his thoughts turn to those subjects in which he himself has taken a more or less conspicuous part. I do not know, therefore, that I need make any apology for laying before you some thoughts that have been passing through my mind on looking back upon the progress that has been made in the metallurgical world, and especially in a retrospect of the rapid advances made by the process to which my name was given thirty-seven years ago.

If we go back to the year 1861, just one-third of a century, we shall find Sheffield by far the largest producer of steel in the world, the greater portion of her annual make of 51,000 tons, realising from £50 to £60 per ton.*

For this purpose the costly bar-iron of Sweden was chiefly employed as the raw material, costing from £15 to £20 per ton ; the conversion of this expensive iron into crude steel occupied about ten days—that is, about two days and nights for the gradual heating of the furnace, in which the cold iron bars had been carefully packed in large stone boxes with a layer of charcoal powder between each bar, in these boxes the metal was retained for six days at a white heat, two days more being required to cool down the

---

* It is stated in the Jurors' Report to the Commissioners of the International Exhibition of 1851, that the production of steel in Sheffield was at that period 51,000 tons annually.

furnace and get out the converted bars. The steel so produced was broken into small pieces and melted in crucibles holding not more than 40 or 50 lb. each, and consuming from 2 to 3 tons of expensive oven coke for each ton of steel so melted. This steel was excellently adapted for the manufacture of knives, and for all other cutting instruments, but its hard and brittle character, as well as its excessively high price, absolutely precluded its use for the thousands of purposes to which steel is now universally applied.

It was under such conditions of the steel trade that, thirty-three years ago, I endeavoured to introduce an entirely novel system of manufacture—so novel, in fact, and so antagonistic to the preconceived notions of practical men, that I was met on all sides with the most stolid incredulity and distrust. Perhaps I ought to make some allowance for this feeling, for I proposed to use as my raw material crude pig-iron costing £3 per ton, instead of the highly purified Swedish bar-iron then used, costing from £15 to £20 per ton. I proposed also to employ *no fuel whatever* in the converting process, which, in my case, occupied only *twenty-five to thirty minutes,* instead of the ten days and nights required by the process then in use; and I further proposed to make from 5 to tons of steel at a single operation, instead of the small separate batches of 40 or 50 lbs., in which all the Sheffield cast steel was at that time made. What, however, appeared still more incredible was the fact that I proposed to make steel bars at £5 or £6 per ton, instead of £50 or £60—the then ruling prices of the trade. One and all of these propositions have long since become well-established commercial facts, and Bessemer cast-steel is now produced without resorting to *any one* of the *expensive* and *laborious* processes practised in making Swedish bar-iron, while the old Sheffield process of converting wrought-iron bars into crude or blister-steel, by ten days' exposure, at a very high temperature, to the action of carbon, is rendered unnecessary. The slow and expensive process of melting 40 or 50 lbs. of steel in separate crucibles is also dispensed with; and in lieu of all these combined processes, from 5 to 10 tons of crude or cast-iron, worth only £3 per ton, is converted into Bessemer cast-steel in thirty minutes, wholly without skilled manipulation, or the employment of fuel; and while still retaining its initial heat, can be at once rolled into railway bars or other required forms.

So great was the departure of my invention from all the preconceived notions and practice of the trade, that no steel manufacturer could be induced to adopt it, in fact the whole steel and iron trade of the kingdom had declared it to be the mere dream of a wild enthusiast; and it was only by building a steelworks of my own in the town of Sheffield, and underselling other manufacturers in the open market, that I was able at last to overcome prejudice and the utter disbelief in the practicability of my invention. But as soon as my works were completed, and I was enabled to throw my cheap steel upon the market, there came a complete panic in the trade, followed by the adoption of my invention at two of the largest works in Sheffield. As an example of the irresistible competition thus established, I may refer to the manufacture of steel railway-wheel tyres, which were at that time selling at £60 per ton. These tyres we put upon the market at £50, but the extent to which even that price was capable of reduction will be readily understood from the fact that tyres made at the present date, by the same process, and by the identical machinery then actually employed, are now sold at £8 per ton with a profit. No sooner were these facts rendered indisputable by the steady commercial working of my process, than it began rapidly to spread throughout England, and thence to every State in Europe. The advantages which my system offered soon attracted the attention of our energetic brethren in the United

States, where it advanced by leaps and bounds, and where it has since culminated, in the year 1892, in the production of no less than 4,160,072 tons, or about *eighty times* the whole production of Sheffield in 1851.*

The visit of the Iron and Steel Institute to America in 1890 was quite a revelation. The development of the iron and steel trade of that country, and the enormous extension of their railroad system, has produced economic changes of vast importance both to them and to us, and demands the serious consideration of all thinking men.

We have it on the undoubted authority of Mr. Abram Hewitt that the annual production of steel by the acid and basic treatment of pig-iron in the Bessemer converter in both Europe and America amounted in 1892 to no less than 10,500,000 tons, about two-fifths only of which was made into rails. Now, taking the average price of rails in 1891 and 1892 in England at £4 10s. per ton, and in the United States and on the Continent of Europe at £5 10s., and adding to this the much higher prices obtained for tyres, axles, cranks, sheets, wire-rods, boiler-plates, forgings, castings, &c., we may fairly assume that the average selling price of the whole of this steel would be £8 per ton, taking one article with another, hence yielding a net amount of 84 millions sterling.

It is a curious fact that high numbers like these do not adequately impress themselves on the minds of many people of undoubted intelligence, and it is not until such figures are broken up as it were, and presented pictorially to the mind's eye, that they are fully understood and appreciated. Thus, if, instead of looking at the eight figures which represent the number of tons, we could have that quantity of steel bodily before us, we should form a very different estimate of its importance. Let us use the mind's eye to assist us, and imagine standing erect before us a plain round column or tower of solid steel 20 feet in diameter and 100 feet high; this, no doubt, would impress us as a very large and heavy mass, and but few persons would be prepared at first to accept the simple fact that the production of Bessemer steel in 1892 would make 1,671 such columns and leave a remainder of 5,535 tons. Yet such is the fact. These tall columns would form a goodly row, and, if placed side by side in a straight line, and in contact with each other, would extend to a distance of 6 miles and 580 yards; indeed there is on an average 5⅓ such columns produced on *every working day in the year*, bringing up each day's production of steel to 33,546 tons, as compared with Sheffield's former production of 51,000 tons *annually*.

We may put this in another way, and imagine a plain cylindrical solid column of 100 feet in diameter, a good idea of which may be formed by a glance at some of the very large gasometers in the Metropolis; then further imagine this gasometer, not as a thin iron shell, but as a ponderous solid mass rising before you to an altitude of 6,684 feet 6 inches, or nearly one mile and a third in height. Such a huge solid mass would be exactly equal to one year's make of Bessemer steel. But even in this form we must draw powerfully on the imagination; for but few persons can in their mind's eye fully realise a huge solid mass of such heavy matter rising to more than sixteen and a-half times the height of the cross of St. Paul's.

A graphic representation of such a column of steel, standing between St. Paul's Cathedral

* We learn from the Bulletin of the American Iron and Steel Association that the output of Bessemer Steel Ingots in the United States in the year 1892 was the largest ever reported, and amounted to not less than 4,160,072 tons.

SOLID STEEL COLUMN
6684 feet 6 inches in height
and
100 feet in diameter
Representing the World's production
of
BESSEMER-STEEL
in the year 1892.

accurately to scale of
1 inch to 1000 feet.

S.T PAUL'S CATHEDRAL.

GASOMETER.
100 feet diameter.
120 feet high.

THE MONUMENT
to fire of London

FIG. 107.

and the Monument erected to commemorate the Great Fire of London, is shown accurately to scale (see Fig. 107), and will aid the mind in more fully realising the magnitude of the ponderous masses annually produced, every pound of which, during the brief period of its conversion into steel, has been raised to such an excessively high temperature as to become as brilliantly incandescent as the poles of the electric arc lamp.

It is this new material, so much stronger and tougher than common iron, that now builds our ships of war and our mercantile marine. Steel forms their boilers, their propeller-shafts, their hulls, their masts and spars, their standing rigging, their cable chains and anchors, and also their guns and armour-plating.

This new material has covered with a network of steel rails the surface of every country in Europe, and in America alone there are no less than 175,000 miles of Bessemer steel rails, binding together its widely-scattered cities, and bringing them within easy commercial contact with each other. Over these long stretches of smooth steel road there ceaselessly run hundreds of thousands of steel wheel-tyres, impelled by hundreds of locomotive engines, which owe their power and endurance to the same ubiquitous material, the great strength and elasticity of which, as compared with common iron, renders it so especially suitable for the construction of our bridges and viaducts, our steam boilers, and our machinery of every description, while its great resistance to wear and abrasion gives it a durability vastly superior to iron. As an example, I may state that every steel rail now in use will bear at least six times the amount of traffic to pass over it that would suffice to wear out an iron rail. This question of durability is one of vast importance, for it has enabled companies to construct lines in localities where the rapid wearing out of iron rails would not profitably permit of their construction. The increased durability of steel will be better realised when we consider that the 175,000 miles of steel railroads now existing in America would have had to be broken up and laid with new rails six times (if the rails had been made of iron) during the period that the steel rails will last in a safe and workable condition.

But to descend from large things to smaller ones, it may be interesting to pass from the almost unrealisable column of solid steel representing the world's yearly production to the average quantity made in every one of the 24 hours comprised in the 313 working days of the year, and thus bring our mass more in accord with some of the tall columns in this metropolis, which are, say, 7 or 8 feet in diameter, and reach 100 feet or more in altitude. It must be remembered that the process of converting crude iron into steel goes on ceaselessly in the converter for the whole twenty-four hours of each day, so that our one hour's production is only one twenty-fourth part of a single day's work; but if all the steel produced in the Bessemer converter in this short interval of time were collected, it would form a solid cylindrical mass of 8 feet in diameter, and 139 feet in height, thus overtopping the Duke of York's column and the Nelson Monument. What a noble portico would twenty-four such columns make, the work of a single day, but yet large enough to dwarf the grand old ruins of Karnak or Thebes.

It may be interesting to put this matter in another form, in order to bring it vividly home to the imagination. A steel ingot of one ton weight is as nearly as possible five cubic feet of solid matter. Let us now imagine a solid square ingot of steel, having a base measuring 50 feet by 50 feet, and standing, say, 400 feet high. This would make a square tower of solid steel, much larger than the clock-tower of the Houses of Parliament

3 c

(which is precisely 40 feet square, and about half as high as this imaginary square tower); in fact, such a tower would only be about four feet below the top of the cross of St. Paul's Cathedral. This tower would contain precisely 1,000,000 cubic feet, and would weigh just 200,000 tons. Now, the Thames Embankment from Westminster Bridge to Blackfriars Bridge, measured down the centre of the roadway, is one mile and a-quarter and a few yards. Let us suppose one of these gigantic towers to stand opposite the Clock Tower, and in a line with the roadway over Westminster Bridge, and a similar one erected at the other end of the Embankment in a line with the roadway passing over Blackfriars Bridge. Let us further imagine fifty other precisely similar towers placed equi-distant between them, thus leaving a space of only 27 yards between each tower. This row of gigantic towers would represent 10,400,000 tons, or just 100,000 tons less than one year's production of Bessemer steel, each of the fifty-two towers being 1,923 tons less than the average weekly production.

We might think of many other object lessons that would be likely to convey to the mind's eye a vivid and realistic picture of the enormous bulk of matter represented by 10,500,000 tons of steel. Let us select one other illustration. Imagine a straight wall 100 miles in length,* 5 feet in thickness, and 20 feet in height. Such a wall would stand on 60½ acres of land. But suppose that this wall, like a gigantic armour-plate, was formed into a circle, and used to surround London; the enclosure so made would extend to Watford on the north side, to Croydon on the south, to Woolwich on the east, and to Richmond on the west. It would, in point of fact, form a circular enclosure of 31¾ miles in diameter, and would embrace an area of 795 square miles. This great wall of London would just be equal to a single year's production of Bessemer steel.

I have thought it would be interesting to give these illustrations of the enormous mass of Bessemer steel that is now annually produced, because its magnitude is more easily conveyed to the mind by such object lessons than in any other way, and it has long been a hobby of mine to convey an idea of large numbers by such illustrations. Some of my old friends will, I doubt not, remember that in 1878 I published a letter in the *Times* entitled "A Billion Dissected," in which I broke up the elements of that measure of numbers in the same way; and in 1882 I dealt, in the *Times*, in a similar manner with our coal output. If this fresh illustration, which has naturally exceptional interest to myself, should bring home to you an idea of the magnitude of modern industrial operations, in respect of a material that bears my name, I shall be much gratified.

As a commercial question it is impossible to form even an approximately correct idea of the value of this material when manufactured into the almost endless variety of useful articles into which it is now made.

As a single instance, I may refer to the manufacture of steel nails. It is an important and well-known fact that a steel nail can be driven into dry hard wood without boring a hole for it. This property of steel nails results in an immense saving of labour, and in the United States, where so many houses are built of wood, it has proved of considerable value. I find from reliable statistics furnished by nail manufacturers, that in 1892 no less than 171,200 tons of unforged nails, and 139,900 tons of steel-wire nails were made in America alone. Medium-sized nails run from 80,000 to 120,000 to the ton, and I have

---

* Or more accurately 99 miles and 2,280 feet in length.

before me some beautifully-formed carpet nails, with large flat heads, of which a single ton of steel will make 3,870,000.

It is an interesting fact that at the International Exhibition of 1862, I exhibited the first steel nails that were ever made. Every form and pattern of nail was shown, large spikes, 6 inches long, weighing only 10 to the pound, or 22,400 to the ton, down to the minute tacks used by upholsterers, and known as gymp tacks, so small that one ton of steel will make more than 14 millions of them.

I well remember how many thousands of people at the Exhibition passed heedlessly by these germs of a new and important industry, apparently without the remotest idea of the future universal employment of steel nails in lieu of iron ones.

Those who have passed through Wolverhampton and the "Black Country" a dozen years ago, must have seen the hundreds of young girls sacrificing all the feminine hopes and aspirations of their young lives, each one toiling from dewy morn to dusky eve, in smoky, grimy smithies, with a pair of iron tongs, holding the red-hot nail in one hand, while with the other she showered upon it blows from the uplifted hammer in such rapid succession as to maintain the incandescence of the iron she was shaping, amid the ceaseless din of her fellow-workers, who, with grimy faces and horny hands, were reeking in the heat and foul air of the nailers' den.

Time in this, as in so many other things, has wrought its wonted change, for to-day the inexorable power of steam, acting on unconscious matter which suffers from neither heat, fatigue, nor moral degradation, now yields from a single machine from 50 to 100 nails per minute, at less cost and of better quality than were ever wrung from human sinews and female degradation. The extent of the change will be better appreciated when it is known that the annual value of *unforged* steel nails now manufactured exceeds ten millions sterling; and I have often felt that if in my whole life I had done no other useful thing than the introduction of unforged steel nails, this one invention would have been a legitimate source of self-congratulation and thankfulness, in so far as it has successfully wiped out so much of this degrading species of slavery from the list of female-employing industries in this country.

The great financier who is constantly dealing with the realised values of many millions would have a very keen appreciation of what £84,000,000 really means, yet I doubt if even the Chancellor of the Exchequer could off-hand give anything like the correct dimensions of a mass of standard gold of that value. It can, however, be easily ascertained with accuracy. Since fifty-seven sovereigns weigh just 1 lb. avoirdupois, the weight of 84,000,000 sovereigns would be 657 tons 17 cwt. 3 qrs. and 16 lbs.; and as the specific gravity of standard gold coin is 17.167, we should have a mass equal to 1374.70 cubic feet, from which we could make a plain cylindrical column of solid gold 5 feet in diameter and 109 feet 5 inches in height, as a representative of the commercial value of the larger column of steel which I have referred to. It is an interesting fact that the statistics published by the *Annales des Mines* for 1893* shows that it would take more than three years' production of all the gold mines in the world to pay in gold for one year's production of Bessemer steel.

* Taken from a paragraph in the *Times*, showing the weight in tons and value in pounds sterling of the world's production of gold in 1893.

In June, 1897, my mother died, and her loss was a blow from which my father never recovered; their happy union had lasted for more than sixty years and he did not long survive her; his own death occurred on the 15th March, 1898.

In the earlier pages of his Autobiography, Sir Henry Bessemer wrote not a little about his father, but the glimpses which he gives us of the elder Bessemer cause regret that he did not say a great deal more; it is true that the interesting part of his career appears to have ended when, as quite a young man, he fled from Paris during the stormy days of the Revolution, leaving behind him almost all the considerable means he had accumulated during his residence in France. It was during the unsettled period before the Revolution that he had made a name and a distinguished position in the French Academy of Sciences. My father makes a reference to a copying and engraving machine invented by the elder Bessemer, which was largely used in the Paris Mint, for reproducing in metal, artists' designs modelled in wax, either in cameo or intaglio. In the Autobiography of James Nasmyth, to which I have already referred, there is an interesting notice of one of these machines. It had been sent from Paris to the London Mint years after my grandfather had returned to this country, and Nasmyth, speaking of it in the highest terms, relates how it was sent from the Mint, in 1830, to Messrs. Maudslay's, for repair, and the work of its repair was entrusted to him. During the prosperous period of my grandfather's life in France, miniatures of himself and of my grandmother were painted by an artist famous at the time, and these portraits were among the objects he saved in his flight from Paris. I have been able to reproduce them here (see Plate L.), and I think that the portraits of the founders of the Bessemer family will not be without interest.

PRINTED AT THE BEDFORD PRESS, 20 AND 21, BEDFORDBURY, STRAND, LONDON, W.C.